Biochemical and Biophysical Studies
of Proteins and Nucleic Acids

Biochemical and Biophysical Studies of Proteins and Nucleic Acids

Edited by:

Tung-Bin Lo, Ph.D.
Dean, School of Natural Sciences, National Taiwan University
Taipei, Taiwan, R.O.C.

Teh-Yung Liu, Ph.D.
Director, Division of Biochemistry and Biophysics, Office of Biologics,
National Center for Drugs and Biologics, Food and Drug Administration
Bethesda, Maryland, U.S.A.

Choh-Hao Li, Ph.D.
Director, Hormone Research Laboratory, University of California
San Francisco, California, U.S.A.

Elsevier
New York • Amsterdam • Oxford

Proceedings of the International Symposium on Biochemical and Biophysical Studies of Proteins and Nucleic Acids, held August 11–13, 1982, in Taipei, Taiwan.

Published by:

Elsevier Science Publishing Co., Inc.
52 Vanderbilt Avenue, New York, New York 10017

Sole distributors outside the USA and Canada:

Elsevier Science Publishers B.V.
P.O. Box 211, 1000 AE Amsterdam. The Netherlands

Library of Congress Cataloging in Publication Data

Main entry under title:

Biochemical and biophysical studies of proteins and nucleic acids.

Collection of papers presented at the Third International Symposium of Proteins and Nucleic Acids, held Aug. 11–13, 1982 at the National Taiwan University.

Includes index.
1. Proteins—Congresses. 2. Nucleic acids—Congresses. 3. Biological chemistry—Congresses. 4. Biophysics—Congresses. I. Lo, Tung-Bin. II. Liu, Teh-Yung, 1932-   . III. Li, Choh-Hao, 1913-   . IV. International Symposium of Proteins and Nucleic Acids (3rd: 1982: National Taiwan University)
QP551.B46   1983   574.19′245   83-20682
ISBN 0-444-00911-6

Manufactured in the United States of America

CONTENTS

v

SESSION IV - BIOCHEMICAL STUDIES OF PROTEINS
Chairmen: R. L. Heinrikson and C. C. Yang

PREFACE

This volume is a collection of invited papers presented at the Third
International Symposium of Proteins and Nucleic Acids held August 11-13, 1982,
at the Auditorium, Institute of Earth Sciences, Academica Sinica, located on
the main campus of the National Taiwan University, Taipei, Taiwan, R.O.C. The
time of the symposium was chosen shortly before the Twelfth International
Congress of Biochemistry in Australia, to provide an opportunity for biochemists
from the United States, Japan, United Kingdom, and Taiwan to meet on topics
in which participants have research programs of mutual interest. The symposium
was organized by Drs. C. H. Li, T. B. Lo, A. N. Schechter, S. I. Chan, C. W.
Wang, K. T. Wang, and T.-Y. Liu. The scope of the symposium included receptors,
membranes, nucleic acid biochemistry and biochemical and biophysical studies
of proteins.

Dr. S. L. Chien, President of the Academica Sinica, opened the conference
by welcoming the participants, especially those who have come from abroad.
He reviewed the status of protein research in the Republic of China and pointed
out the importance of protein and nucleic acid research as the fundamentals of
genetic engineering, which is one of the fields of concentration in the
national research projects. Dr. C. M. Chang, Chairman of the National Science
Council, traced the history of the practice of biochemistry as a profession on
the island of Taiwan during the last fifty years and acknowledged the contribution of Dr. C. H. Li, the Chairman of the Symposium, in guiding the development of modern basic biochemical research on this island since the late fifties.
Dr. Chang echoed his government's determination to promote and foster recombinant DNA engineering programs and expressed the hope that the symposium would
provide many valuable and stimulating new ideas.

The symposium could not have taken place without the support of Drs. C. M.
Chang and S. L. Chien. The editors of this volume wish to thank Dr. R. L.
Heinrikson for being rapporteur for the meeting. Thanks are due to Drs. Kung-
Tsung Wang and Chi-Wu Wang for their dedicated service in making the symposium
a success.

The Editors

INVITED PARTICIPANTS

Gilbert Ashwell
Laboratory of Biochemistry
    and Metabolism
National Institutes of Health
Building 10, Room 9N105
Bethesda, MD 20205, USA

Sunney I. Chan
California Institute of Technology
Noyes Laboratory, 127-72
Pasadena, CA 91125, USA

Yee-Hsiung Chen
Department of Biochemical Sciences
National Taiwan University
Taipei, Taiwan, ROC

Elliot Elson
Department of Biological Chemistry
Washington University
St. Louis, MO 63110, USA

Robert L. Heinrikson
Department of Biochemistry
University of Chicago
Chicago, IL 60637, USA

Chien Ho
Department of Biological Sciences
Carnegie Mellon University
4400 Fifth Avenue
Pittsburgh, PA 15213, USA

Alice S. Huang
Department of Microbiology and
    Molecular Genetics
Children's Hospital Medical Center
Harvard Medical School
300 Longwood Avenue
Boston, MA 02115, USA

Choh Hao Li
Hormone Research Laboratory
University of California
San Francisco, CA 94143, USA

Fore-Lien Huang
Department of Zoology
National Taiwan University
Taipei, Taiwan, ROC

Paul P. Hung
Genetic Division
Bethesda Research Laboratory, Inc.
P. O. Box 6009
Gaithersburg, MD 20877, USA

Lou-Sing Kan
Division of Biophysics
School of Hygiene and Public Health
Johns Hopkins University
Baltimore, MD 21205, USA

Hiroshi Kawauchi
School of Fisheries Sciences
Kitasato University
Kesen-gun, Iwate
022-01, JAPAN

Chung-Yen Lai
Roche Institute of Molecular
    Biology
Nutley, NJ 07110, USA

Chuan-Pu Lee
Department of Biochemistry
Wayne State University
School of Medicine
540 E. Canfield St.
Detroit, MI 48201, USA

Yuan Chuan Lee
Department of Biology
Johns Hopkins University
Baltimore, MD 21218, USA

Sidney Pestka
Roche Institute of Molecular
    Biology
Nutley, NJ 07110, USA

Shutsung Liao
Department of Biochemistry
Ben May Laboratory
University of Chicago
950 E. 59th Street, Box 424
Chicago, IL 60637, USA

Jong-Yau Lin
Institute of Biochemistry
College of Medicine
National Taiwan University
Taipei, Taiwan, ROC

Chen-Seng Liu
Institute of Biological Chemistry
Academia Sinica
Taipei, Taiwan, ROC

Teh-Yung Liu
Division of Biochemistry
    and Biophysics
Office of Biologics
8800 Rockville Pike
Bethesda, MD 20205, USA

Tung-Bin Lo
Institute of Biochemical Sciences
National Taiwan University
Taipei, Taiwan, ROC

Richard Mathies
Department of Chemistry
University of California
Berkeley, CA 94720 USA

Richard Perham
Department of Biochemistry
University of Cambridge
Tennis Court Road
Cambridge, England CB2N 1QW

Alan N. Schechter
Laboratory of Chemical Biology
National Institute of Arthritis,
    Digestive Diseases, and Kidney
National Institutes of Health
Bethesda, MD 20205, USA

Jordan Tang
Laboratory of Protein Research
Oklahoma Medical Research Foundation
Oklahoma City, OK 73104, USA

Kung-Tsung Wang
Institute of Biological Chemistry
Academia Sinica
Taipei, Taiwan, ROC

Ray J. Wu
Section of Biochemistry
Molecular and Cell Biology
Cornell University
Ithaca, NY 14853, USA

Haruaki Yajima
Faculty of Pharmaceutical Science
Kyoto University, Sakyo-Ku
Kyoto, JAPAN

Chen-Chang Yang
Institute of Molecular Biology
National Tsinhua University
Hsinchu, Taiwan, ROC

K. Yasunobu
Department of Biochemistry and
    Biophysics
University of Hawaii
Honolulu, Hawaii 96822, USA

Biochemical and Biophysical Studies
of Proteins and Nucleic Acids

BIOLOGICALLY ACTIVE PEPTIDES: AN OVERVIEW

CHOH HAO LI
Hormone Research Laboratory, University of California, San Francisco, California

In 1882, exactly 100 years ago, Curtius[1] obtained benzoyldiglycine by condensing hippuric acid with glycine. This achievement represents the first synthesis
of a peptide. We may take this Symposium to celebrate a century of peptide
chemistry.[2] Fifty-three years later, Harrington and Mead[3] described the
synthesis of glutathione, a tripeptide γ-glutamylcysteinylglycine. In 1953,
du Vigneaud et al.[4] reported the synthesis of nonapeptide oxytocin. In the same
year Bricas and Fromageot[5] published a comprehensive review on "Naturally
Occurring Peptides" and concluded that there are only *six* peptides with known
structures, namely glutathione, pteroyltriglutamic acid, polymyxin D, D-
glutamyl-polypeptide, β-alanyl-L-histidine (carnosine) and β-alanyl-L-1-
methylhistidine (anserine). There are now over 3,000 peptides and proteins with
known amino acid sequences.

It is generally agreed that substances with a molecular weight less than
10,000 are peptides. I would like to suggest here for convenience of discussion
the following classifications: peptides, molecular weight < 10,000; polypeptides,
molecular weight 10,000-25,000 and proteins, molecular weight >25,000.

The primary structures of all peptide, polypeptide and protein hormones from
the pituitary, pancreas and parathyroid glands have been known. In addition,
many peptide hormones from hypothalamus and gastrointestinal tissue have been
isolated and their structures elucidated. Some gastrointestinal peptides and
their biological functions are shown in Table 1. Biologically active peptides
have also been obtained from mammalian brains (Table 2). Indeed, biologically
active peptides occur in all living matter. Some examples for the universality
of biologically active peptides are given in Table 3. It is of particular
interest to note that frog skins[6] produce a number of peptides similar to that
from mammalian brain (Table 4).

As another example, an 11-amino acids neuropeptide from the freshwater
codenterate hydra (*Hydra attenuata*) has recently been isolated and sequenced
with the primary structure[7] pGlu-Pro-Pro-Gly-Gly-Ser-Lys-Val-Ile-Leu-Phe-OH.
It was named head activator because it controls head-specific growth and differentiation processes. A peptide of identical amino acid sequence to the
head activator has also been isolated from human hypothalamus, bovine hypothalamus and rat intestine.[8] The function of the "head activator" peptide in
mammals is yet to be investigated.

TABLE 1

SOME GASTROINTESTINAL PEPTIDES

Peptide	Amino Acid Residues	Action on Gastrointestinal Tract
Secretin	27	Increases pancreatic water and electrolyte secretion
Gastrin	17	Increases gastric acid secretion
Cholecystokinin-pancreozymin	33	Increases gall bladder contraction and pancreatic enzyme secretion
Vasoactive intestinal peptide	28	Secretin-like action and causes intestinal secretion
Gastric inhibitory polypeptide	43	Inhibits gastric secretion and emptying, increases insulin secretion
Motilin	22	Increases motor activity of gastrointestinal tract
Urogastrone	53	Inhibits gastric acid secretion
Caerulein	10	Like cholecystokinin but 10-fold more potent

It is important to obtain biologically active peptides from tissues of more than one species. These are *natural* analogs. The best case is the hormone calcitonin.[9] Salmon calcitonin is known to be more potent than the human hormone as a hypocalcemic agent. On the other hand, salmon calcitonin does not cause hypocalcemic effect in the salmon. In addition, bovine thyrotropin is three times more potent than the human hormone.[10] In opiate-receptor binding assay using rat brain membrane preparations, ostrich β-endorphin is seven times more active than the human homologue.[11]

The biological profile of a biologically active peptide isolated from one species is not necessarily identical to that from another species. We have

TABLE 2

BIOLOGICALLY ACTIVE PEPTIDES FROM MAMMALIAN BRAIN

Substance P	Oxytocin
Neurotensin	Vasopressin
Leu-enkephalin	Thyroliberin (TRH)
Met-enkephalin	Gonadoliberin (GnRH)
Cholecystokinin (CCK-8)	Somatostatin (GH-RIH)

```
                              5                    10
Human:    H-Ser-Tyr-Ser-Met-Glu-His-Phe-Arg-Trp-Gly-

Turkey:

Ostrich:


                             15                   20
         Lys-Pro-Val-Gly-Lys-Lys-Arg-Arg-Pro-Val-

                              Arg-Arg-Lys         Ile

                              Arg


                             25                   30
         Lys-Val-Tyr-Pro-Asn-Gly-Ala-Glu-Asp-Glu-

                                   Ser-Val

                                   Val-Gln-Glu


                             35                   39
         Ser-Ala-Glu-Ala-Phe-Pro-Leu-Glu-Phe-OH

         Glu-Gln-Ala-Ser-Tyr      Val

         Thr-Ser        Gly
```

Fig. 1. Amino acid sequence of ACTH from human, turkey and ostrich pituitary glands.

TABLE 5

RELATIVE STEROIDOGENIC POTENCY OF ACTH FROM

HUMAN, TURKEY AND OSTRICH PITUITARY GLANDS

ACTH	Corticosterone Production in Cortical Cells	Aldosterone Production in Capsular Cells
Human	1.0	1.0
Turkey	3.5	1.7
Ostrich	1.7	4.8

4

TABLE 3

UNIVERSALITY OF BIOLOGICALLY ACTIVE PEPTIDES (SOME EXAMPLES)

Peptide	Source	Number of AA Residues	Activity
Adipokinetic Hormone (AKH)	Locust	10	Fat mobilizing
Melittin	Bee venom	26	Erythrocyte lysis
Phalloidin	Mushroom	7	Poisonous
Muramyldipeptide	Mycobacterial cell wall	2	Immune response enhancing
Cyclosporin A	Fungus	11	Suppressing immune rejection

recently isolated corticotropin (ACTH) from turkey[12] and ostrich[13] pituitary glands and their primary structures are shown in Fig. 1. When compared with the human hormone, 7-12 amino acid residues have been replaced by different residues beyond residue position 14. Bioassay results reveal that the three ACTHs have different relative steroidogenic potency. As shown in Table 5, the turkey hormone is the most potent for corticosterone production whereas the ostrich hormone is the most potent for aldosterone production when compared with the human hormone.

TABLE 4

SOME BIOLOGICALLY ACTIVE PEPTIDES FROM FROG SKIN AND MAMMALIAN BRAIN

Skin (frog)	Brain (mammalian)
Dermorphin:	Enkephalin
Tyr-D-Ala-Phe-Gly-Tyr-Pro-Ser[7]-NH$_2$	
Bombesin:	Bombesin-like peptide
Pyr-Gln-Arg-Leu-Gly-Asn-Gln-Trp-Ala-Val-Gly-His-Leu-Met[14]-NH$_2$	
Xenopsin:	Neurotensin
Pyr-Gly-Lys-Arg-Pro-Trp-Ile-Leu[8]-OH	
Caerulein:	CCK-8,CCK-4,gastin-6
Pyr-Gln-Asp-Tyr(SO$_3$H)-Thr-Gly-Trp-Met-Asp-Phe[10]-NH$_2$	

TABLE 6

SYNTHESIS OF COMPLEX PEPTIDES WITH FULL BIOLOGICAL ACTIVITY

Synthetic Peptide	Number of Amino Acids	Methods Employed
Insulin (Human)	51	Solution
Gastrin (Human)	34	Solution
Pancreatic trypsin inhibitor (bovine)	56	Solid phase
β-Lipotropin (ovine)	91	Solid phase
Pancreatic ribonuclease A (bovine)	124	Solution

Remarkable advances have been made in the synthesis of biologically-active peptides since the first synthesis of the hormone oxytocin in 1953.[4] The introduction of the solid-phase method by Merrifield[14] in 1963 and repetitive methods in solution have made possible the total synthesis of some complex peptides in recent years. Each synthesis presents a formidable and unpredictable challenge. Only a few of these syntheses[15] have provided target peptides and polypeptides with full biological activity and proof of homogeneity by several chemical and physical characteristics (see Table 6).

Synthetic analogs have been shown to possess higher biological activity when compared with the parent molecule. For instance, Manning et al.[16] obtained a synthetic analog of arginine-vasopressin, [1-deamino, 4-D-valine, 8-D-arginine]-vasopressin, which possesses an anti-diuretic to pressor activity ratio 125,000:1 as compared with a ratio of 1:1 for the parent hormone. Arimura et al.[17] reported that des-$(Gly-NH_2^{10})$-[D-Leu^6]-gonadoliberin ethylamide elicited 54- to 82-fold greater in $vivo$ effects than gonadoliberin. Yamashiro et al.[18] synthesized two methionine-enkephalin analogs, [Thr^2, Thz^5]- and [D-Met^2, Thz^5]-enkephinamides, and showed that the two are 4.2-4.8 times more potent than morphine when injected intravenously whereas Met- or Leu-enkephalin does not exhibit analgesia by the mouse tail-flick assay. In the β-endorphin series, replacement of a single amino acid residue is position 27 by tryptophin, i.e., [Trp^{27}]-$β_h$-endorphin, causes nearly 4-fold increase of analgesic potency of the parent molecule.[19]

Recent studies reveal that natural or synthetic peptide segments of certain peptide and polypeptide hormones act as an inhibitor to parent hormones. A peptide with a sequence identical to residues 7-38 of the ACTH structure (see

6

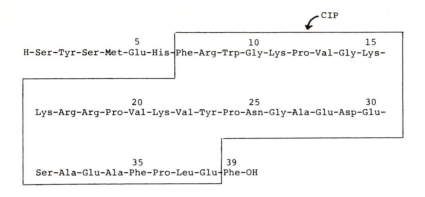

Fig. 2. Amino acid sequence of α_h-ACTH and CIP.

Fig. 2) has been isolated from human pituitary glands[20] and is capable of inhibiting ACTH-stimulated corticosterone production in the isolated rat adrenal cells. It was named corticotropin-inhibiting peptide (CIP). A segment of β-endorphin (see Fig. 3), β_c-EP-(6-31), has been synthesized[21] and shown to inhibit β-endorphin-induced analgesia in mice by the tail-flick test.[22]

The placental hormone human choriogonadotropin (HCG) and the pituitary hormones [lutropin (LH) and follitropin (FSH)] consist of two nonidentical subunits, α and β, which are held firmly together by noncovalent bonds. The

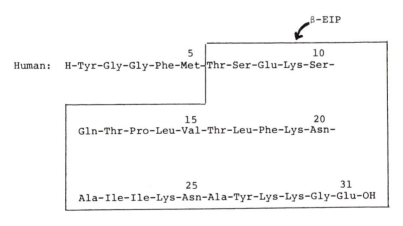

Fig. 3. Amino acid sequence of β_h-endorphin and β-EIP

isolated subunits exhibit minimal biological activity. In the Leydig cell *in vitro* assay, β-subunits of HCG and LH were able to inhibit the biological activity of human choriogonadotropin.[23]

These striking observations that a segment of a peptide or polypeptide hormone acts as an inhibitor to the parent hormone may have physiological significance in metabolic functions of the hormone.

Since 1956, a number of industrially-produced synthetic peptide pharmaceuticals have been introduced including oxytocin, ACTH-(1-24), TRH, salmon calcitonin and human ACTH (Table 7). In addition, lysine-vasopressin, pentagastrin, GnRH, human β-endorphin and thymosin α_1 have been produced in large scale for clinical studies (Table 8). Other biologically active peptides may have possible use in man. Some examples are given in Table 9.

One of the major problems for peptide pharmaceuticals is that most of them are not orally active. They are injectable drugs. In order to overcome this difficulty, synthetic analogs may be found to render them orally active. Alternatively, it may be possible to make biologically-active peptides orally active with artificial liposome encapsulation.

For large scale production of biologically active peptides by either the solid-phase or solution method, molecules with more than 30 amino acid residues will have difficulties in obtaining pure products. Fortunately, the revolutionary recombinant DNA technique to produce polypeptides and proteins in bacteria has now become available. By this technique, it is possible to produce industrially large quantities of human growth hormone (191 amino acid residues) and human leukocyte interferon (166 amino acid residues) for clinical studies. However, it should be noted that it is not possible to use the recombinant DNA technique to synthesize peptides containing uncommon amino acids such as D-amino acids, N-methyl amino acids, etc., as well as peptides

TABLE 7

INDUSTRIALLY-PRODUCED SYNTHETIC PEPTIDE PHARMACEUTICALS

Synthetic Peptide	Number of Amino Acids	Date of Introduction
Oxytocin	9	1956
ACTH-(1-24)	24	1964
Thyroliberin (TRH)	3	1974
Calcitonin (salmon)	32	1974
ACTH (human)	39	1977

8

TABLE 8

INDUSTRIALLY-PRODUCED SYNTHETIC PEPTIDES IN CLINICAL STUDIES

Synthetic Peptide	Number of Amino Acids	Date of Introduction
Lysine-vasopressin	9	1961
Pentagastrin	5	1961
Gonadoliberin (GnRH)	10	1975
β-Endorphin (human)	31	1980
Thymosin α_1	28	1980

with blocked N^{α}-NH_2 groups. In addition, new biologically active peptides, poly-peptides and proteins cannot be discovered by genetic engineering in bacteria or yeast and they have come from biochemical and chemical studies of biological materials.

REFERENCES

1. Curtius, T. and Prakt, J. (1882) Chem., 24, 239-240.
2. Wieland, T. (1981) in Perspectives in Peptide Chemistry, Eberle, A., et al., eds., Karger, Basd, pp. 1-13.
3. Harrington, C. R. and Mead, T. H. (1935) Biochem. J., 29, 1602-1611.
4. du Vigneaud, V., et al. (1953) J. Am. Chem. Soc., 75, 4879-4880.
5. Bricas, E. and Fromageot, C. (1953) Adv. Prot. Chem., 8, 1-125.

TABLE 9

PEPTIDES MAY HAVE POSSIBLE USE IN MAN

Peptide	Number of Amino Acids	Function
Monellin	94	Sweet tasting
Anthopleurin-A	49	Cardiotonic effect
MCD-peptide (mast cell degranulating)	22	Countering rheumatic diseases
Cyclosporin A	11	Suppressing immune rejection
GnRH antigonist	10	Antifertility
δ-Peptide	9	Sleep stimulator
Arg-vasopressin analog	9	Stimulate memory and learning
Bleomycin	6	Inhibit cancer growth

6. Erspamer, V. and Melchiorri, P. (1980) in Excerpta Medica Intern. Congr. Series, 495, 185-200.
7. Schaller, H. C. and Bodenmüller, H. (1981) Proc. Natl. Acad. Sci. USA, 78, 7000-7004
8. Bodenmüller, H. and Schaller, H. C. (1981) Nature, 293, 579-580.
9. Copp, D. H. (1976) in Handbook of Physiology, 7, 431-442.
10. Sairam, M. R. and Li, C. H. (1977) Can. J. Biochem., 55, 747-754.
11. Yamashiro, D., Hammonds, R. G., Jr., and Li, C. H. (1982) Int. J. Pept. Prot. Res., 19, 251-253.
12. Chang, W-C., Chung, D., and Li, C. H. (1980) Int. J. Pept. Prot. Res., 15, 261-270.
13. Li, C. H., et al. (1978) Biochem. Biophys. Res. Commun., 81, 900-906.
14. Merrifield, R. B. (1964) Biochemistry, 3, 1385-1390.
15. Meienhofer, J. (1980) in Excerpta Medica Int. Congr. Series, 495, 3-18.
16. Manning, M., et al. (1973) J. Med. Chem., 16, 975-978.
17. Arimura, A., et al. (1974) Endocrinology, 95, 1174-1177.
18. Yamashiro, D., Tseng, L-F., and Li, C. H. (1977) Biochem. Biophys. Res. Commun., 78, 1124-1129.
19. Li, C. H., Yamashiro, D., and Nicolas, P. (1982) Proc. Natl. Acad. Sci. USA, 79, 1042-1044.
20. Li, C. H., et al. (1978) Proc. Natl. Acad. Sci. USA, 75, 4306-4309.
21. Li, C. H., et al. (1978) Int. J. Pept. Prot. Res., 11, 154-158.
22. Lee, N. M., et al. (1980) Proc. Natl. Acad. Sci. USA, 77, 5525-5526.
23. Moudgal, N. R. and Li, C. H. (1982) Proc. Natl. Acad. Sci. USA, 79, 2500-2503.

Biochemical and Biophysical Studies of Proteins and Nucleic Acids,
Lo, Liu, and Li, eds.

THE HUMAN INTERFERONS - PROTEIN PURIFICATION, EXPRESSION IN BACTERIA, AND BIOLOGICAL PROPERTIES

SIDNEY PESTKA
Roche Institute of Molecular Biology
Nutley, New Jersey 07110

ABSTRACT

In the past few years, we have isolated and purified several of the human
interferons by developing and applying new techniques of high performance
liquid chromatography. Although the purification of fibroblast interferon
yielded a single protein, purification of interferon produced by leukocytes
yielded a family of proteins. Partial amino acid sequences for several of
these interferons were obtained and the sequence of two leukocyte interferon
species (α_2 and β_1) was determined with less than 200 µg of each. These amino
acid sequences provided essential information for expressing the interferons
in recombinant DNA vectors. In addition, the amino acid sequences of IFL-α_2
and β_1 yielded the surprising result that these interferons isolated from
leukocytes were ten amino acids shorter than expected from the DNA sequence of
the respective recombinants.

DNA recombinants for human leukocyte and fibroblast interferons were constructed, identified, and expressed in *Escherichia coli*. Recombinant leukocyte
interferons were purified with monoclonal antibodies to leukocyte interferon.
They are currently in clinical trial. With the availability of pure recombinant A interferon, we have been able to crystallize this interferon in several
forms. The interferon DNA recombinants were used to construct new synthetic
interferon species whose biological properties have proved to be quite interesting. Study of the biological properties of the wide variety of natural,
recombinant, and synthetic species suggest that various receptors for interferon may exist and trigger distinct activities through different mechanisms.

INTRODUCTION

Isaacs and Lindenmann[1,2] described interferon in 1957. A similar phenomenon
was described by Nagano and Kojima[3] in independent studies. Since that time,
many properties of interferon were established (see references 4-20 for reviews
and general compendia). It was established that interferon is generally not
virus specific. Interferon produced by a cell protects similar cells from
infection by many different viruses. Some viruses, however, are more sensitive

to inhibition by interferon than others. The original concept that the interferons are highly species specific in their activity has been modified over the past few years. Nevertheless, each interferon shows a characteristic species activity profile. For example, human leukocyte interferon exhibits a high degree of antiviral activity in bovine or porcine cell cultures, whereas fibroblast interferon is hardly active and immune interferon shows no activity on these cells.

The interferons are proteins with antiviral activities produced by various cells. Even within a species, several interferons are produced with biologically, chemically, and physically distinct antiviral activities. Until the last several years, the lack of reliable definitive data on composition and structure of the interferons was due to the failure to obtain significant amounts of any pure interferon for analysis. However, progress in the purification of several interferons has provided sufficient material for chemical analysis as well as biological and physical characterization. The amino acid compositions as well as amino acid sequences for several human interferons have been reported. Furthermore, with the successful construction and identification of DNA recombinants containing the human interferon-coding sequences and genes, knowledge of the primary structures of several of the human interferons has emerged rapidly.

ACTION AND ASSAY

Unlike antibodies that react with and neutralize viruses directly, interferon interacts with the host cell to induce resistance to viruses. Interferon exerts its effect by rendering cells incapable of supporting virus multiplication. Ordinarily lytic viruses cause infected cells to lyse and, in the process, produce large numbers of progeny viruses. Since interferon prevents both cellular lysis produced by these viruses as well as viral replication, interferon can be measured by determining either the extent of inhibition of cell lysis (cytopathic effect) or of new virus formation. The cytopathic effect inhibition assay[21-24] is a convenient assay based on this ability of interferon to make cells resistant to lysis or destruction by viruses. Detailed methodology for these and other assays have been described.[18,19] In practice, samples are titered relative to a standard preparation of interferon to determine the quantity of interferon present. The unit of activity is a value arbitrarily established by international agreement. Reference preparations of human interferon are listed in Table 1.

TABLE 1

CLASSES OF HUMAN INTERFERONS

Class	Abbreviations	Usual Production Conditions	Reference Preparations	Source
Fibroblast	IFF, IFN-β	fibroblasts and poly(I)·poly(C)	GO23-902-527	NIAID
Leukocyte	IFL, IFN-α	leukocytes and virus (NDV or Sendai)	69/19 (MRC Research Standard B)	NIBSC
			GO23-901-527 (NIH)	NIAID
Immune	IFI, IFN-γ	leukocytes and mitogen	in preparation	-

All the above international reference standards are recognized by the World Health Organization (WHO). The leukocyte interferon preparation GO23-901-527 is a reference preparation that has been standardized with reference to leukocyte interferon preparation 69/19 which is considered the WHO international standard. NIAID, Research Resources Branch, National Institute of Allergy and Infectious Diseases, National Institutes of Health, Bethesda, Maryland, USA; NIBSC, The International Laboratory for Biological Standards, National Institute for Biological Standards and Control, Holly Hill, Hampstead, London, UK.

CLASSES OF INTERFERON

Several different classes of interferons have now been identified. Although this discussion is limited to the human interferons, the general considerations are applicable to interferons from other sources. Interferons have been classified (Table 1) as fibroblast (F), leukocyte (L), and immune (I) classes with F, L, and I designating these individual classes, respectively.[25] In a nomenclature recently proposed,[26] these have been also termed α, β, and γ classes, respectively. The terms fibroblast, leukocyte, and immune have historical as well as biological significance. These are the species that are predominantly synthesized by fibroblasts, buffy coat leukocytes, and T-lymphocytes. The three classes represent protein molecules of different structure. As described below, many individual members of the leukocyte interferon class have been isolated and defined. So far, however, only one definitive member of the fibroblast interferon class has been isolated and characterized by several groups,[27-30] although there have been reports that fibroblast interferon may consist of multiple species.[31,32] Work is in progress on the purification of immune interferon in many laboratories.[33-38] The isolation and expression of an immune interferon-DNA recombinant[39,40] provides a step toward elucidation of the structure of a representative of this class. Results from several

14

laboratories suggest that the natural immune interferon class may consist of only a single species. The relationships of members within and between classes will be discussed in context with their structures.

PURIFICATION

Human leukocyte interferon. Since its discovery, many attempts were made to purify the interferons with little success until recent years. In fact, interferon used in experiments as well as in initial human clinical trials was essentially a crude protein fraction less than one percent of which by weight consisted of interferon. Because of the use of such crude interferon-containing material, it was not clear what activities of these preparations were, indeed, due inherently to the interferon present and what activities were due to the numerous other contaminating proteins. Accordingly, it was essential to obtain pure interferon to determine what activities were an inherent part of the interferon molecule as well as to establish their chemical composition and structure. As will be pointed out later, some of the activities ascribed to interferon turned out to be exhibited by the pure forms, but other properties were not demonstrable with pure preparations and were therefore, due in whole or in part to one of the contaminants in the crude preparations.

We began purification of interferon from human leukocytes in 1977. This interferon was produced by incubating human white blood cells with Newcastle disease virus or Sendai virus for 6 to 24 hours[41-44] by a combination of techniques that were previously reported.[45,46] The antiviral activity was found in the cell culture medium after overnight incubation of the leukocytes. We substituted milk casein for human or bovine serum in the culture medium as had been described.[45] The use of casein, a single protein, instead of serum which contains many different and uncharacterized proteins simplified the initial concentration and purification steps. We used leukocytes from normal donors as well as from patients with chronic myelogenous leukemia. These leukemic cells make large amounts of human leukocyte interferon when induced with Newcastle disease virus or Sendai virus.[42,47-49]

Because classical techniques for protein purification were not remarkably successful in isolation of the human interferons, we applied high-performance liquid chromatography (HPLC) to the purification of proteins. Udenfriend, Stein and co-workers[50-52] had developed sensitive fluorescent techniques for detection of amino acids and peptides and had achieved the separation of pep-tides by reverse-phase HPLC. However, separation of proteins had not yet been accomplished. In the early experiments, there was uniform failure to achieve

LEUKOCYTES, NDV
MEDIUM, CASEIN

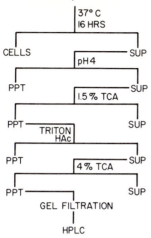

Fig. 1. Flow chart of initial steps in purification of leukocyte interferon. Cells and debris were removed by low speed centrifugation from the medium containing interferon. Casein was used as a serum substitute. By acidification of the medium to pH 4 with hydrochloric acid, the bulk of the casein, which precipitated, was removed from the interferon, which remained in solution. The interferon was concentrated by two steps, involving precipitation with trichloroacetic acid. The concentrated solution containing relatively crude interferon was separated into components of different sizes by gel filtration on Sephadex G-100 in the presence of 4 M urea. Details of these procedures have been described.[44,54,55]

any purification of proteins with reverse-phase HPLC because the interferon activity was continually lost. At that time, increasing ethanol concentration was used to elute proteins and this was tried for interferon and other proteins without success. It was necessary to use a more nonpolar solvent to elute interferon. Although there was initial hesitancy to use n-propanol above 20% (v/v) and other organic solvents because of the limited solubility of proteins in such solvents, it was found that n-propanol gradients effectively eluted interferon and other proteins without noticeable precipitation of the proteins at the concentrations employed.[53-56] Furthermore, by changing the pH of the elution buffer, a completely different separation could be achieved during elution of the same reverse-phase column with n-propanol. As subsequently demonstrated with fibroblast interferon,[30] a large number of different columns and solvent systems could be used to effect resolution of proteins. By applying normal-phase chromatography with a diol silica column in between the two reverse-phase columns, it was possible to use three sequential HPLC steps to purify human leukocyte interferon to homogeneity. Sufficient amounts were purified in high yield for initial chemical characterization of the protein and for determination of amino acid composition. The amino acid composition of the human leukocyte interferon species γ_2 was the first reported for any purified interferon.[54]

The initial steps included selective precipitations and gel filtration (Fig. 1) followed by HPLC. The HPLC steps were reverse-phase chromatography

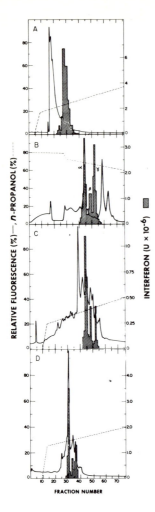

Fig. 2. High-performance liquid chromatography of interferon. (A) Chromatography on LiChrosorb RP-8 at pH 7.5. The
fluorometer scale was set to 100 and 2%
of the column effluent was directed to
the flourescamine monitoring system.
(B) Chromatography on LiChrosorb diol
at pH 7.5. The fluorometer scale was
30 and 2% of the column effluent was
directed to the fluorescamine monitoring
system. (C) Chromatography on LiChrosorb
RP-8 at pH 4.0. The fluorometer scale
was 1 and 5% of the column effluent was
directed to the fluorescamine monitoring
system. (D) Rechromatography on
LiChrosorb RP-8. The conditions were
similar to those of Step C. Several
preparations carried through Step C were
pooled (13 x 10^6 units) and applied to
the last column. The gradations on the
abscissa correspond to the end of the
fractions. (Taken from reference 54).

at pH 7.5 on LiChrosorb RP-8, normal partition chromatography on LiChrosorb
Diol, and reverse-phase chromatography at pH 4.0 on LiChrosorb RP-8. Gradients
of *n*-propanol were used for elution of interferon from these columns (Fig. 2).
The overall purification was about 80,000-fold and the specific activity of
pure interferon was 2-4 x 10^8 units/mg.[54] Interferon prepared by this procedure
yielded a single band of molecular weight 17,500 on polyacrylamide gel electrophoresis. The antiviral activity was associated with the single protein
band.[53] A summary of the purification of interferon prepared from leukocytes
obtained from patients with chronic myelogenous leukemia is shown in Table 2.

TABLE 2

PURIFICATION OF LEUKOCYTE INTERFERON FROM CML CELLS

Step	Units Recovered (x 10^{-6})	Protein Recovered (mg)	Specific Activity (units/mg)	Degree of Purification	Percentage Recovery
1. Incubation medium	800	20,000	4×10^4	1	100
2. pH 4 supernatant	800	4,000	2×10^5	5	100
3. 1.5% trichloroacetic acid precipitate	780	2,000	3.9×10^5	9.8	97
4. Triton X-100/acetic acid supernatant	760	510	1.5×10^6	37.5	95
5. 4% trichloroacetic acid precipitate	810	350	2.3×10^6	57.5	100
6. Sephadex G-100	660	130	5.1×10^6	128	82
7. LiChrosorb RP-8, pH 7.5	510	26	2×10^7	500	64
8. LiChrosorb Diol					
Peak α	149	5	3×10^7	750	
Peak β	148	3	5×10^7	1,250	
Peak γ	139	1.7	8×10^7	2,000	
Total	436				54
9. LiChrosorb RP-8, pH 4.0					
Peak α_1	9	0.035	2.6×10^8	6,500	
Peak α_2	26	0.065	4.0×10^8	10,000	
Peak β_2	30	0.075	4.0×10^8	10,000	
Peak β_3	13	0.032	4.0×10^8	10,000	
Peak γ_1	15	0.058	2.6×10^8	6,500	
Peak γ_2	31	0.077	4.0×10^8	10,000	
Peak γ_3	26	0.074	3.5×10^8	8,750	
Peak γ_4	3.5	0.010	3.5×10^8	8,750	
Peak γ_5	4.5	0.050	0.9×10^8	2,250	
Total	158.0				20

Activity was determined on bovine MDBK cells. Protein was measured with bovine serum albumin as standard. The data shown for α_1 and α_2 represent results after chromatography. Data taken from Rubinstein et al.[55]

Several reports had previously described high-performance liquid chromatography of proteins, mainly on ion exhcange and size exclusion columns.[57,58] However, those systems were either not commercially available or had a low capacity. With proper choice of eluent and pore size, octyl and octadecyl silica, could be used for high resolution reverse-phase HPLC of both peptides and proteins. Accordingly, with n-propanol as eluent, the use of LiChrosorb RP-8 (octyl silica) column for protein fractionation was a major factor in the success of the purification (Fig. 2, panels A, C, and D). In addition, LiChrosorb Diol, which is chemically similar to glycophase resins that have been used for exclusion chromatography of proteins, was introduced as a support for normal partition chromatography of proteins (Fig. 2B). High recoveries of interferon activity were obtained in each chromatographic step (Table 2), a requirement when small amounts of initial starting material are present. Although the initial experiments were performed with leukocytes from normal donors,[53,54] it was found that leukocytes from patients with chronic myelogenous leukemia (CML), who were undergoing leukapheresis to lower their peripheral white blood cell counts, were a rich source of interferon that appeared to be essentially identical to the human leukocyte interferon purified from leukocytes from normal donors.[49] Although the protein profiles were almost identical, the activity profiles showed that the amount of activity under peak γ was lower in preparations from leukemic cells compared to normal leukocytes.[49] However, even from normal leukocytes, the ratio of peaks α, β, and γ varied from one preparation to another. Human lymphoblastoid interferon produced by suspension cultures of Namalva cells was purified by a combination of immunoaffinity chromatography and other methods by Zoon et al.[59] The amino acid composition of human leukocyte interferon purified by HPLC as described above[54] shows similarity with human lymphoblastoid interferon[59] and one of the types of mouse L-cell interferon.[60]

During the purification of leukocyte interferon, it became evident that multiple species existed (Fig. 2; Table 2). Human leukocyte interferon is heterogeneous and several bands containing antiviral activity ranging in molecular weight from 15,000 to 21,000 are observed on SDS-polyacrylamide gel electrophoresis.[61] Heterogeneity of human leukocyte interferon has also been observed by isoelectric focusing[62] and several types of chromatographic procedures.[55,63-66] As noted above, our initial work with high-performance liquid chromatography revealed three major groups of interferon species which were labeled α, β, and γ according to their order of elution from a LiChrosorb Diol (polar-bonded phase) column.[54,55] These groups were further resolved into several homogeneous components. Although others had reported heterogeneity in crude human leukocyte interferon preparations,[63-66] this was not thought to be due to amino acid sequence heterogeneity. In fact, it had been reported by a number of groups that leukocyte interferon contained carbohydrate and that heterogeneity was due to differences in carbohydrate content of the protein.[67-70] Thus, the well-established heterogeneity of human leukocyte interferon was attributed to differences in the degree of glycosylation. However, five purified species of leukocyte interferon examined contained no detectable carbohydrate.[55] We still have not examined all the purified leukocyte interferon species isolated. However, the amino sugar content of each species analyzed (α_2, β_1, β_2, β_3, and γ_3) was determined to be much less than one residue of either glucosamine or galactosamine per molecule of interferon.[55] We therefore conclude that, contrary to general dogma, human leukocyte interferon is largely devoid of carbohydrate. Nevertheless, we would expect that some species exhibiting molecular weights significantly in excess of 19,000 would contain carbohydrate. In fact, it appears we have now detected one natural species of interferon that contains some carbohydrate (to be published). In general, these high molecular weight species represent a small fraction of the natural human leukocyte interferons. Because peptide mapping and sequencing revealed significant structural differences among the species, we concluded that leukocyte interferon represents a family of homologous proteins.

By analogous procedures, additional leukocyte interferon species were isolated from cultured myeloblasts.[71,72] Since our initial purification, other reports[50,73-75] have also described multiple species of leukocyte interferon. Allen and Fantes[73] also found no carbohydrate on the species of leukocyte interferon they purified. The dogma that interferons are glycoproteins has been so universal that leukocyte interferons are still called glycoproteins despite the data to the contrary. However, there appear to be some minor species of leukocyte interferon that are glycosylated.

Since only relatively small amounts of each species were isolated in these early experiments, it was difficult to obtain information about their amino acid sequence. The first determinations of the amino acid sequences of the amino-terminal ends of human interferons was by Zoon et al.[76] for a leukocyte interferon, and Knight et al.[77] for human fibroblast interferon. A few months later, Levy et al.[78] reported the amino-terminal sequence of another human leukocyte interferon; and Friesen et al.,[30] Stein et al.,[29] and Okamura et al.,[79] the amino-terminal sequence of human fibroblast interferon. It was clear that all the amino acid sequences obtained for human fibroblast interferon were

20

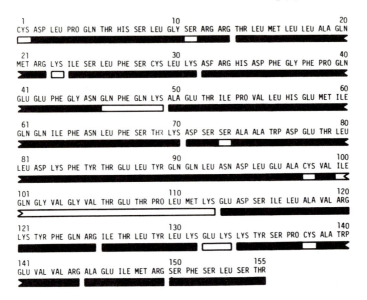

HUMAN LEUKOCYTE INTERFERON α_2,β_1

```
1                          10                          20
CYS ASP LEU PRO GLN THR HIS SER LEU GLY SER ARG ARG THR LEU MET LEU LEU ALA GLN

21                         30                          40
MET ARG LYS ILE SER LEU PHE SER CYS LEU LYS ASP ARG HIS ASP PHE GLY PHE PRO GLN

41                         50                          60
GLU GLU PHE GLY ASN GLN PHE GLN LYS ALA GLU THR ILE PRO VAL LEU HIS GLU MET ILE

61                         70                          80
GLN GLN ILE PHE ASN LEU PHE SER THR LYS ASP SER SER ALA ALA TRP ASP GLU THR LEU

81                         90                          100
LEU ASP LYS PHE TYR THR GLU LEU TYR GLN GLN LEU ASN ASP LEU GLU ALA CYS VAL ILE

101                        110                         120
GLN GLY VAL GLY VAL THR GLU THR PRO LEU MET LYS GLU ASP SER ILE LEU ALA VAL ARG

121                        130                         140
LYS TYR PHE GLN ARG ILE THR LEU TYR LEU LYS GLU LYS LYS TYR SER PRO CYS ALA TRP

141                        150         155
GLU VAL VAL ARG ALA GLU ILE MET ARG SER PHE SER LEU SER THR
```

Fig. 3. Sequence of human leukocyte interferon α_2 and β_1. The bars under the sequence represent the tryptic peptides isolated and sequenced. Solid bars represent the sequence determined by Edman degradation of the peptides. Unfilled bars represent sequences that were consistent with the composition of the peptides and/or sequence assignments derived from the DNA of the corresponding recombinant. Details of these experiments have been reported.[80,81] The molecular weight of the protein as shown is 18,035.

identical and that we had all purified and sequenced the same protein. However, it was striking that the amino-terminal sequence of the human leukocyte interferon species α_1, α_2, and β_1[78] was different from that of the leukocyte interferon reported by Zoon et al.[76] There were two differences in the sequence. Because the differences involved amino acids that would not be expected to cause errors, we felt that both sequences were correct and that the differences dramatically confirmed the fact that the leukocyte interferons consist of a family of closely related proteins. We subsequently isolated a human inter-

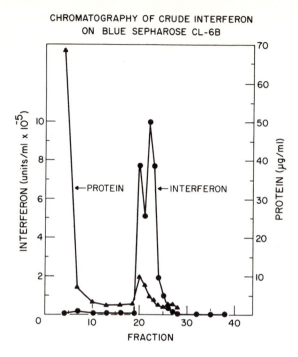

CHROMATOGRAPHY OF CRUDE INTERFERON
ON BLUE SEPHAROSE CL-6B

Fig. 4. Blue-Sepharose chromatography of crude human fibroblast interferon. Fractions 1–9 are eluted with 30% ethylene glycol, whereas the remaining fractions are eluted with 50% ethylene glycol. The volume per fraction is 25 ml. ●——●, interferon; ▲——▲, protein. (Taken from reference 89.)

feron DNA recombinant (see below), the coding sequence of which was virtually identical to the sequence of our pure proteins.

Thereafter, Levy et al.[80] and Shively et al.[81] reported amino acid sequences of three species of human leukocyte interferon. So far, the sequences of two leukocyte interferons were determined almost completely (Fig. 3). Additional sequences were reported by Zoon et al.[82] Allen and Fantes[73] reported the sequences of tryptic fragments obtained from a mixture of several leukocyte interferon species. All these sequences are sufficiently different to establish very clearly the concept of a family of closely related proteins. As will be described below, the first clone of human leukocyte interferon isolated in my laboratory was almost identical in sequence to human leukocyte interferons α_2 and β_1 (Fig. 3). The α_2 and β_1 species, however, were ten amino acids shorter

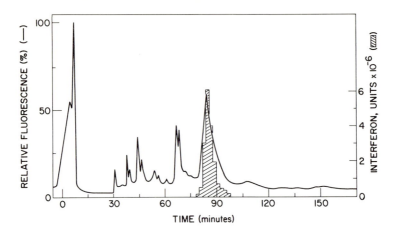

Fig. 5. High-performance liquid chromatography of human fibroblast interferon. About 10^6 units of interferon were applied to an RP-8 column. A step gradient of increasing n-propanol at constant pH (4.2) was pumped at 22 ml/hr. Interferon was eluted in a broad peak toward the end of the 32% n-propanol step. A portion (3%) of the column effluent was monitored with fluorescamine for determination of protein. (Taken from reference 29.)

than expected from the DNA sequence. It should be noted that during this time several groups[60,83-87] reported the purification of mouse interferons, and some amino acid sequences were reported.[88]

Human fibroblast interferon. Several laboratories reported the purification and partial structural analysis of human fibroblast interferon with the use of SDS-polyacrylamide gel electrophoresis as the last step in the purification.[27,28] To obtain a product salt- and solvent-free, we developed a simple two-step purification of fibroblast interferon.[29,89] The first step in the purification from the crude interferon-containing medium involved Blue-Sepharose chromatography, a procedure described previously.[65,77] The second step involved high-performance liquid chromatography on octyl silica. The amino acid composition and 19 residues of the amino-terminal sequence of human fibroblast interferon were determined.[29] The sequence was identical to the first 13 residues reported by Knight et al.[77] and Okamura et al.,[79] and the first ten residues reported by Friesen et al.[30].

The Blue-Sepharose step (Fig. 4) provides a high purification factor, but results in a dilute solution of interferon in 50% ethylene glycol. The final product is then obtained in concentrated form free of ethylene glycol and buffer salts by HPLC (Fig. 5). Because only volatile eluents are used, the interferon may be recovered salt- and solvent-free simply by evaporation. The specific activity of the purified protein is 3×10^8 units/mg of protein, similar to that of the purified human leukocyte interferon species. Comparison of the first 19 amino acids of human fibroblast interferon with mouse interferons A and B[88] reveals four identical amino acids at positions 3, 6, 11, and 18. It is therefore likely that these proteins are related and have a common ancestral origin. However, the amino-terminal sequences of human leukocyte and fibroblast interferons have very little homology.

Unlike human leukocyte interferon, which has been isolated as several different species,[54,55,71-75] only a single human fibroblast interferon species has so far been isolated. Although little or no homology exists between the amino-terminal sequence of human fibroblast and human leukocyte interferon through positions 1-19, because both interferons appear to bind to the same cell receptor[90] and have similar activities, some homology between the two in the area determining the active site is likely. Some homology is evident from comparison of the DNA sequences coding for these interferons.

CLONING AND EXPRESSION OF HUMAN INTERFERONS IN BACTERIA

Although a great deal of research has been directed toward an understanding of the molecular mechanism of interferon induction and activity, progress had been slow due largely to the small amounts of pure interferon protein available for study. Recombinant DNA technology offers an opportunity to produce a large amount of both the interferon proteins and the interferon gene sequences for use in studying many of these issues.

Isolating human interferon DNA sequences was a formidable task since it meant preparing DNA recombinants from cellular messenger RNA (mRNA) that was present at a low level. This had never been accomplished for a protein whose structure was unknown. In addition, to properly reconstruct the DNA recombinants for expression of natural interferon, it is useful to know the partial amino acid sequence of the proteins, particularly, the NH_2- and COOH-terminal ends. Without this information, synthesis of natural human interferon in bacteria would not have been possible. Thus, purification and determination of the structure of the human interferons (see above) assisted us and others in these efforts.

The approach that we took to isolate DNA recombinants containing the human interferon sequences involved a number of procedures. First, it was

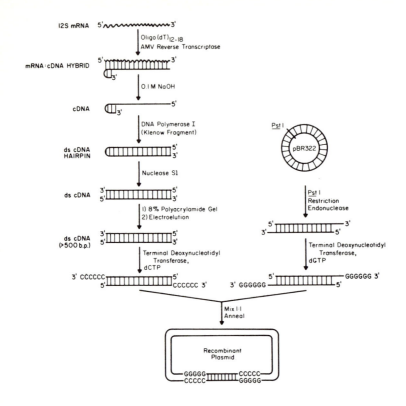

Fig. 6. Preparation of interferon DNA recombinants from messenger RNA.

necessary to isolate and measure the interferon mRNA. This was accomplished several years earlier when interferon mRNA was translated in cell-free extracts[91,92] and in frog oocytes.[93-96] The next step was to prepare sufficient mRNA from cells synthesizing interferon. This was accomplished for both fibroblast[93-98] and leukocyte interferon mRNA.[42,97,98]

A cDNA library was prepared[99-103] from mRNA isolated from human leukocytes synthesizing interferon (Fig. 6). Partially purified mRNA from induced cells was used as a template for cDNA synthesis.[99] The dC-tailed double-stranded DNA obtained was hybridized to dG-tailed pBR322 cleaved at the *PstI* restriction nuclease site. These were introduced into *Escherichia coli* χ1776 by transformation. Tetracycline-resistant ampicillin-sensitive transformants were

CLONE SCREENING: HYBRIDIZATION

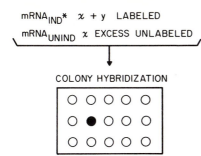

mRNA$_{IND}$* $x + y$ LABELED

mRNA$_{UNIND}$ x EXCESS UNLABELED

COLONY HYBRIDIZATION

Fig. 7. Schematic outline of hybridization procedure. In the presence of
excess unlabeled mRNA from uninduced
cells, [^{32}P]mRNA from induced cells
binds preferentially to sequences
specific for the induced cells. The
induced-specific sequences include
interferon sequences as well as others
that are induced concomitantly with
interferon.

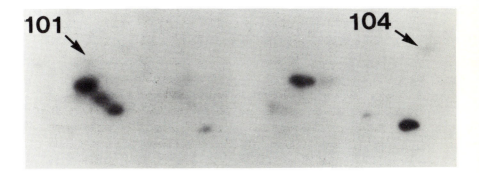

Fig. 8. Autoradiogram of colony hybridization procedure. The colonies could
be classified into three groups by intensity of autoradiograph: intense, less
intense (from barely visible to moderately intense), and no intensity (not
visible). Colonies 101 and 104 were shown subsequently to contain human fibroblast and leukocyte interferon sequences, respectively. Data from Maeda
et al.[99]

26

CLONE SCREENING: BINDING ACTIVE mRNA

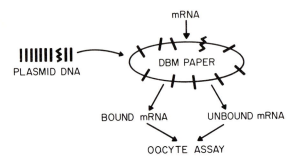

Fig. 9. Schematic illustration of screening of recombinants with DBM paper. Plasmid DNA isolated from single colonies or from several pooled colonies is covalently linked to DBM paper.[105] To detect a recombinant containing an interferon sequence (zigzag plasmid DNA in the illustration), mRNA prepared from cells synthesizing interferon is hybridized to the DNA on the filter as described.[105] If a recombinant plasmid bound to the filter contains a sequence homologous to the interferon mRNA, bound mRNA yields active interferon after translation in *Xenopus laevis* oocytes.

obtained that provided a large group of cDNA recombinants representing DNA copies of all the mRNAs in the cell.

It was then necessary to devise a procedure by which we could isolate and specifically identify recombinants containing interferon-specific DNA. We chose the following approach. First, we would screen all recombinants for their ability to bind to mRNA from cells synthesizing interferon (induced cells), but not to mRNA from cells not producing interferon (uninduced cells) (Fig. 7). Individual transformant colonies were screened by colony hybridization for the presence of induced-specific sequences with ^{32}P-labeled interferon mRNA as a probe. In the presence of excess mRNA from uninduced cells (Fig. 7), only cloned sequences that were representative of mRNA sequences existing in induced cells should be evident on hybridization. Two classes of positive colonies were seen (Fig. 8). One class, comprising 2%-3% of the total, hybridized very strongly to ^{32}P]mRNA and exhibited dense spots on autoradiography. The second class, which consisted of 10% of the total, hybridized to ^{32}P]mRNA in various degrees, but significantly less strongly than the first class. The remainder of the colonies were negative. By this initial screening, 80%-90% of the transformants were eliminated from further screening. This

TABLE 3

ASSAY OF PLASMID DNAs BY mRNA HYBRIDIZATION TO DNA-DBM PAPER

Plasmid	Type of Colony	Interferon Activity
Pools K1-K6	+++	<2 (<2); <2 (<2)
Pools K7-K9; K11; K12	+	<2 (<2); <2 (<2)
Pool K10	+	16 (<2) 128 (16); 128 (<2)
97	+	4 (<2)
98-100	+	<2 (<2)
101	+	48 (<2)
102	+	<2 (<2)
104	+	48 (<2)
105, 106	+	<2 (<2)

The symbols +++ and + are indications of the autoradiographic densities of the colonies by colony hybridization. Interferon activity is expressed relative to a reference standard for samples assayed by injection of eluted mRNA into *Xenopus laevis* oocytes. Values in parentheses are control values for mRNA bount to pBR322-DBM paper. The numbers represent the numbers of the pools or of the individual colonies. Data from Maeda et al.[99]

screening approach[99,104] allowed us to discard about 90% of the recombinants which did not contain an interferon DNA sequence.

We then needed a method to identify the recombinants containing the interferon DNA sequences among the remaining 10%. To do this, we pooled the recombinants in groups of ten and determined whether any pools would specifically bind interferon mRNA[99,105] (Fig. 9). Specifically, the positive colonies were examined for the presence of interferon-specific sequences by an assay which depends upon hybridization of interferon mRNA specifically to plasmid DNA. Plasmid DNA from ten recombinants was isolated, cleaved with *Hind*III restriction endonuclease, denatured, and covalently bound to diazobenzyloxylmethyl (DBM) paper.[99,105] One microgram of purified mRNA from induced cells was hybridized to each filter. Unhybridized mRNA was removed by washing. The specifically hybridized mRNA was eluted and translated in *Xenopus laevis* oocytes. To measure levels of mRNA nonspecifically bound to the filters, plasmids pBR322 and pβG1, which contains the gene for rabbit β-globin,[106] were used as negative controls. These procedures are schematically illustrated in Fig. 9. Once a positive pool was identified, in order to identify the specific interferon cDNA clone, the nine individual colonies were grown, the plasmid DNAs were prepared, and each individual DNA was examined by mRNA hybridization as

28

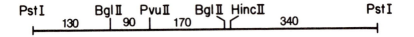

Fig. 10. Restriction map of the double-stranded cDNA insert of leukocyte interferon clone #104. The cDNA insert, which is illustrated as a horizontal line, can be removed from the pBR322 genome by digestion with *PstI* restriction endonuclease. Vertical lines represent sites on the DNA fragment where the restriction enzymes *PstI*, *BglII*, *PvuII*, and *HincII* produce double-stranded breaks. The numbers above the horizontal line indicate the approximate map distance in nucleotide base pairs between restriction sites. Data from Maeda et al.[99]

above (Fig. 9). Of the nine colonies in positive pool K10 (Table 3), two bound interferon mRNA well above background levels. By these procedures, two recombinants were identified: plasmid 104 (p104) contained most of the coding sequence for human leukocyte interferon, and plasmid 101 (p101) contained the sequence for human fibroblast interferon.[99] Within a few months of each other, several groups reported the cloning of human interferons.[99,107-112] In fact, Taniguchi et al.[107] was the first by several months to report the cloning of any interferon.

Figure 10 shows the restriction map of the inserted cDNA in plasmid 104. The DNA fragment between the two *BglII* sites was subcloned into phage M13[113] for DNA sequence analysis by the dideoxynucleotide termination method.[114,115] It and other fragments were also sequenced by the method of Maxam and Gilbert.[116] The DNA sequence was determined[99] and found to correspond to the available amino acid sequence determined for purified human leukocyte interferon.[80] The other clone, plasmid 101, was shown to contain the sequence for human fibroblast interferon. By both restriction mapping and DNA sequencing, it was found that plasmid 101 contains at least 90% of the coding sequence for human fibroblast interferon. Thus, the nucleotide sequences coding for human leukocyte and fibroblast interferons were identified. Paradoxically, the initial screening of the strong positive colonies demonstrated that they contained no interferon-specific sequences. We have not yet characterized these cDNA sequences. Thus, as noted above, comparison of DNA sequence data with amino acid sequence data showed conclusively that plasmid 104 contains part of the human leukocyte interferon sequence. DNA sequence data indicated that a termination triplet is located 15 codons from the right $Bg\ell$II site (Fig. 10). Therefore, of the 780-base pair insert of plasmid 104 (p104), a 430-base pair sequence which is located upstream from the termination codon codes for the interferon. Since the molecular weight of human leukocyte interferon is about 19,000, the 430-base pair sequence is too short to code for the total sequence of interferon. The DNA sequence data also showed that the insert in plasmid 104 does not have a sequence corresponding to the amino-terminal amino acids. Therefore, the cDNA insert in p104 contains the sequence for most of the amino acids, but not for the amino-terminal end of interferon. Comparison of the restriction map of the p104 insert with the data published by Nagata et al.[111] indicated that the p104 insert has one additional $Bg\ell$II site. Other differences as well are evident in the restriction map. These observations support the hypothesis that human leukocyte interferon mRNA consists of a population of several distinct species. The purification of multiple species of human leukocyte interferon by high-performance liquid chromatography agrees as well with these results.

Because p104 contains more than 80% of the sequence for human leukocyte interferon, this DNA was used as a probe for finding a full-length copy of the interferon cDNA sequence for expression of human leukocyte interferon in *Escherichia coli* (see below). In addition, this DNA (p104) was used to isolate human genomic DNA sequences of the human leukocyte interferons. So far, we have isolated at least 12 distinct human genomic sequences for human leukocyte interferon from human DNA[20,117] (Table 4). With the human fibroblast cDNA (p101), a single genomic clone from the human DNA library[118] was found corresponding to human fibroblast interferon.[117] None of these genomic clones was found to contain an intervening sequence as is present in many genes from higher organisms. Two of the genomic fragments (G8 and G93) contain two distinct tandem leukocyte interferon genes.

The coding regions of the leukocyte interferon genes that were isolated in our laboratory and others[99,111,117,119-128] have shown that the interferon genes comprise a family of homologous proteins. As shown in Fig. 11, these proteins are highly related to each other and yet differ from each other in amino acid sequences. Thus, it appears that the diversity in human leukocyte interferon is a result of distinct genes representing each expressed human leukocyte interferon sequence. The cloned human leukocyte A interferon sequence corresponds to our natural leukocyte interferons α_2 and β_1.

Expression of human leukocyte interferon in *E. coli*. As noted above, p104 contained most of the sequence for human leukocyte interferon, but did not

TABLE 4

FRAGMENTS OBTAINED BY CLEAVAGE OF THE GENOMIC DNAs WITH EcoRI ENZYME

Genomic DNAs	Class	EcoRI Fragments
G8	Leukocyte	3.3, 3.1, 2.9, 2.2, 1.4, 1.3
G48	Leukocyte	14, 2.2, 1.4, 0.6
G55	Leukocyte	7.5, 2.7, 2.3, 2.0, 1.7, 0.9, 0.7, 0.4, <0.1
G57, 73, 89	Leukocyte	9.0, 3.2, 1.9, 0.8, 0.7, 0.5, 0.4, 0.3
G68, 77A	Leukocyte	8.5, 4.0, 3.5, 2.5, 1.3, 0.4, 0.2
G76	Leukocyte	10, 2.2, 1.8, 1.7, 1.2
G77B	Leukocyte	9.0, 3.2, 2.3, 1.4, 0.7
G91	Leukocyte	6.5, 5.0, 2.7
G93	Leukocyte	7.4, 7.2, 2.6, 1.4, 0.1
G83	Fibroblast	6.0, 4.5, 2.8, 1.9, 1.2, 0.7, 0.6
	Immune	7.3, 3.3, 2.6, 1.0

The numbers indicate the size of the fragments in kilobases. The fragments hybridizing with the appropriate cDNA probes are underlined. Data from Maeda et al.[117,128] and from studies on the immune interferon genomic sequence (unpublished) data.

contain the coding sequence for the amino terminus of the protein. Accordingly, p104 was used to screen additional cDNA recombinants. A number of cDNA recombinants (A, B, C, D, E, F, G, H) which hybridized to the unique 260-base pair BgℓII restriction fragment of p104 were identified. Most contained interferon DNA sequences of sufficient size to code for the entire human leukocyte interferon protein. One (clone E) appeared to be a transcribed pseudogene since a single thymidylic acid residue was inserted in the coding region and since there were two in-phase termination codons.[123] The DNA sequences of these and other recombinants are shown in Fig. 12. Although there is great homology within the coding sequence, the homology among the sequences precipitously decreases to the 5'- and 3'-side of the coding sequences. It is likely that the sequence corresponding to IFLrL is also a pseudogene because this sequence contains an in-phase termination codon at positions S20 (Fig. 11) or 118-120 (Fig. 12).[128]

The first recombinant isolated in our laboratory was leukocyte A interferon;[99] the others were subsequently identified.[117,123] The entire PstI insert of the full-length leukocyte A interferon recombinant was sequenced and the results shown in Fig. 12.[119] Our protein sequence information (see above) permitted us to determine the correct translational reading frame and allowed

```
            S1                          S10                           S20        S23
A    MET ALA LEU THR PHE ALA LEU LEU VAL ALA LEU VAL LEU SER CYS SER SER VAL GLY
B                        TYR MET             VAL             TYR         PHE SER     LEU
C            SER     SER         MET     VAL             TYR         ILE             LEU
D        SER PRO             MET VAL VAL                                             LEU
F            SER     SER         MET     VAL             TYR         ILE             LEU
G*
H                PRO     SER MET MET         VAL             TYR         ILE         LEU
K        ARG SER SER     SER     MET VAL VAL                 TYR         ILE         LEU
L            SER     SER         MET VAL                     TYR         ILE END     LEU

     1                          10                          20
A    CYS ASP LEU PRO GLN THR HIS SER LEU GLY SER ARG ARG THR LEU MET LEU LEU ALA GLN MET ARG LYS ILE SER
B                                        ASN         ALA     ILE                     ARG
C                                        ASN         ALA     ILE     GLY         GLY ARG
D            GLU                 ASP ASN                                         SER ARG
F                                ASN         ALA     ILE                         GLY ARG
G*
H        ASN     SER             ASN ASN                             MET         ARG
H        ARG ASN                 ALA     ILE                         GLY ARG
K                        THR     ARG ASN     ALA     ILE     GLY     GLY ARG
L

                30                      40                      70
A    LEU PHE SER CYS LEU LYS ASP ARG HIS ASP PHE GLY PHE PRO GLN GLU GLU PHE --- GLY ASN GLN PHE GLN LYS
B    PRO                                     GLU                     ASP ASP LYS
C    PRO                             ARG ILE                         ASP
D    PRO SER             MET                                         ASP
F    PRO                                                             ASP
G                                            GLU                     ASP
H    PRO                         GLU     ARG             GLU         ASP HIS
K    PRO                                 ARG ILE                     ASP
L    PRO

                60                      70
A    ALA GLU THR ILE PRO VAL LEU HIS GLU MET ILE GLN GLN ILE PHE ASN LEU PHE SER THR LYS ASP SER SER ALA
B        GLN ALA         SER                 THR                             GLU
C        GLN ALA         SER                 THR
D        PRO ALA         SER             LEU                 THR
F        GLN ALA         SER                     THR
G        GLN ALA         SER                     THR
H        GLN ALA         SER             MET     THR                     ASN
K    THR GLN ALA         SER                     THR             GLU
L        GLN ALA         SER                     THR             GLU

                80                      90                      100
A    ALA TRP ASP GLU THR LEU LEU ASP LYS PHE TYR THR GLU LEU TYR GLN GLN LEU ASN ASP LEU GLU ALA CYS VAL
B        LEU                     GLU     ILE         ASP                             VAL LEU CYS
C            GLU GLN SER     GLU         SER
D                ASP                     CYS
F    THR     GLU GLN SER     GLU         SER         ASN                 MET
G    THR                                                                         MET
H                        GLU         ILE         PHE     MET
K        GLU GLN SER     GLU         SER     ILE
L        GLU GLN SER     GLU         SER ILE

                110                     120
A    ILE GLN GLY VAL GLY VAL THR GLU THR PRO LEU MET LYS GLU ASP SER ILE LEU ALA VAL ARG LYS TYR PHE GLN
B    ASP     GLU             ILE     SER         TYR
C            GLU             GLU             ASN
D    MET     GLU GLU ARG     GLY             ASN VAL             LYS             ARG
F            GLU             GLU             ASN VAL             LYS
G    MET     GLU             GLU ASP         ASN VAL     THR
H            GLU             GLU             ASN
K            GLU             GLU             ASN     PHE
L            GLU             GLU             ASN

                130                     140                     150
A    ARG ILE THR LEU TYR LEU LYS GLU LYS LYS TYR SER PRO CYS ALA TRP GLU VAL VAL ARG ALA GLU ILE MET ARG
B                    THR                             SER
C                    ILE     ARG
D                    THR
F                    THR
G                    THR
H                    MET
K                    MET
L                    ILE     ARG

                160                     166
A    SER PHE SER LEU SER THR ASN LEU GLN GLU SER LEU ARG SER LYS GLU
B                    ILE             LYS ARG     LYS
C    LEU     PHE                     LYS ARG     ARG     ASP
D    LEU                             ARG             ARG
F            LYS ILE PHE             ARG             ARG
G            ALA                     ARG             ARG
H        PHE                         LYS ARG     ARG
K        PHE                         LYS ARG     ARG     ASP
L    LEU     PHE                     LYS ARG     ARG     ASP
```

Fig. 11. Summary of amino acid sequences of human leukocyte interferons. The amino acid sequences of leukocyte interferons A, B, C, D, F, G, H, K, and L derived from the respective DNA sequences (see Fig. 12) are shown. The entire amino acid sequence of IFLrA is given. The corresponding residues of B, C, D, F, G, H, K, and L are shown only where they differ from that of the IFLrA sequence. A gap is introduced in the IFLrA sequence between residues 43 and 44 to provide homology with the other recombinants. Residues S1 to S23 represent the precursor peptide sequence; 1 to 166, the sequence of the mature interferons. The IFLrG sequence begins at residue 34 because only a partial length clone of this recombinant was isolated.

```
                                        50                                    100
A  TGAGCCTAAA CCTTAGGCTC ACCCATTTCA ACCAGTCTAG CAGCATCTGC AACATCTACA ATG GCCTTGA CCTTTGCTTT ACTGGTGGCC CTCCTGGTGC
B                          T  T CTC                              C T        T TA      A              AG
C            CAAGGT TAT CA CTC AGTAG CTA GCAATAT TG C ACATCC       C T     T      TA      G G
D    C G GTC AG        C  GA CC   A T      T   G                   C E          A T    G G
F                                                        ACATCC    C T     T                  G G
G
H      A GG TTCAGT T A C  TCA        C C      T GGG TCC            A   C       A A      GG
K   CT  AGCC C TGG TCAAGT TA  CACCTC GGTAG CTA GT ATAT TG C A ATCC   CG T   T      A T G G    A
L   CATAGGCCG AG CAAGGT TAT CA CTC AGTAG CTA GCAATAT TG C ACATCC    C T     T      TA      G G

                                        150                                   200
A  TCAGCTGCAA GTCAAGCTGC TCTGTGGGCT GTGATCTGCC TCAAACCCAC AGCCTGGGTA GCAGGAGGAC CTTGATGCTC CTGGCACAGA TGAGGAAAAT
B    A    TT A    C                       G T        A G       A          A    C A G
C    A  A CT T    C                       G          C  AT G       A    G A G A G
D                  C                     C G G           A A               A    C G
F    A  A CI T    C                       G          A   AT G       A          A G A G
G
H                        C              A  T         AA   A       T       A    A G
K    A  A C T     C                       G          C  AT G       A    G   A G A G
L    A  A C T A   C                       G        C  C  AT G       A    G A  A G A G

                                        250                                   300
A  CTCTCTTTTC TCCTGCTTGA AGGACAGACA TGACTTTGGA TTTCCCCAGG AGGAGTT--- TGGCAACCAG TTCCAAAAGG CTGAAACCAT CCCTGTCCTC
B    C          C              A  C         TGA  AT A       G       T
C    C  C            T CC  A C              TGA            G    C G   T
D    C C     TC   T                         TGA            G    CC G   T
F    C       C                     C   A     TGA            G    C G   T
G                                    T        TGA            G    C G   T
H    C          C                   A TGA     GA C G   T
K    C          C       A CA   C AG          TGA   C        G A  C G   T
L    C          C       T CC  A C           TGA            G A  C G   T

                                        350                                   400
A  CATGAGATGA TCCAGCAGAT CTTCAATCTC TTCAGCACAA AGGACTCATC TGCTGCTTGG GATGAGACCC TCCTAGACAA ATTCTACACT GAACTCTACC
B             C          C                                   AC G      A    T C      T
D    C        C          C    T C     A T          A       AC  G      G C       TA
F             C                                       A   AC  A     A  T    T        T T
G             C                                       A              A  T         T T
H      G      C                        A                          A      T      T
K             C                       G                       AC G      A    T C      T
L             C                       G                       AC  G     A    T C      A T

                                        450                                   500
A  AGCAGCTGAA TGACCTGGAA GCCTGTGTGA TACAGGGGGT GGGGGTGACA GAGACTCCCC TGATGAAGGA GGACTCCATT CTGGCTGTGA GGAAATACTT
B              T CTGTGTG AT   AA       T   T         T C       C
C    A             A          A  T   GA          T          T C
D      T           A       G  A A A   GG    A          T T        C T       A
F    A        C    A          C  T    A   T         T T        C T
G                  A G       A  T A   GA   C   T     T T    T  C   A       A
H    AA           A       A   T   GA            T          C
K    A            A       A   T   GA            T   T C
L    A            A       A   T   GA            T          C

                                        550                                   600
A  CCAAAGAATC ACTCTCTATC TGAAAGAGAA GAAATACAGC CCTTGTGCCT GGGAGGTTGT CAGAGCAGAA ATCATGAGAT CTTTTTCTTT GTCAACAAAC
B              A     C                                              C C      A   TC
C              T    A T   G                                         CC C G   T
D    G              C                                               CC C       A
F              T    C                          T                    C C      A   A TT
G    T      C    C              A                C C    A  G
H              T    TG                                              C C   T
K              T    A TG                                            C C   T
L              T    A T    G                                        CC C G   T

                                        650                                   700
A  TTGCAAGAAA GTTTAAGAAG TAAGGAATGA AAACTGGTTC AACATGGAAA TGATTTTCAT TGATTCGTAT GCCAGCTCAC CTTTTTATGA TCTGCCATTT
B       A     A GAG              GC    A       C            A C AA  C AG   TCT   AC  G  GT TGC C
C       A     A G                   C     CC G   C AA  C ATT T     AC  C G    GT CTTCCA
D       A     A G        A TT      C      A CA  CT    C A CA   G    GC  C    AT CTGTCA
F    T        A G  G       C GTT CA  CATG A  T GATC G AT GAC AATACA C  AGT CACN   C ATGAC   TCTG CA
G       A     A G  G               C        C A  CA   TT   AC C TGAG  TCTG CG
H       A     A G  G  T            T        C AA  C AT T    AC  C  T    T
K       A     A G  G  T            T        C A AA GC AT T    AC  C   GT CTTCCA
L       A     A G  G  T            C     CC G   C AA  C ATT T    AC  C  GT CTTCCA

                                        750                                   800
A  CAAAGACTCA TGTTTCTGCT ATGACCATGA CACGATTTAA ATCTTTTCAA ATGTTTTTAG GAGTATTAAT CAACATTGTA TTCAGCTCTT AAGGCACTAG
B  TTC A GA C CT G T CTG CCA AACCAT GCTATGAATT GAA CAAATG TGTCAAG GT TTTC GG G GTTA GCAAC A  CTG TCA GCT T TGG
C  TTC A GA T CAC     ATA CC G C C GTT A C     AT   CA    TGT   CAGC AGTGTAA GA AGTGTCGTGT A ACCTGTGC G CACTAGT
D  TTC A GA T CTCAC CCTG C ATAAC AT G  C  GCTG   AAAC G T T AC A TT A A     T  TT AC AT C A A  A T A   TTATT TT
F  TC AGACTC AT CTCCTA TAAC ACC C ATGAG G  TCAAAA TTT CA A C  TT C G G GTA AGGA ACA C A GTTTA C  GT CAGGC C
G  TC  ATA T T  AT CTGCTA TATC ATGAC TTGAG G  TCAAAA TTT CAAACG  TT C CACG GT A G ACACT CTTTAG  G C CAGGGACT
H            CT C A AAC CACCA  A T TGAATCA#  T TCCAAATG T T CAGG   TGT  AG G   T G GT T AC T TG AG GCACT G CC
K  TTC A GA T CAC     ATA CC GA GTT A C     AT  CCA    TGT   CAG  AGTGT A GA AGGATCGTGT    ACCTGTGC G CACTAGT
L  TTC A GA T CAC     ATA CC G C T GTT  A C     AT    CA   TGT   CAGC AGTGTAA GA AGTGTCGTGT  AACCTGTGC G CACTAGT

                                        850                                   900
A  TCCCTTACAG AGGACCATGC TGACTGATCC ATTATCTATT TAAATATTTT TAAAATATTA TTTATTTAAC TATTTATAAA ACAACTTATT TTTGTTCATA
B  CA TAGT CC TTACAG   A CC TGCTGAT GGATCTAT C ATCTAT A   T    C T A  TAG T A CTAC   GG GACTTAA    AG T  GT C
C  C TT ACAGA T AC AT CT GATG CTCTG T C  T   GTTTA A A   T TT  A T A  A A A TAG T   TC TGAG C GGT A  T
D  GTT A  T A C TCATG       ACCT T CA  TG GGT  G GT    AAAC ATGTTCC     A T ACTCA A  AAA  a
F  AGTCCTTTA CA TG CCA   CTGATAGA TC AT  C   TC   GA A T T TAT   A TA T G AT A ATT  TTTTG CCA  G AA AT
G  AGT   TT C  ATGATCAT GCTGAC   T   C TCTA   T TCG CA C TTG CG T  AC AT T   AT T  TT T ATG   CA GT
H  TTACAGAT    CC TTC A  T CC TT CA C A TA  TT A A     TTTAT     AC   TTA TA TA T TATTT  TA  G AA ATCAT
K  CT A AGA T AC AA
L  C TT A AGA T AC AT CT GATG CTCTG  C    T   GTTTA A A   T TT A T A A A TAG T   TC TGAG C GC T A   T

                                        950
A  TTACGTCATG TGCACCTTTG CACAGTGGTT AATGTAATAA AATATGTTCT TTGTATTTGG TAAAAAAAA
B  A  TTAT  T ATGTGAAC T TTTACA T  G  TGTG   C A AACATG   C TA ATT   TT TTTT
C  G GGT A        AACAA A A TGTTC TCA   T TAGCC   T    TAAT   CC T  CA  T   TTTT
D
F  GTGTA TT  A  TTG GT T  T TCAAAA T     T ATC T ATAT   AG  CAATA ATT ATTTTCTTT
G  TTTATG T AGTTT AGT TTGT GTTAA T  AAC A  T G T  G  GGTC  A AT       TTTGCT
H  GAGTAC T  T ACATTG GGT T AT  AACA   A GT CT TCATAT  AG CCAATA ATT A TTTCCTT
K
L  G GGT A   A  ATA    TT TTCAA A TT  CC AT T  T
```

Fig. 12. Summary of the nucleotide sequences of human leukocyte interferon DNA recombinants. The nucleotide sequences of DNA recombinants corresponding to leukocyte interferons A, B, C, D, F, G, H, K, and L are shown. The entire nucleotide sequence of the DNA corresponding to IFLrA is shown. The corresponding residues of B, C, D, F, G, H, K, and L are presented only where they differ from that of the A sequence. A gap of three nucleotides is introduced at positions 258-260 of the A sequence to provide homology with the other recombinants. The boxes designate the ATG-initiation and TGA-termination codons. Because only a partial length clone of the recombinant corresponding to IFLrG was isolated, the G sequence begins at residue 229. K and L are genomic clones; the rest are cDNA recombinants. The data summarize the sequences given in several reports.[99,119,123,128]

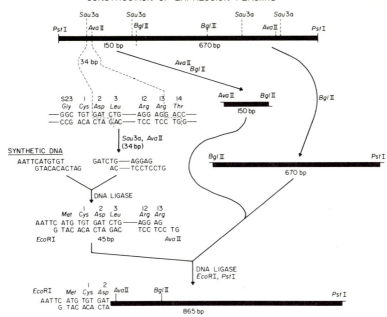

CONSTRUCTION OF EXPRESSION PLASMID

Fig. 13. Construction of expression plasmid for human leukocyte interferon. A cDNA recombinant containing the full coding sequence for leukocyte A interferon was isolated. The entire cDNA insert was removed by the restriction endonuclease *PstI*. The following fragments as shown were isolated: *Sau3a-AvaII* (34 bp); *AvaII-BglII* (150 bp); *BglII-PstI* (670 bp). The small *Sau3a-AvaII* fragment corresponds to the NH$_2$-terminus of mature leukocyte interferon containing the nucleotides coding for amino acids #2 to #13. A synthetic DNA linker was synthesized that restored the TGT codon for *Cys*#1, introduced a methionine initiation codon (ATG) just to the 5'-side of the *Cys* codon, and contained *EcoRI* and *Sau3a* restriction sites. This synthetic DNA fragment was ligated to the 34-bp *Sau3a-AvaII* fragment and the resultant fragment ligated to the remaining *AvaII-BglII* (150 bp) and *BglII-PstI* (670 bp) fragments as shown. The reconstructed 865-bp expression fragment (*EcoRI-PstI*) contained the coding sequence for a mature human leukocyte interferon with an additional methionine at the NH$_2$-terminal end. This was ligated to the tryptophan promoter-operator (Fig. 14). Data taken from Goeddel et al. [119]

us to predict the entire human leukocyte interferon sequence coded by this recombinant (IFLrA). The first ATG-translational initiation codon is found 60 nucleotides from the 5'-end of the sequence and is followed 188 codons later by a TGA-termination triplet; there are 335 untranslated nucleotides at the 3'-end followed by a poly(A) sequence. The signal peptide is 23 amino acids long. The 165-amino acid polypeptide constituting the mature human leukocyte interferon (IFLrA) has a calculated molecular weight of 19,241.

The precursor form of the recombinant (IFLr-preA) was expressed in *Escherichia coli* by ligating the 1,000-base pair *PstI* insert into the *PstI* site of plasmid pBR322 containing part of the *E. coli*-tryptophan operon extending from the promoter-operator region through the *trp* E gene, with deletion of a portion of the *trp* leader and *trp* E-coding regions.[119] The mature leukocyte interferon IFLrA was expressed directly by reconstruction of the recombinant. The leader sequence of IFLrA was removed and an ATG-translational initiation codon was placed immediately preceding the codon for amino acid 1 (cysteine) of the mature leukocyte interferon IFLrA (Fig. 13). Next, a 300-base pair *EcoRI* fragment was constructed which contains the *E. coli-trp* promoter-operator, and the *trp* leader ribosome-binding site, but stops short of the ATG sequence for initiation of translation of the leader peptide. This DNA fragment was then attached to the reconstructed human leukocyte interferon preceding the ATG codon (Fig. 14). Clone IFLrA-25, in which the *trp* promoter was inserted in the desired orientation, yielded high levels of activity: about 2×10^8 units of interferon per liter of culture. The IFLrA protein produced in *E. coli* behaves similar to authentic human leukocyte interferon. It is neutralized by antiserum to human leukocyte interferon. Furthermore, the IFLrA protein binds to monoclonal antibodies specific for human leukocyte interferon and has been purified to homogeneity with the use of monoclonal antibodies.[129,130]

Expression of human fibroblast interferon in *E. coli*. Similar to the studies above with human leukocyte interferon, a bacterial clone containing fibroblast interferon DNA was identified. Unlike the multiple species of leukocyte interferon, experiments indicated that there is only a single human fibroblast interferon gene corresponding to this recombinant.[117] To express mature human fibroblast interferon (IFF) directly in *E. coli*, a series of plasmids which placed the synthesis of the 166-amino acid polypeptide under *trp* promoter control (Fig. 14) were constructed.[112] These constructions utilized synthetic deoxyribooligonucleotides which primed the synthesis of double-stranded IFF DNA beginning precisely with the coding sequence of mature IFF. The fibroblast interferon produced in bacteria is similar to authentic human fibroblast interferon by several criteria. It contains the same amino acid sequences that have

REGULATION OF INTERFERON EXPRESSION

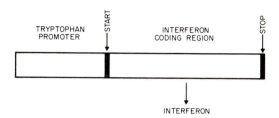

Fig. 14. Regulation of interferon expression in E. coli. The tryptophan promoter-operator containing the ribosome-binding site was ligated to the interferon-coding sequence that was reconstructed to express mature leukocyte interferon (Fig. 13). Leukocyte interferon (IFLrA) expression was thus regulated by the trp promoter-operator.

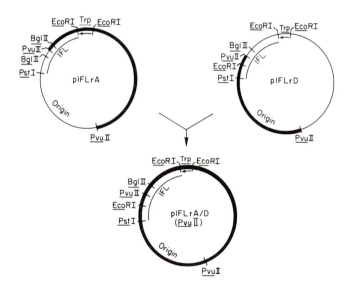

Fig. 15. Schematic illustration of construction of the pIFLrA/D (PvuII) recombinant. The interferon-coding sequence is illustrated by IFL. By ligating the appropriate PvuII fragments together, the IFLrA/D hybrid was constructed. By ligating the complementary fragments, the IFLrD/A hybrid was constructed

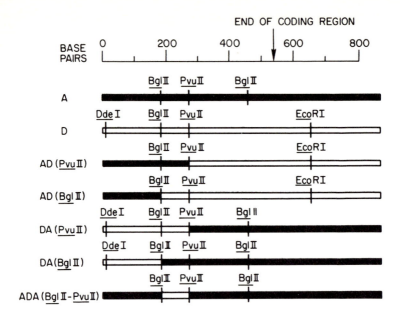

Fig. 16. Schematic illustration and restriction maps of the coding regions of the mature proteins of pIFLrA, pIFLrD and hybrids constructed from these two molecules.

been identified in human fibroblast interferon. It behaves like authentic fibroblast interferon when its antiviral activity is compared on human and bovine cells. However, fibroblast interferon made in bacteria is not glyco-sylated, whereas that made by human cells contains carbohydrate.[27,30]

   Hybrid interferons. With the isolation of multiple leukocyte cDNA and genomic recombinants, it was possible to consider the construction of hybrid molecules. Those recombinants that have restriction endonuclease sites in common can be simply cut and religated to the respective complementary segments. In this way, a series of hybrid leukocyte interferon molecules were pre-pared.[131-133] For example, IFLrA (165 amino acids) and IFLrD (166 amino acids) could be ligated at common $BglII$ and $PvuII$ restriction endonuclease sites giving rise to hybrid molecules (Figs. 15, 16) IFLrA$_{1-62}$/D$_{64-166}$ $(BglII)$, IFLrA$_{1-91}$/D$_{93-166}$ $(PvuII)$, IFLrD$_{1-63}$/A$_{63-165}$ $(BglII)$, IFLrD$_{1-92}$/A$_{92-165}$ $(PvuII)$ and IFLrA$_{1-62}$/D$_{64-92}$/A$_{92-165}$. The subscripts represent the amino acids of the respective mature interferons.

Since IFLrD exhibits greater antiviral activity on bovine MDBK cells than on human AG-1732 cells, whereas the activity of IFLrA is approximately similar on both cells, it was of interest to examine the antiviral activity of the hybrid molecules on these cells. The antiviral activity of the hybrid interferons on these cells seems to correlate with the amino-terminal half of the molecule: the antiviral activity of the IFLrA/D hybrids resembled the activity of IFLrA, whereas the antiviral activity of IFLrD/A hybrids resembled the activity of IFLrD. It is of interest that some of the hybrid molecules exhibit new activities not exhibited by either of the parent molecules. For example, the hybrid IFLrA/D ($Bg\ell$II) is fairly active on mouse cells, whereas neither of the parent molecules are. Changing of a single amino acid of IFLrA seems to be able to alter the activity of the molecule. Because an astronomical number of new interferon molecules can be generated by making hybrids, amino acid replacements, and other alterations, it may be possible to find molecules with potent antiviral activity but devoid of antiproliferative activity or certain side effects; or to tailor the molecule to exhibit the properties desired, such as high antitumor activity.

MONOCLONAL ANTIBODIES

Antibodies have proven to be invaluable for characterization, quantitative analysis, and purification of proteins. Until recently, quantitation of interferon could only be determined by relatively complex and time-consuming antiviral assays. In addition, purification of the interferons was still quite difficult. Our success in the purification of the human leukocyte interferons made it feasible to attempt the production of interferon-specific monoclonal antibodies by the technique of Köhler and Milstein.[134] Accordingly, with our purified human leukocyte interferons, we obtained 13 monoclonal antibodies to human leukocyte interferon.[135] Secher and Burke also reported a monoclonal antibody to human leukocyte interferon.[136] The individual monoclonal antibodies exhibit different patterns of binding to purified leukocyte interferon species that are consistent with the structural multiplicity of the human leukocyte interferons. These antibodies have been useful for the purification of the interferons as well as for rapid and quantitative determination of interferon with very high sensitivity by radio- and enzyme-immunoassays.[137] In these assays, they did not cross-react with human fibroblast or immune interferons.

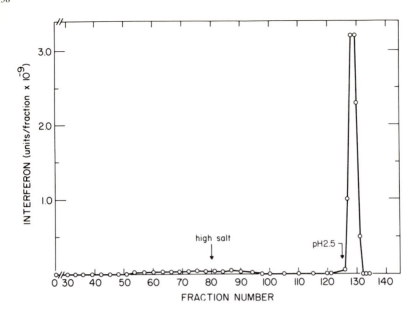

Fig. 17. Purification of interferon by monoclonal antibody immunoabsorbant column chromatography. A suspension of crude *E. coli* extract prepared from bacteria synthesizing IFLrA was loaded onto an immunoabsorbant column (monoclonal antibody LI-8 attached to Affi-Gel 10). The column was then washed sequentially with 25 mM Tris·HCl, 0.01% thiodiglycol, 10 μM phenylmethylsulfonyl fluoride (fractions 30-80), 0.5 M NaCl, 25 mM Tris·HCl, pH 7.5, 0.2% Triton X-100 (fractions 81-116), 0.15 M NaCl, 0.1% Triton X-100 (fractions 117-124), and then eluted with 0.2 N acetic acid, 0.15 M NaCl, 0.1% Triton X-100, pH 2.5 (fractions 125-140). Almost all the proteins applied to the column were found in the flow-through fraction. Interferon was measured by radioimmunoassay. One radioimmunoassay unit corresponds to about 0.7 antiviral unit. Data taken from Staehelin et al.[129]

Purification of recombinant human leukocyte interferon. Monoclonal antibodies to the human leukocyte interferons were used to purify recombinant human leukocyte A interferon (IFLrA) produced in bacteria.[129,130] *E. coli* containing the human leukocyte interferon, IFLrA, were broken. Unbroken cells and cellular debris were removed by centrifugation. The IFLrA and soluble bacterial proteins remain in the cell lysate. Nucleic acids (DNA and RNA) that make the lysate viscous and thus difficult to handle easily are precipitated by combination with polymin P. The soluble proteins remaining in the lysate can then be concentrated, or without concentration, passed directly through a column containing a monoclonal antibody to human leukocyte interferon (Fig. 17). The

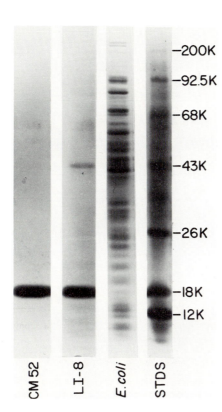

Fig. 18. Sodium dodecyl-sulfate polyacrylamide gel electrophoresis (SDS-
PAGE) of purified interferon. The gel
was stained with 2.5% Coomassie Brilliant Blue. Approximately 20 µg of
interferon purified by monoclonal antibody column LI-8 were subjected to
electrophoresis. Stds, represents the
standard molecular weight markers; E.
coli, represents total proteins from
Escherichia coli S30; LI-8, represents
the fraction after passage through the
monoclonal antibody column; and CM-52,
represents the purified interferon after
passage through the CM-52 column. Data
from Staehelin et al.[129,130]

HUMAN FIBROBLAST INTERFERON GENE

```
                                                                    ↓
AAT CGT AAA GAA GGA CAT CTC ATA TAA ATA GGC CAT ACC CAT GGA GAA AGG ACA TTC TAA CTG CAA CCT TTC GA/
                           ‾‾‾‾‾‾‾‾‾‾‾
                              -100
                                                                  S1
                                                                  Met Thr Asn Lys Cys Leu Leu Gln
GCC TTT GCT CTG GCA CAA CAG GTA GTA GGC GAC ACT GTT CGT GTT GTC AAC ATG ACC AAC AAG TGT CTC CTC CAA
                                                                  ‾‾‾
Ile Ala Leu Leu Leu Cys Phe Phe Thr Thr Ala Leu Ser Met Ser Tyr Asn Leu Leu Gly Phe Leu Gln Arg Ser
ATT GCT CTC CTG TTG TGC TTC TTC ACT ACA GCT CTT TCC ATG AGC TAC AAC TTG CTT GGA TTC CTA CAA AGA AGC
                                                    1   ‾‾‾
                                                        25
Ser Asn Phe Gln Cys Gln Lys Leu Leu Trp Gln Leu Asn Gly Arg Leu Glu Tyr Cys Leu Lys Asp Arg Met Asn
AGC AAT TTT CAG TGT CAG AAG CTC CTG TGG CAA TTG AAT GGG AGG CTT GAA TAC TGC CTC AAG GAC AGG ATG AAC
100
                                      50
Phe Asp Ile Pro Glu Glu Ile Lys Gln Leu Gln Gln Phe Gln Lys Glu Asp Ala Ala Leu Thr Ile Tyr Glu Met
TTT GAC ATC CCT GAG GAG ATT AAG CAG CTG CAG CAG TTC CAG AAG GAG GAC GCC GCA TTG ACC ATC TAT GAG ATG
                            200
                                         75
Leu Gln Asn Ile Phe Ala Ile Phe Arg Gln Asp Ser Ser Ser Thr Gly Trp Asn Glu Thr Ile Val Glu Asn Leu
CTC CAG AAC ATC TTT GCT ATT TTC AGA CAA GAT TCA TCT AGC ACT GGC TGG AAT GAG ACT ATT GTT GAG AAC CTC
                                                          ‾‾‾‾‾‾‾‾‾‾‾
                                                      300
Leu Ala Asn Val Tyr His Gln Ile Asn His Leu Lys Thr Val Leu Glu Glu Lys Leu Glu Lys Glu Asp Phe Thr
CTG GCT AAT GTC TAT CAT CAG ATA AAC CAT CTG AAG ACA GTC CTG GAA GAA AAA CTG GAG AAA GAA GAT TTC ACC
Arg Gly Lys Leu Met Ser Ser Leu His Leu Lys Arg Tyr Tyr Gly Arg Ile Leu His Tyr Leu Lys Ala Lys Glu
AGG GGA AAA CTC ATG AGC AGT CTG CAC CTG AAA AGA TAT TAT GGG AGG ATT CTG CAT TAC CTG AAG GCC AAG GAG
400
                        150
Tyr Ser His Cys Ala Trp Thr Ile Val Arg Val Glu Ile Leu Arg Asn Phe Tyr Phe Ile Asn Arg Leu Thr Gly
TAC AGT CAC TGT GCC TGG ACC ATA GTC AGA GTG GAA ATC CTA AGG AAC TTT TAC TTC ATT AAC AGA CTT ACA GGT
          166                           500
Tyr Leu Arg Asn END
TAC CTC CGA AAC TGA AGA TCT CCT AGC CTG TGC CTC TGG GAC TGG ACA ATT GCT TCA AGC ATT CTT CAA CCA GCA
              ‾‾‾                                 600
GAT GCT GTT TAA GTG ACT GAT GGC TAA TGT ACT GCA TAT GAA AGG ACA CTA GAA GAT TTT GAA ATT TTT ATT AAA
                                                                                    ↓
TTA TGA GTT ATT TTT ATT TAT TTA AAT TTT ATT TTG GAA AAT AAA TTA TTT TTG GTG CAA AAG TCA ACA TGG CAG
700                                        ‾‾‾‾‾‾‾
```

Fig. 19. DNA sequence of fibroblast interferon gene. The nucleotide sequence of the fibroblast interferon gene is shown. The amino acid sequence predicted from the DNA sequence is also shown. The Asn-Glu-Thr sequence outlined is the likely glycosylation site. Underlined areas include the Goldberg-Hogness box (TATAAATA) and the polyadenylation consensus sequence (AATAAA). The *Met* (S1) initiation codon, *Met*#1 codon, and the termination codon TGA are also outlined by boxes. Vertical arrows indicate the putative cap and polyadenylation sites. The sequence of the precursor peptide is designated by S1 to S21. Data from Maeda et al.[117,128]

antibodies bind only the interferon; all the other components and proteins pass through the column. After washing the column, the IFLrA bound to the column is removed by elution with an acidic solution.[129,130] A virtually pure interferon solution is eluted from the monoclonal antibody column (Fig. 18). The column is then washed and neutralized so that it can be used repeatedly. The interferon solution is concentrated by passage over a column of carboxymethyl-cellulose.[129,130] The activity of the purified interferon made in bacteria is comparable to that of the same human leukocyte interferon species synthesized by human cells.

TABLE 5

AMINO ACID COMPOSITION OF HUMAN RECOMBINANT INTERFERONS

	IFLrA	IFLrB	IFLrC	IFLrD	IFLrF	IFLrH	IFLrK	IFLrL	IFF	IFI
ASN	4	4	6	6	6	9	5	6	12	10
ASP	8	12	8	11	7	7	7	8	5	10
THR	10	6	7	9	8	8	8	8	7	5
SER	14	15	14	13	14	13	13	13	9	11
GLN	12	13	14	10	14	13	13	14	11	9
GLU	14	15	15	15	15	15	17	15	13	9
PRO	5	4	5	6	5	4	5	5	1	2
GLY	5	2	5	3	5	2	3	4	6	5
ALA	8	9	9	9	9	9	9	9	6	8
CYS	4	4	4	5	4	4	4	4	3	2
VAL	7	7	7	7	8	7	7	7	5	8
MET	5	4	4	6	5	9	5	4	4	8
ILE	8	10	10	7	9	7	8	11	11	4
LEU	21	22	20	22	18	17	19	19	24	7
TYR	5	5	4	4	3	4	4	4	10	10
PHE	10	10	9	8	11	12	12	9	9	5
HIS	3	3	3	3	3	3	4	3	5	2
LYS	11	10	7	8	10	10	8	7	11	20
ARG	9	10	13	12	10	11	13	14	11	8
TRP	2	1	2	2	2	2	2	2	3	1
Term.	UGA	UGA	UGA	UAA	UGA	UGA	UGA	UGA	UGA	UAA
Total	165	166	166	166	166	166	166	166	166	146
Molecular Weight	19,219	19,472	19,384	19,392	19,308	19,719	19,683	19,497	20,004	17,126

The amino acid compositions and molecular weights of recombinant leukocyte interferons IFLrA through IFLrK are taken from the corresponding sequences reported.[117,119,123,128] The composition and molecular weight of fibroblast interferon (IFF) is taken from the sequences shown for the cDNA[107-110,112] and genomic[117] recombinants.

The composition and molecular weight of immune interferon (IFI) is taken from the cDNA recombinants.[39,40] The molecular weights given in the table are for the primary polypeptide chains only.

HUMAN IMMUNE INTERFERON cDNA

```
                 S1                              S10                            S20
                 Met Lys Tyr Thr Ser Tyr Ile Leu Ala Phe Gln Leu Cys Ile Val Leu Gly Ser Leu Gly
AC TTC TTT GGC TTA ATT CTC TCG GAA ACG ATG AAA TAT ACA AGT TAT ATC TTG GCT TTT CAG CTC TGC ATC GTT TTG GGT TCT CTT GGC

 1                          10                         20                                      30
Cys Tyr Cys Gln Asp Pro Tyr Val Lys Glu Ala Glu Asn Leu Lys Lys Tyr Phe Asn Ala Gly His Ser Asp Val Ala Asp Asn Gly Thr
TGT TAC TGC CAG GAC CCA TAT GTA AAA GAA GCA GAA AAC CTT AAG AAA TAT TTT AAT GCA GGT CAT TCA GAT GTA GCG GAT AAT GGA ACT
                                             100

         40                           50                              60
Leu Phe Leu Gly Ile Leu Lys Asn Trp Lys Glu Glu Ser Asp Arg Lys Ile Met Gln Ser Gln Ile Val Ser Phe Tyr Phe Lys Leu Phe
CTT TTC TTA GGC ATT TTG AAG AAT TGG AAA GAG GAG AGT GAC AGA AAA ATA ATG CAG AGC CAA ATT GTC TCC TTT TAC TTC AAA CTT TTT
                                                   200

             70                         80                                90
Lys Asn Phe Lys Asp Asp Gln Ser Ile Gln Lys Ser Val Glu Thr Ile Lys Glu Asp Met Asn Val Lys Phe Phe Asn Ser Asn Lys Lys
AAA CGA GAT GAC TTC GAA AAG CTG ACT AAT TAT TCG GTA ACT GAC TTG AAT GTC CAA CGC AAA GCA ATA CAT GAA CTC ATC CAA GTG ATG
                                                         400

             100                          110                               120
Lys Arg Asp Asp Phe Glu Lys Leu Thr Asn Tyr Ser Val Thr Asp Leu Asn Val Gln Arg Lys Ala Ile His Glu Leu Ile Gln Val Met
AAA CGA GAT GAC TTC GAA AAG CTG ACT AAT TAT TCG GTA ACT GAC TTG AAT GTC CAA CGC AAA GCA ATA CAT GAA CTC ATC CAA GTG ATG
                                                         400

             130                          140                  146
Ala Glu Leu Ser Pro Ala Ala Lys Thr Gly Lys Arg Lys Arg Ser Gln Met Leu Phe Arg Gly Arg Arg Ala Ser Gln END
GCT GAA CTG TCG CCA GCA GCT AAA ACA GGG AAG CGA AAA AGG AGT CAG ATG CTG TTT CGA GGT CGA AGA GCA TCC CAG TAA TGG TTG TCC
                                                                                                        500

TGC CTG CAA TAT TTG AAT TTT AAT TCT ATT TAT TAA TAT TTA ACA TTA TTT ATA TGG GGA ATA TAT TTT TAG ACT CAT CAA TCA
                                                                                                              600

AAT AAG TAT TTA TAA TAG CAA CTT TTG TGT AAT GAA AAT GAA TAT CTA TTA ATA TAT GTA TTA TTT ATA ATT CCT ATA TCC TGT GAC TGT

CTC ACT TAA TCC TTT GTT TTC TGA CTA ATT AGG CAA GGC TAT GTG ATT ACA AGG CTT TAT CTC AGG GCC CAA CTA GGC AGC CAA CCT AAG
   700

CAA GAT CCC ATG GGT TGT GTG TTT ATT TCA CTT GAT GAT ACA ATG AAC ACT TAT AAG TGA AGT GAT ACT ATC CAG TTA CTG CCG GTT TGA
                800

AAA TAT GCC TGC AAT CTG AGC CAG TGC TTT AAT GGC ATG TCA GAC AGA ACT TGA ATG TGT CAG GTG ACC CTG ATG AAA ACA TAG CAT CTC
                900

AGG AGA TTT CAT GCC TGG TGC TTC CAA ATA TTG TTG ACA ACT GTG ACT GTA CCC AAA TGG AAA GTA ACT CAT TTG TTA AAA TTA TCA ATA
                                                  1000

TCT AAT ATA TAT GAA TAA AGT G
                        1072
```

Fig. 20. Nucleotide sequence of an immune interferon cDNA recombinant. The amino acid sequence of immune interferon predicted from the DNA sequence is shown. Two potential glycosylation sites are outlined in boxes: Asn-Gly-Thr and Asn-Tyr-Ser at positions 28 and 100, respectively. The codons representing the initiation *Met* (S1), *Cys*#1 (the beginning of the mature interferon), and the termination signal TAA are also outlined by boxes. The sequence of the precursor peptide is designated by S1 to S20. Data taken from Devos et al.[40]

Since repeated use of the monoclonal antibody columns is possible, these affinity columns provide a convenient method for preparing homogeneous human leukocyte interferon from bacterial fermentations. Interferon prepared by modifications of these procedures is now being used in clinical trials in humans. Several biological activities of this purified recombinant interferon have been determined. The recombinant IFLrA exhibits antiviral activity and antiproliferative activity comparable to crude and purified natural leukocyte interferons. IFLrA also stimulates natural killer cell activity. With the

HUMAN RECOMBINANT LEUKOCYTE K INTERFERON (IFLrK)

AG AAA GCA AAA ACA GAC ATA GAA AGT AAA ACT AGG CAT TTA GAA AAT GGA AAT TAG TAT GTT CAC TAT TTA AGA CCT
                                                                                              -100

ATG CAC AGA GCA AAG TCT CCA GAA AAC CTA GAG CCA CTG GTT CAA GTT ACC CAC CTC AGG TAG CCT AGT GAT ATT

              S1   S10   S20
              Met Ala Arg Ser Phe Ser Leu Leu Met Val Val Leu Val Leu Ser Tyr Lys Ser Ile Cys Ser
TGC AAA ATC CCA ATG GCC CGG TCC TTT TCT TTA CTG ATG GTC GTG CTG GTA CTC AGC TAC AAA TCC ATC TGC TCT

   S23 1
Leu Gly Cys Asp Leu Pro Gln Thr His Ser Leu Arg Asn Arg Arg Ala Leu Ile Leu Leu Ala Gln Met Gly Arg
CTG GGC TGT GAT CTG CCT CAG ACC CAC AGC CTG CGT AAT AGG AGG GCC TTG ATA CTC CTG GCA CAA ATG GGA AGA
                                                             100

   25
Ile Ser Pro Phe Ser Cys Leu Lys Asp Arg His Glu Phe Arg Phe Pro Glu Glu Glu Phe Asp Gly His Gln Phe
ATC TCT CCT TTC TCC TGC TTG AAG GAC AGA CAT GAA TTC AGA TTC CCA GAG GAG GAG TTT GAT GGC CAC CAG TTC
                                                                                     200

   50
Gln Lys Thr Gln Ala Ile Ser Val Leu His Glu Met Ile Gln Gln Thr Phe Asn Leu Phe Ser Thr Glu Asp Ser
CAG AAG ACT CAA GCC ATC TCT GTC CTC CAT GAG ATG ATC CAG CAG ACC TTC AAT CTC TTC AGC ACA GAG GAC TCA

   75
Ser Ala Ala Trp Glu Gln Ser Leu Leu Glu Lys Phe Ser Thr Glu Leu Tyr Gln Gln Leu Asn Asp Leu Glu Ala
TCT GCT GCT TGG GAA CAG AGC CTC CTA GAA AAA TTT TCC ACT GAA CTT TAC CAG CAA CTG AAT GAC CTG GAA GCA
                                          300

   100
Cys Val Ile Gln Glu Val Gly Val Glu Glu Thr Pro Leu Met Asn Glu Asp Phe Ile Leu Ala Val Arg Lys Tyr
TGT GTG ATA CAG GAG GTT GGG GTG GAA GAG ACT CCC CTG ATG AAT GAG GAC TTC ATC CTG GCT GTG AGG AAA TAC
                                                        400

   125
Phe Gln Arg Ile Thr Leu Tyr Leu Met Glu Lys Lys Tyr Ser Pro Cys Ala Trp Glu Val Val Arg Ala Glu Ile
TTC CAA AGA ATC ACT CTT TAT CTA ATG GAG AAG AAA TAC AGC CCT TGT GCC TGG GAG GTT GTC AGA GCA GAA ATC
                                                                                 500

   150   166
Met Arg Ser Phe Ser Phe Ser Thr Asn Leu Gln Lys Ser Arg Leu Arg Arg Lys Asp END
ATG AGA TCC TTC TCT TTT TCA ACA AAC TTG CAA AAA AGA TTA AGG AGG AAG GAT TGA AAA CTG GTT CAT CAT GGA

AAT GAT TCT CAT TGA CTA ATG CAT CAT CTC ACA CTT TCA TGA GTT CTT CCA TTT CAA AGA CTC ACT TCT ATA ACC
                600

ACC ACA AGT TGA ATC AAA ATT TCC AAA TGT TTT CAG GAG TGT TAA GAA GCA TCG TGT TTA CCT GTG CAG GCA CTA
                                               700

GTC CTT TAC AGA TGA CCA

Fig. 21. DNA sequence of leukocyte interferon gene IFLrK. The nucleotide
sequence of the leukocyte interferon gene IFLrK as well as the amino acid
sequence predicted from the DNA are shown. The underlined area represents the
Goldberg-Hogness box (TATTTA). The codons representing the initiation Met (S1),
Cys#1, and the termination signal (TGA) are outlined by boxes. The sequence
of the precursor peptide is designated by residues S1 to S23. Data from
Maeda et al.[117,128]

eventual availability of large amounts of homogeneous IFLrA, extensive clinical
trials, biological studies, and determination of its structure are achievable.

STRUCTURE AND AMINO ACID COMPOSITION OF THE INTERFERONS

The primary sequences of several recombinant and natural interferons were
determined. A summary of their amino acid sequences are shown in Figs. 3, 11,
19, 20, and 21. The amino acid compositions of these interferons are summarized in Table 5 as well as their molecular weights.

44

RECOMBINANT HUMAN LEUKOCYTE A INTERFERON (IFLrA)

Fig. 22. Schematic structure of human leukocyte A interferon.

It is evident from the sequences of the interferons that all the recombinant
leukocyte interferons (except for IFLrA) have a predicted amino acid sequence
of 166 amino acids from the cysteine amino terminus to the first in phase
termination codon. IFLrA contains 165 amino acids (Fig. 11). In addition,
fibroblast interferon (Fig. 19) consists of 166 amino acids as predicted from
the DNA sequence of the fibroblast interferon gene[117] or cDNA
recombinants.[107-110,112] However, the primary sequence of immune interferon
(IFI) consists of 146 amino acids (Fig. 20). Neither the fibroblast (Fig. 19)
nor the leukocyte (Fig. 21) interferon genes contain any intervening sequence.
As noted above, fibroblast interferon produced by human cells contains carbohydrate and the protein contains the sequence Asn-Glu-Thr (Fig. 19). None of
the leukocyte interferon sequences (A, B, C, D, F, G, K, or L) except for IFLrH

contains an asparagine glycosylation site. In the case of IFLrH, there are two potential glycosylation sequences: Asn[2]-Leu-Ser and Asn[72]-Ser-Ser. Thus, although no carbohydrate was detected in the purified human leukocyte interferon species examined,[55,73] it is conceivable that some minor species of leukocyte interferon may be glycosylated. The nucleotide sequence of a human immune interferon cDNA recombinant and its corresponding protein is shown in Fig. 20. There are two likely glycosylation sites, one at position 28 and the other at position 100.[39,40]

When the amino acid sequence of the recombinant leukocyte A interferon (IFLrA) was compared to the amino acid sequence of the corresponding protein produced by human buffy coat cells, it was surprising to find that the native interferon was 10 amino acids shorter than IFLrA at the COOH-terminal end.[80] Apparently, there is some processing event during secretion or during production of these interferons.

Disulfide linkages in IFLrA were determined to reside between Cys#1-Cys#98 and Cys#29-Cys#138.[138] It has been suggested[139] that only the Cys#29-Cys#138 disulfide is necessary for activity. A schematic structure of the interferon molecule is represented in Fig. 22. Since these molecules are strongly hydrophobic, it is likely that the protein contains a central hydrophobic core.

A number of recombinant and natural leukocyte interferons have been isolated. Table 6 presents a summary correlating the various interferon preparations with one another.

BIOLOGICAL PROPERTIES OF HUMAN INTERFERONS

Numerous cellular-regulatory properties have been attributed to interferon. The ability to inhibit viral growth was the first recognized biological activity of interferon and was the property for which the protein was named. Later the ability to inhibit cell multiplication, i.e., the antiproliferative activity, was detected as well as immunomodulatory activities.[140-142] Effective administration of interferon as a therapeutic agent requires a thorough knowledge of its biological properties. With pure human leukocyte interferons from both natural sources as well as from bacteria expressing the human interferons, we have examined several of the biological activities attributed to interferon. We have found that the growth-inhibitory activity of leukocyte interferon purifies with the antiviral activity.[143,144] In addition, enhancement of natural killer cell activity is a property of homogeneous leukocyte interferon.[145,146] Furthermore, homogeneous human leukocyte interferon inhibits hemopoietic colony formation of human bone marrow cells.[147]

TABLE 6

CORRELATION OF NATURAL AND RECOMBINANT HUMAN LEUKOCYTE INTERFERONS

Recombinant	Natural	
	Leukocytes	Myeloblasts
A	α_1	(a)
B	α_2	b_1
C	β_1	b_2
D	β_2	b_3
F	β_3	c_1
G	γ_1	c_2
H	γ_2	c_3
K	γ_3	d_1
L	γ_4	d_2
	γ_5	

The recombinant leukocyte interferon sequences are from Maeda et al.[117,128] and Goeddel et al.[119,123] Natural leukocyte interferons from leukocytes and myeloblasts are from Rubinstein et al.[55] and from Hobbs and Pestka,[72] respectively. As indicated, recombinant A corresponds to natural α_2 and β_1; and recombinant D, to natural γ_3. None of the interferons from myeloblasts have been definitively related to the other natural or recombinant species. The "a" species from myeloblasts have not been purified to homogeneity.

The antiviral activity of the purified interferons was determined on human and bovine cells.[55] The results of these studies are summarized in Table 7. The relative antiviral activity of the various human leukocyte interferon species differs on human cells. Species γ_3 and γ_5 exhibit low activity on human cells, but are active on bovine cells. The human leukocyte interferons are all active on bovine cells, but human fibroblast and immune interferons exhibit low and no activity, respectively, on these cells. All these leukocyte interferons inhibited growth of lymphoblastoid cells, but relative to the anti-viral activity they showed a wide range of activities (Table 7). It should be noted that the antiproliferative activity is a function of the human cell line tested. Species that inhibit lymphoblastoid cell growth may be relatively inactive in blocking growth of human astrocytoma cells.[148] Purified preparations of human fibroblast interferon also exhibited antiproliferative activity.[17,149] Similar results were reported with purified mouse interferons.[83,85]

TABLE 7

BIOLOGICAL ACTIVITIES OF HUMAN INTERFERON SPECIES

Species	Relative anti-viral activity		Relative anti-proliferative activity	NK activity
	Human Cells	Bovine Cells		
Fibroblast	3	0.01	6	++
Leukocyte				
α_1	3	3	13	+
α_2	3	4	7	+
β_1	4	3	+	+
β_2	2	4	62	++
β_3	3	4	3	++
γ_1	2	3	8	++
γ_2	2	4	29	+
γ_3	0.2	4	10	+
γ_4	4	4	3	+
γ_5	0.02	1	46	+

The relative antiviral and antiproliferative activities for the various interferons are shown. Multiplication of the antiviral activity by 10^8 yields the specific activity of that interferon in units/mg.[29,55] The antiproliferative activity is given relative to a fixed number of antiviral units on human Daudi lymphoblastoid cells.[144] The natural killer cell activity is taken from the data of Ortaldo, Mantovani, Hobbs, Rubinstein, Pestka, and Herberman (in preparation.

Because these two biological activities purify together and because the nomogeneous species exhibit both activities, it is evident that the antiproliferative and the antiviral activities are both intrinsic properties of these human leukocyte interferons. However, the ratio of the growth-inhibitory to antiviral activity was not constant among the species. Similar results were found with interferons purified from a myelocytic cell line and with recombinant interferons.[17,150] All species of human leukocyte interferon stimulate natural killer cell activity to one degree or another.[145,146,151,152] Therefore, antiproliferative activity and stimulation of killer cell activity as well as antiviral activity are intrinsic activities of the pure interferon molecules.

These studies as well as the considerations in preceding sections allow us to conclude that the individual members of the family of human leukocyte interferon proteins differ in structure and in their biological activities. Our

results indicate that the molecules exhibiting maximal antiviral effects differ
from those exhibiting maximal antiproliferative activities. Thus, it will be
necessary to evaluate all the biological activities with each of the natural
and recombinant leukocyte interferons. Furthermore, it should be possible to
modulate selectively these functions by appropriate alterations in their
structures.

Activities of hybrid interferons, specific molecular activity, and multiple interferon receptors

In comparing the antiviral and antiproliferative activities of several hybrid
interferons on cells from several species, paradoxical results were obtained.[133]
To enable a clear comparison of different interferons on different cell types,
we thus defined "specific molecular activity" as the molecules/cell necessary
to yield a specific effect. This parameter is more meaningful than determing
effects of interferon on the basis of antiviral units because standards are un-
necessary and one can accurately measure effects across species where standards
are inaccurate and their use uncertain. Antiviral assays relative to a heterologous reference standard can be quite confusing and provide erroneous conclusions[133] when cells of different species are compared, because the titer of
the standards vary from cell to cell independently of the interferon being
assayed. By measuring the antiviral activity of IFLrA, IFLrD, and several
hybrid interferon molecules on cells from various species, the specific molecular activities were determined for these interferons (Table 8).

IFLrA was highly active on human, bovine and feline cells, but active on
mouse and rat cells only at over 6,000- and 120,000-fold, respectively, the
molecular/cellular ratio active on human cells. IFLrD was most active on bovine
and feline cells, slightly active on human cells, but much less active on mouse
and rat cells. IFLrA/D (*Bgl*) was highly active on all but the rat cell lines
examined; IFLrA/D (*Pvu*) was active on all except the rat cell lines, but
exhibited less specific molecular activity than IFLrA/D (*Bgl*) on most of these
cells. IFLrD/A (*Bgl*) exhibited antiviral activity on bovine and feline cells,
minimal activity on human cells, but none on mouse or rat cells. IFLrD/A (*Pvu*)
was most active on bovine and feline cells, slightly on human cells, barely on
mouse cells, but not on rat cells. IFLrA/D/A was most active on bovine and
feline cells, about 1/20th as active on human cells, and about 1/600th as active
on mouse cells.

The most active species inhibited the cytopathic effect 50% at a concentration of 500 to 5,000 molecules/cell. Within a factor of ten, all the species
showed a similar specific molecular activity on bovine cells; and most were

TABLE 8

MOLECULES/CELL FOR 50% INHIBITION OF VIRAL CYTOPATHIC
EFFECT FOR VARIOUS CELL LINES

Interferon	Cell Line				
	AG-1732	MDBK	L-Cells	Felung	Rat C6
A	4,900	1,100	3.1×10^7	8,700	6.1×10^8
D	360,000	6,600	4.3×10^6	3,700	1.1×10^6
A/D ($Bg\ell$)	5,500	1,200	3,300	450	210,000
A/D (Pvu)	4,100	2,400	48,000	1,800	$>1.7 \times 10^6$
D/A ($Bg\ell$)	1.5×10^6	1,800	$>7.2 \times 10^7$	42,000	$>3.8 \times 10^7$
D/A (Pvu)	590,000	5,400	7.1×10^6	12,000	$>3.0 \times 10^7$
A/D/A	36,000	2,000	1.1×10^6	1,500	$>1.8 \times 10^7$

The molecular weights of the various species are as follows: IFLrA, 19,219; IFLrD, 19,392; IFLrA/D ($Bg\ell$), 19,395; IFLrA/D (Pvu), 19,427; IFLrD/A ($Bg\ell$), 19,216; IFLrD/A (Pvu), 19,184; IFLrA/D/A, 19,205. The number of cells in the microtiter well was determined at the time of addition of cells: AG-1732, 1.2×10^4; MDBK, 2.7×10^4; L-cells, 2.5×10^4; Felung, 1.8×10^4; Rat C6, 5.4×10^4. Data from Rehberg et al.[133]

significantly active on feline cells. The seven interferons exhibited a range of specific molecular activities on human cells over a factor of 100-fold. Only IFLrA/D ($Bg\ell$) and IFLrA/D (Pvu) showed activity on mouse cells at 200,000 molecules per cell or less.

The data suggest that decrease in antiviral activity on human cells appears to be mediated by determinants on the amino-terminal portion of the leukocyte interferon molecule. It is clear that the hybrids designated $IFLrD_{1-63}/A_{63-165}$ ($Bg\ell$II) and $IFLrD_{1-92}/A_{92-165}$ (PvuII) as well as IFLrD show a much lower activity on human AG-1732 cells than does IFLrA (Table 8). When the three amino acid substitutions were introduced into the native IFLrA molecule (as accomplished by construction of the A/D/A hybrid), the specific molecular activity on human cells was reduced about ten-fold. A/D hybrids of both types have approximately equal activity on both human, bovine and feline cell lines. Thus, the amino-terminal half of IFLrD is responsible for the structural changes that result in a marked reduction in antiviral activity on human cells.

When the antiviral activity of these interferons was examined on L-cells, greatest activity was found in all species containing some or all of the carboxy-terminal 103 amino acids of IFLrD: IFLrD, $IFLrA_{1-62}/D_{64-166}$ ($Bg\ell$II),

TABLE 9

ANTIPROLIFERATIVE ACTIVITY: CONCENTRATIONS OF
INTERFERON AT 50% INHIBITION OF CONTROL CELL GROWTH

Interferon	Molecules/cell	Antiviral units/ml
A	13,000	23
D	450,000	12
A/D ($Bg\ell$)	9,300	21
A/D (Pvu)	30,000	62
D/A ($Bg\ell$)	910,000	4
D/A (Pvu)	4,200,000	73
A/D/A	60,000	17

The values in the table represent the concentrations of the interferons that inhibited growth of Daudi cells 50%. Antiviral units were determined on AG-1732 cells. Molecules/cell were calculated from the cell number present at the time of addition of the interferons. Data taken from Rehberg et al.[133]

IFLrA$_{1-91}$/D$_{93-166}$ (PvuII), IFLrD$_{1-92}$/A$_{92-165}$ (PvuII), and

IFLrA$_{1-62}$/D$_{64-92}$/A$_{92-166}$ ($Bg\ell$II-PvuII). It appears that when the three amino acids specific for IFLrD from the central portion of the molecule (IFLrA/D/A) were introduced into IFLrA, the molecule showed slight activity on mouse cells. Similarly, when the 74 carboxy-terminal amino acids of IFLrD replaced those of IFLrA (IFLrA/D, Pvu), the molecule exhibited significant activity on mouse cells. When both these alterations were introduced, the molecule, IFLrA/D ($Bg\ell$), was very active on mouse cells.

Because IFLrA exhibited antigrowth activity,[150] the antiproliferative activities of these interferons were determined on Daudi lymphoblastoid cells. The quantity of each interferon that produced 50% inhibition of growth of the cells is summarized in Table 9. It is ev ident that the ratio of the quantity required for 50% inhibition of viral cytopathic effect (antiviral activity) to the quantity required for antigrowth activity (Table 10) varied over a range of 12-fold. In the case of IFLrA/D (Pvu) and IFLrD/A (Pvu), about 7- to 8-fold more interferon was required for 50% inhibition of growth than 50% inhibition of viral cytopathic effect.

This disparity between antiviral and antiproliferative activities indicates that these two activities are mediated by different mechanisms. A similar suggestion was made previously when it was observed that the individual puri-fied human leukocyte interferons exhibited different ratios of antiviral to

TABLE 10

RATIO OF SPECIFIC MOLECULAR ACTIVITIES OF INTERFERONS FOR
ANTIPROLIFERATIVE AND ANTIVIRAL ACTIVITY ON HUMAN CELLS

Interferon	AP/AV
A	2.7
D	1.3
A/D ($Bg\ell$)	1.7
A/D (Pvu)	7.3
D/A ($Bg\ell$)	0.6
D/A (Pvu)	7.1
A/D/A	1.7

The specific molecular antiproliferative activity (molecules/cell) for inhibition of growth of human lymphoblastoid Daudi cells (Table 8), AP, was divided
by the specific molecular antiviral activity on human AG-1732 fibroblasts, AV.
The ratio AP/AV is given in the table. Data from Rehberg et al.[133]

antiproliferative activity.[143-144] Since the effects of interferon are mediated
by a number of different mechanisms,[150,153-158] it is not surprising that some
of the effects can be dissociated. Thus, after comparison of the antiproliferative and antiviral activities of human leukocyte, fibroblast, and immune interferons, Eife et al.[159] observed that the antiproliferative:antiviral activity
ratios differed significantly for the three interferons. Effects of interferon on lytic viruses such as vesicular stomatitis virus were dissociated from
those on Moloney murine leukemia virus.[160] These results indicate that many
of the effects of the interferons can be dissociated and are due to different
molecular mechanisms. The individual interferons can apparently turn on
several pathways to different degrees. The existence of at least two distinct
interferon receptors that respond differentially to each of the interferons
could account for the differences in the molecular antiproliferative:antiviral
activity ratios. It is conceivable, in fact, that a family of interferon
receptors may exist. Interferon receptors with distinct binding constants for
the individual interferons could be demonstrated by direct and competitive
binding studies. In fact, with the use of our purified IFLrA, Branca and
Baglioni[90] have reported differences in the binding of immune interferon to
cells from the binding of leukocyte and fibroblast interferons. If it is
assumed that the ability of an interferon to exhibit antiviral activity reflects
the ability of that interferon to interact with the appropriate receptors for
interferon, then the conformational changes alter the binding of interferon to
these receptors.

52

Fig. 23. Prismatic crystal of recombinant human leukocyte A interferon. Taken from Miller et al.[164]

Since IFLrA/D ($Bg\ell$) is highly active on mouse cells, it should prove useful in developing model systems for evaluating the effectiveness of interferon *in vivo* as an antiviral and antitumor agent. In preliminary studies, IFLrA/D ($Bg\ell$) has been shown to exhibit antiviral activity in mice.[161,162] Studies have been initiated in nude mice carrying human tumors to evaluate several of the recombinant and hybrid leukocyte interferons for *in vivo* antitumor activity (N. Kaplan and M. Grunberg, personal communication) as well as *in vitro* antiproliferative activity.[163] In contrast to its high activity on mouse cells, IFLrA/D ($Bg\ell$) was only slightly active on rat cells.

It should be noted that, although IFLrA showed activity on mouse cells when very high concentrations of the purified IFLrA were used, antiviral activity on mouse cells was demonstrable only at 3×10^7 molecules/cell, 10,000-fold the concentration of IFLrA/D ($Bg\ell$) necessary for the same effect. This result indicates that the species barrier, when it does exist, is only relative; and that the mouse receptors with low affinity for human interferon IFLrA do bind IFLrA when it is present at high concentrations. Similarly, at very high concentrations, IFLrA exhibits activity on rat cells.

CRYSTALLIZATION OF RECOMBINANT INTERFERONS

With the purification of IFLrA made in bacteria, for the first time large amounts of pure interferon were available for study of its structure as well as for initiating studies in humans. Crystals of the IFLrA[164] were prepared

(Fig. 23). Such crystals should permit the eventual determination of the secondary and tertiary structure of this interferon by X-ray crystallography.

CLINICAL USE OF INTERFERONS

Because of their antiviral and antiproliferative activities as well as other activities, interferons have been considered to have potential therapeutic applications in treating viral diseases and cancer. Prior to our purification of human leukocyte interferon, impure interferon preparations were available and only in relatively small amounts. Despite these limitations, Cantell et al.[165,166] prepared concentrated preparations of crude human leukocyte interferon for initial clinical studies. With these preparations, a number of investigators reported that these leukocyte interferon concentrates exhibited some effectiveness in treating a number of viral diseases in humans.[167] Strander[168] and later others[169,170] reported some positive results of these preparations in treating human cancer. All these groups reported some human tumors responding to these preparations to one extent or another. However, dosage was limited by the total amount of material available as well as by the impurity of the preparation. Furthermore, since impure preparations of interferon were used, it was not certain that the effects were due to one of the interferons in the preparation, to some other component in the crude concentrates, or to some combination of both factors.

With the synthesis of human interferon in bacteria, the supply problem was solved in that we could now make virtually unlimited amounts of the material. With the purification of the recombinant human leukocyte interferon, dosage was no longer limited by impurities present. Accordingly, after appropriate safety and toxicity tests, on January 15, 1981, recombinant human leukocyte A interferon (IFLrA) prepared at Hoffmann-La Roche was the first pure human interferon used in a patient. For the first time there will be sufficient interferon available so that the clinical trials will not be limited by the availability of material.

It must be emphasized that we are entering new vistas in the therapy of viral diseases and cancer. The rapid maturation of this fundamental research into clinical application will be one of most rewarding for science as well as for people. A report of this initial clinical study with IFLrA has appeared.[171]

CONCLUDING COMMENTS

The use of interferon in clinical trials in humans was achieved by the rapid transfer of technology from laboratory-level basic science to appropriate pharmaceutical areas. Although interferon has yet to prove itself worthy of a permanent position in our pharmaceutical armamentarium, there is a cautious optimism prevailing for its use as an antiviral and/or an antitumor agent. Clinical trials are progressing on schedule and results will be forthcoming continually in the months ahead.

ACKNOWLEDGEMENTS

I would like to acknowledge the able and important contributions of many collaborators and colleagues who provided the substance and sustenance for the work described in this review. Although their specific contributions are cited in the appropriate references, I would like to thank the following individuals for their special efforts: Larry Brink, Ralph L. Cavalieri, Nancy Chang, Tsu-Rong Chiang, Marian Evinger, Philip C. Familletti, Heinz-Jürgen Friesen, David V. Goeddel, Jordan U. Gutterman, Ronald B. Herberman, Jeane P. Hester, Eileen G. Hoal, Donna S. Hobbs, Bruce Kelder, Michael J. Kramer, Hsiang-fu Kung, Chun-Yen Lai, Warren P. Levy, Shuichiro Maeda, Russell McCandliss, James L. McInnes, David L. Miller, John A. Moschera, John R. Ortaldo, Edward Rehberg, Sara Rubinstein, Menachem Rubinstein, John E. Shively, Alan Sloma, Theophil Staehelin, and Stanley Stein as well as members of the Biopolymer, Immunotherapy, and Microbiology Departments of Hoffmann-La Roche who assisted in these efforts. In addition, I would like to thank John Burns, Sidney Udenfriend, and Herbert Weissbach for their support, encouragement and advice.

REFERENCES

1. Isaacs, A. and Lindenmann, J. (1957) Proc. R. Soc. London Ser. B, 147, 258-267.
2. Isaacs, A., Lindenmann, J., and Valentine, R. C. (1957) Proc. R. Soc. London Ser. B, 147, 268-273.
3. Nagano, Y. and Kojima, Y. (1958) C. R. Seances Soc. Biol. Ses Fil., 152, 1627-1629.
4. Ng, M. H. and Vilček, J. (1972) Adv. Protein Chem., 26, 173-241.
5. Colby, C. and Morgan, M. J. (1971) Annu. Rev. Microbiol., 25, 333-360.
6. De Clercq, E. and Merigan, T. C. (1970) Annu. Rev. Med., 21, 17-46.
7. Ho, M. and Armstrong, J. A. (1975) Annu. Rev. Microbiol., 29, 131-161.
8. Tex. Rep. Biol. Med., (1977) 35, 1-573.
9. Pestka, S. (1978) in Dimensions in Health Research: Search for the Medicines of Tomorrow, Weissbach, H. and Kunz, R. M., eds., New York: Academic Press, pp. 29-56.
10. Stinebring, W. F. and Chapple, P. J., eds. (1978) Human Interferon: Production and Clinical Use, New York: Plenum Press.

11. Finter, N. B., ed. (1973) Interferons and Interferon Inducers, New York: American Elsevier Publishing Co.
12. Khan, A., Hill, N. O., and Dorn, G. L., eds. (1980) Interferon: Properties and Clinical Uses, Dallas, Texas: Leland Fikes Foundation Press.
13. Stewart, W. E., II, ed. (1979) The Interferon System, New York: Springer-Verlag.
14. Annals of the New York Academy of Science (1980) 350, 1-641.
15. Yabrov, A. A., ed. (1980) Interferon and Nonspecific Resistance, New York: Human Sciences Press.
16. Pestka, S., et al. (1980) in Polypeptide Hormones, Beers, R. F., Jr., and Bassett, E. G., eds., New York: Raven Press, pp. 33-48.
17. Pestka, S., et al. (1981) in Cellular Responses to Molecular Modulators, Mozes, L. W., et al., eds., New York: Academic Press, pp. 455-489.
18. Pestka, S., ed. (1981) Methods in Enzymology, Vol. 78, Part A, New York: Academic Press, 632 pp.
19. Pestka, S., ed. (1981) Methods in Enzymology, Vol. 79, Part B, New York: Academic Press, 677 pp.
20. Pestka, S., et al. (1981) in Recombinant DNA, Walton, A. G., ed., Amsterdam: Elsevier-North Holland, pp. 51-74.
21. Armstrong, J. A. (1981) Methods Enzymol., 78, 381-387.
22. Johnston, M. D., Finter, N. B., and Young, P. A. (1981) Methods Enzymol., 78, 394-399.
23. Familletti, P. C., Rubinstein, S., and Pestka, S. (1981) Methods Enzymol., 78, 387-394.
24. Rubinstein, S., Familletti, P. C., and Pestka, S. (1981) J. Virol., 37, 755-758.
25. Pestka, S. and Baron, S. (1981) Methods Enzymol., 78, 3-14.
26. Interferon Nomenclature (1980) Nature (London), 286, 110.
27. Knight, E., Jr. (1976) Proc. Natl. Acad. Sci. USA, 73, 520-523.
28. Berthold, W., Tan, C., and Tan, Y. H. (1978) J. Biol. Chem., 253, 5206-5212.
29. Stein, S., et al. (1980) Proc. Natl. Acad. Sci. USA, 77, 5716-5719.
30. Friesen, H.-J., et al. (1981) Arch. Biochem. Biophys., 206, 432-450.
31. Sehgal, P. B. and Sagar, A. D. (1980) Nature, 288, 95-97.
32. Weissenbach, J., et al. (1980) Proc. Natl. Acad. Sci. USA, 77, 7152-7156.
33. Yip, Y. K., et al. (1981) Proc. Natl. Acad. Sci. USA, 78, 1601-1605.
34. Langford, M. P., et al. (1979) Infect. Immun., 26, 36-41.
35. Von Wussow, P., et al. (1982) J. Interferon Res., 2, 11-20.
36. de Ley, M., et al. (1980) Eur. J. Immunol., 10, 877-883.
37. Wolfe, R., et al. in preparation.
38. Braude, I. A., personal communication.
39. Gray, P. W., et al. (1982) Nature, 295, 503-508.
40. Devos, R., et al. (1982) Nucleic Acids Res., 10, 2487-2501.
41. Familletti, P. C. and Pestka, S. (1981) Antimicrob. Agents Chemother., 20, 1-4.
42. Familletti, P. C., McCandliss, R., and Pestka, S. (1981) Antimicrob. Agents Chemother., 20, 5-9.
43. Waldman, A. A. et al. (1981) Methods Enzymol., 78, 39-44.
44. Hershberg, R. D., et al. (1981) Methods Enzymol., 78, 45-48.
45. Cantell, K. and Tovell, D. R. (1971) Appl. Microbiol., 22, 625-628.
46. Wheelock, E. F. (1966) J. Bacteriol., 92, 1415-1421.
47. Lee, S. H. S., van Rooyen, C. E., and Ozere, R. L. (1969) Cancer Res., 29, 645-652.
48. Hadházy, C. Y., et al. (1967) Acta Microbiol. Acad. Sci. Hung., 14, 391-397.
49. Rubinstein, M., et al. (1979) in Peptides: Structure and Biological Function, Function, Gross, E. and Meienhofer, J., eds., Rockford, Illinois: Pierce Chemical Co., pp. 99-103.
50. Udenfriend, S., et al. (1972) Science, 178, 871-872.

51. Stein, S., et al. (1973) Arch. Biochem. Biophys., 155, 203-212.
52. Böhlen, P., et al. (1975) Anal. Biochem., 67, 438-445.
53. Rubinstein, M., et al. (1978) Science, 202, 1289-1290.
54. Rubinstein, M., et al. (1979) Proc. Natl. Acad. Sci. USA, 76, 640-644.
55. Rubinstein, M., et al. (1981) Arch. Biochem. Biophys., 210, 307-318.
56. Rubinstein, M. and Pestka, S. (1981) Methods Enzymol., 78, 464-472.
57. Regnier, F. E. and Noel, R. (1976) J. Chromatogr. Sci., 14, 316-320.
58. Chang, S.-H., Noel, R., and Regnier, F. E. (1976) Anal. Chem., 48, 1839-1845.
59. Zoon, K. C., et al. (1979) Proc. Natl. Acad. Sci. USA, 76, 5601-5605.
60. Cabrer, B., et al. (1979) J. Biol. Chem., 254, 3681-3684.
61. Stewart, W. E., II (1974) Virology, 61, 80-86.
62. Stewart, W. E., II, et al. (1977) Proc. Natl. Acad. Sci. USA, 74, 4200-4204.
63. Törma, E. T. and Paucker, K. (1976) J. Biol. Chem., 251, 4810-4816.
64. Chen, J. K., et al. (1976) J. Virol., 19, 425-434.
65. Jankowski, W. J., et al. (1976) Biochemistry, 15, 5182-5187.
66. Brob, P. M. and Chadha, K. C. (1979) Biochemistry, 18, 5782-5786.
67. Bose, S., et al. (1976) J. Biol. Chem., 251, 1659-1662.
68. Bridgen, P. J., et al. (1977) J. Biol. Chem., 252, 6585-6587.
69. Stewart, W. E., II, et al. (1977) Proc. Natl. Acad. Sci. USA, 74, 4200-4204.
70. Bose, S. and Hickman, J. (1977) J. Biol. Chem., 252, 8336-8337.
71. Hobbs, D. S., et al. (1981) Methods Enzymol., 78, 472-481.
72. Hobbs, D. S. and Pestka, S. (1982) J. Biol. Chem., 257, 4071-4076.
73. Allen, G. and Fantes, K. H. (1980) Nature (London), 287, 408-411.
74. Zoon, K. C. (1981) Methods Enzymol., 78, 457-464.
75. Berg, K. and Heron, I. (1981) Methods Enzymol., 78, 487-499.
76. Zoon, K. C., et al. (1980) Science, 207, 527-528.
77. Knight, E., Jr., et al. (1980) Science, 207, 525-526.
78. Levy, W. P., et al. (1980) Proc. Natl. Acad. Sci. USA, 77, 5102-5104.
79. Okamura, H., et al. (1980) Biochemistry, 19, 3831-3835.
80. Levy, W. P., et al. (1981) Proc. Natl. Acad. Sci. USA, 78, 6186-6190.
81. Shively, J. E., et al. (1982) Anal. Biochem., 126, 318-326.
82. Zoon, K. (1981) in The Biology of the Interferon System (De Maeyer, E., Galasso, G., and Schellekens, H., eds., Amsterdam: Elsevier-North Holland, pp. 47-55.
83. De Maeyer-Guignard, J., et al. (1978) Nature (London), 271, 622-625.
84. De Maeyer-Guignard, J. (1981) Methods Enzymol., 78, 513-522.
85. Iwakura, Y., Yonehara, S., and Kawade, Y. (1978) J. Biol. Chem., 253, 5074-5079.
86. Kawade, Y., et al. (1981) Methods Enzymol., 78, 522-535.
87. Kawakita, M., et al. (1978) J. Biol. Chem., 253, 598-602.
88. Taira, H., et al. (1980) Science, 207, 528-530.
89. Kenny, C., Moschera, J. A., and Stein, S. (1981) Methods Enzymol., 78, 435-447.
90. Branca, A. A. and Baglioni, C. (1981) Nature (London), 294, 768-770.
91. Pestka, S., et al. (1975) Proc. Natl. Acad. Sci. USA, 72, 3898-3901.
92. Thang, M. N., et al. (1975) Proc. Natl. Acad. Sci. USA, 72, 3975-3977.
93. Cavalieri, R. L., et al. (1977) Proc. Natl. Acad. Sci. USA, 74, 3287-3291.
94. Cavalieri, R. L., et al. (1977) Proc. Natl. Acad. Sci. USA, 74, 4415-4419.
95. Cavalieri, R. L. and Pestka, S. (1977) Tex. Rep. Biol. Med., 35, 117-123.
96. Reynolds, F. H., Jr., Premkumar, E., and Pitha, P. M. (1975) Proc. Natl. Acad. Sci., 72, 4881-4885.
97. McCandliss, R., Sloma, A., and Pestka, S. (1981) Methods Enzymol., 79, 51-59.
98. Sloma, A., McCandliss, R., and Pestka, S. (1981) Methods Enzymol., 79, 68-71.
99. Maeda, S., et al. (1980) Proc. Natl. Acad. Sci. USA, 77, 7010-7013; (1981) 78, 4648.
100. Pestka, S. (1981) Methods Enzymol., 79, 599-601.

101. McCandliss, R., Sloma, A., and Takeshima, H. (1981) Methods Enzymol., 79, 601-607.
102. Maeda, S. (1981) Methods Enzymol., 79, 607-611.
103. Maeda, S. (1981) Methods Enzymol., 79, 611-613.
104. Maeda, S., Gross, M., and Pestka, S. (1981) Methods Enzymol., 79, 613-618.
105. McCandliss, R., Sloma, A., and Pestka, S. (1981) Methods Enzymol., 79, 618-622.
106. Maniatis, T., et al. (1976) Cell, 8, 163-182.
107. Taniguchi, T., et al. (1979) Proc. Jpn. Acad., B, 55, 464-469.
108. Tanigushi, T., et al. (1980) Gene, 10, 11-15.
109. Derynck, R., et al. (1980) Nature (London), 285, 542-547.
110. Houghton, M., et al. (1980) Nucleic Acids Res., 8, 1913-1931.
111. Nagata, S., et al. (1980) Nature (London), 284, 316-320.
112. Goeddel, D. V., et al. (1980) Nucleic Acids Res., 8, 4057-4074.
113. Messing, J., Crea, R., and Seeburg, P. H. (1981) Nucleic Acids Res., 9, 309-321.
114. Sanger, F., Nicklen, S., and Coulson, A. R. (1977) Proc. Natl. Acad. Sci. USA, 74, 5463-5467.
115. Smith, A. J. H. (1980) Methods Enzymol., 65, 560-579.
116. Maxam, A. M. and Gilbert, W. (1977) Proc. Natl. Acad. Sci. USA, 74, 560-564.
117. Maeda, S., et al. (1981) in Developmental Biology Using Purified Genes, Brown, D., and Fox, C. F., eds., New York: Academic Press, pp. 85-96.
118. Lawn, R. M., et al. (1978) Cell, 15, 1157-1174.
119. Goeddel, D. V., et al. (1980) Nature (London), 287, 411-416.
120. Streuli, M., Nagata, S., and Weissmann, C. (1980) Science, 209, 1343-1347.
121. Mantei, N., et al. (1980) Gene, 10, 1-10.
122. Lawn, R. M., et al. (1981) Proc. Natl. Acad. Sci. USA, 78, 5435-5439.
123. Goeddel, D. V., et al. (1981) Nature (London), 290, 20-26.
124. Lawn, R. M., et al. (1981) Science, 212, 1159-1162.
125. Nagata, S., Mantei, N., and Weissmann, C. (1980) Nature, 287, 401-408.
126. Brack, C., et al. (1981) Gene, 15, 379-394.
127. Ullrich, A., et al. (1982) Cell, in press.
128. Maeda, S., et al., in preparation.
129. Staehelin, T., et al. (1981) J. Biol. Chem., 256, 9750-9754.
130. Staehelin, T., et al. (1981) Methods Enzymol., 78, 505-512.
131. Streuli, M., et al. (1981) Proc. Natl. Acad. Sci. USA, 78, 2848-2852.
132. Weck, P. K., et al. (1981) Nucleic Acids Res., 9, 6153-6166
133. Rehberg, E., et al. (1982) J. Biol. Chem., 257, 11497-11502.
134. Köhler, G. and Milstein, C. (1975) Nature (London), 256, 495-497.
135. Staehelin, T., et al. (1981) Proc. Natl. Acad. Sci. USA, 78, 1848-1852.
136. Secher, D. S. and Burke, D. C. (1980) Nature (London), 285, 446-450.
137. Staehelin, T., et al. (1981) Methods Enzymol., 79, 589-595.
138. Wetzel, R. (1981) Nature (London), 289, 606-607.
139. Wetzel, R., et al. (1982) J. Cell. Biochem., Supplement 6, 89.
140. Paucker, K., Cantell, K., and Henle, W. (1962 Virology, 17, 324-334.
141. Gresser, I. (1977) Tex. Rep. Biol. Med., 35, 394-398.
142. Johnson, H. M. and Baron, S. (1977) Pharmacol. Ther., 1, 349-367.
143. Evinger, M., Rubinstein, M., and Pestka, S. (1980) Ann. NY Acad. Sci., 350, 399-404.
144. Evinger, M., Rubinstein, M., and Pestka, S. (1981) Arch. Biochem. Biophys., 210, 319-329.
145. Herberman, R. B., et al. (1980) Ann. NY Acad. Sci., 350, 63-71.
146. Herberman, R. B., et al. (1981) J. Clin. Immunol., 1, 149-153.
147. Verma, D. S., et al. in preparation.
148. Kaplan, N. O. and Slimmer, C. (1981) in Cellular Responses to Molecular Modulators, Mozes, L. W., et al., eds., New York: Academic Press, pp. 443-453.

58

149. Knight, E., Jr. (1976) Nature (London), 262, 302-303.
150. Evinger, M., Maeda, S., and Pestka, S. (1981) J. Biol. Chem., 256, 2113-2114.
151. Ortaldo, J. R., et al. in preparation.
152. Herberman, R. B., et al. (1982) Cell. Immunol., 67, 160-167.
153. Lengyel, P. (1981) Methods Enzymol., 79, 135-148.
154. Lengyel, P. and Pestka, S. (1980) in Gene Families of Collagen and Other Proteins, Prockop, D. C. and Champe, P. C., eds., Amsterdam: Elsevier-North Holland, pp. 121-126.
155. Maheshwari, R. K. and Friedman, R. M. (1981) Methods Enzymol., 79, 451-458.
156. Sreevalsan, T., Lee, E., and Friedman, R. M. (1981) Methods Enzymol., 79, 342-349.
157. Revel, M., et al. (1981) Methods Enzymol., 79, 149-161.
158. Kerr, I. M. and Brown, R. E. (1978) Proc. Natl. Acad. Sci. USA, 75, 256-260.
159. Eife, R., et al. (1981) J. Immunol. Methods, 47, 339-347.
160. Epstein, D. A., et al. (1981) Eur. J. Biochem., 118, 9-15.
161. Kramer, M. J., et al. (1981) Abstract, Second Annual International Congress for Interferon Research, (1981) San Francisco, California, October 21-23.
162. Weck, P. K., et al. (1982) Infect. Immun., 35, 660-665.
163. Makover, S. D., Telep, E., and Wright, R. B. (1981) Abstract, Second Annual International Congress for Interferon Research, San Francisco, California, October 21-23.
164. Miller, D. L., Kung, H.-F., and Pestka, S. (1982) Science, 215, 689-690.
165. Cantell, K., et al. (1981) Methods Enzymol., 78, 29-38.
166. Cantell, K., Hirvonen, S., and Koistinen, V. (1981) Methods Enzymol., 78, 499-505.
167. Pollard, R. B. and Merigan, T. C. (1978) Pharmacol. Ther. (Part A), 2, 783-811.
168. Strander, H., et al. (1973) J. Natl. Cancer Inst., 51, 733-742.
169. Merigan, T. C., et al. (1978) N. Engl. J. Med., 298, 981-987.
170. Gutterman, J. U., et al. (1980) Ann. Intern. Med., 93, 399-406.
171. Gutterman, J. U., et al. (1982) Ann. Intern. Med., 96, 549-556.

INTRACELLULAR DISSOCIATION OF RECEPTOR-LIGAND COMPLEX IN RAT HEPATOCYTES[*]

GILBERT ASHWELL
National Institutes of Health,
Bethesda, Maryland 20205

Previous studies on the endocytosis and degradation of ^{125}I-asialo-orosomucoid by freshly isolated hepatocytes was monitored as a function of time at 37°C. Experimental values were determined for the rates of internalization, dissociation of the receptor-ligand complex and degradation of the labeled ligand. Compartmental analysis and computer modeling revealed that the data were compatible with dissociation of ligand from receptor preceding ligand degradation. Subsequent studies in primary monolayer cultures of rat hepatocytes suggested that the internalization process was adequately defined by a dual endocytotic pathway. Rate coefficients for each of the transitions that constitute the pathways were computed. Subcellular fractionation on Percoll gradients revealed that, prior to localization in lysosomes, ^{125}I-asialo-orosomucoid resided in a fraction of lower buoyant density than plasma membranes. Neither ammonium chloride (20 mM) nor leupeptin (0.1 mg/ml) affected ligand binding or internalization of prebound ligand. However, both reagents inhibited degradation by greater than 95%. Of the two, only ammonium chloride inhibited receptor-ligand dissociation and the accumulation of ligand in a pre-lysosomal, low-buoyancy fraction. In contrast, exposure of cells to leupeptin led to accumulation of ligand within lysosomes. These results are interpreted in terms of a pH-mediated dissociation of ligand-receptor complex within a non-lysosomal endocytic vesicle.

---

[*] A full text of this contribution appears in T.-Y. Liu, S. Sakakibara, A. N. Schechter, K. Yagi, H. Yajima, and K. T. Yasunobu (Eds.) Biophysical Studies of Macromolecules: Frontiers in Biochemical and Biophysical Studies of Macromolecules. Elsevier-North Holland, New York, 1983.

VIRAL GLYCOPROTEINS AT THE CELL SURFACE

A. S. HUANG, E. J. O'ROURKE, AND L. M. LITTLE
Department of Microbiology and Molecular Genetics, Harvard Medical School,
Boston, Massachusetts; and Division of Infectious Diseases, Children's
Hospital Medical Center, Boston, Massachusetts

INTRODUCTION

Vesicular stomatitis virus (VSV) is a large RNA enveloped virus.[1,2] In the
laboratory, the virus grows in a wide variety of cells and completes its one-
step growth within 8 hr.[3] It synthesizes a vast amount of virus-specific RNA
(2-10 pg/cell[4]) and proteins (> 90%), concomitantly with the shutdown of host
RNA and protein synthesis. During its maturation two virus-coded proteins
associate with the plasma membrane. The M or matrix protein remains attached
to the cytoplasmic side of the bilayer and the G or glycoprotein traverses
the bilayer and extends extracellularly.

The G protein has been extensively studied. Its complete sequence of 511
amino acids has been determined from the nucleotide sequence of its mRNA
(1665 nucleotides[5]). The G mRNA has been copied into DNA, cloned in bacterial
plasmids, and expressed as part of SV40 DNA vectors in eucaryotic cells (Rose,
personal communication). G protein is synthesized in the rough endoplasmic
reticulum, and then transported via the Golgi to the plasma membrane.[6,7] This
process takes about 20 min. Knowing the number of methionine residues in the
G protein and the specific activity of ^{35}S-methionine, we have calculated that
each cell is capable of synthesizing at a minimum 500,000 to 1,000,000 molecules
of G every 20 min. During infection there are 3 major species of G protein:
an underglycosylated, immature, intracellular form, a fully glycosylated plasma
membrane form and a shed form which is missing the hydrophobic amino acids
at the carboxy end.[8,7]

ANTIGENIC MODULATION AND DEGRADATION

Because so much is known about the VSV glycoprotein, studies on antigenic
modulation of G protein at the cell surface by specific antisera were initiated.
When infected baby hamster kidney (BHK) cells were exposed to antibody in the
absence of complement, there was a rapid degradation of the G protein (Fig. 1).
This degradation was dependent on antibody concentration and the stable
association of G protein with the plasma membrane. Absence of G protein at
the infected cell surface did not lead to this rapid degradation. Complete
degradation of the G protein to acid insoluble material could be obtained at
very high antibody concentrations.

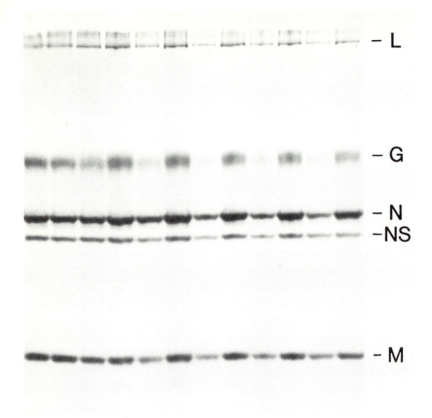

$- L$

$- G$

$- N$

$-NS$

$- M$

A A' B B' C C' D D' E E' F F'

Fig. 1. Kinetics of antibody induced degradation of VSV proteins. VSV-infected cells at 3 hr post infection were labeled with ^{35}S-methionine for 20 min. Then the radioactive incorporation was stopped by the addition of excess cold methionine. The sample was divided in half and incubated with either anti-VSV or normal rabbit serum at a dilution of 1:10. Aliquots were removed at the indicated times during the chase period. Cytoplasmic extracts were made[3] and analyzed by polyacrylamide gel electrophoresis.[20] Lanes A-F contained anti-VSV serum and Lanes A'-F' contained normal serum. A,A', 0 time; B,B', 15 min; C,C', 30 min; D,D', 45 min; E,E', 60 min; F,F', 90 min.

Because of the absence of extracellular degradation products, the mechanism of the degradation was probed by the use of protease inhibitors. Table 1 shows that the lysosomotropic agents chloroquine and ammonium chloride, as well as a serine-specific protease inhibitor, phenylmethylsulfonyl chloride, failed to

TABLE 1

FAILURE OF THREE PROTEOLYTIC INHIBITORS TO PREVENT DEGRATION OF THE G PROTEIN[*]

Inhibitor	Percent remaining G protein with	
	Anti-VSV	Normal serum
phenylmethyl Sulfonyl fluoride (1 mM)		
+	30	85
−	31	95
chloroquine (100 µM)		
+	50	90
−	49	91
ammonium chloride (20 mM)		
+	41	87
−	31	84

[*] VSV-infected cells were incubated at 2-1/2 hr after infection with or without inhibitor for 30 min prior to pulse labeling with ^{35}S-methionine for another 30 min, after which anti-VSV antibody or normal rabbit serum was added at a final concentration of 1:10. Inhibitor was present throughout the whole period. At the end of the 60 min. chase, cultures were harvested. Cytoplasmic extracts were made and analyzed by polyacrylamide gel electrophoresis, as shown for Fig. 1. The amount of G protein was quantitated by densitometer scanning of the radioautograph. The results are expressed as a percentage of G protein at the end of the chase period compared to the total amount of radioactive G protein.

prevent the degradation. Other inhibitors, such as leupeptin, pepstatin, chymostatin, and EP 475, also did not inhibit degradation. These findings point to the ATP-dependent protease pathway,[9] in contrast to the lysosomal one,[10] as another important degradative pathway in eucaryotic cells.

Previous studies on antigenic modulation showed that the kinetics and site of degradation varied widely for different cells and different cell surface antigens.[11-15] Because VSV is ubiquitous, it was used to infect fibroblasts as well as lymphocytes and other cells with the purpose of determining whether different cells handle identical antigens in the same way or differently during antigenic modulation. Preliminary results indicate that cells vary greatly in their ability to degrade the same antigen-antibody complex.

PROBING HUMAN TUMOR ANTIGENS

Another approach for studying cell surface glycoproteins utilizing VSV is to grow the virus in tumor cells and then to examine the progeny for tumor-specific antigens. When VSV buds out of cells to form infectious progeny, the particles will often pick up other glycoproteins, especially virus-coded ones, aside from its own. These progeny have been called pseudotypes, phenotypic mixtures or mosaics.[16] It is thought that only certain types of transmembrane glycoproteins are incorporated into virions and that, perhaps, the VSV M protein, which interacts with the cytoplasmic portion of the G protein, determines the selectivity.

Irrespective of the mechanism of formation of mosaic virions, two populations of VSV, one grown in HeLa cells and the other in BHK cells, were studied in regard to their surface components. When the virions were purified by differential and rate zonal centrifugation[17] and their proteins stained after polyacrylamide gel electrophoresis, the VSV(HeLa) preparation showed more heterogeneity than the VSV(BHK) preparation (Fig. 2). The 5 VSV structural proteins are indicated as L, G, N, NS and M (Fig. 2). VSV(HeLa) contained the same proteins, but the G band was quite faint and there were multiple additional bands indicative of cellular constituents.

Because VSV(HeLa) showed an unusual pattern of neutralization with antisera made against VSV, anti-HeLa cell sera were used to see if cellular determinants could be detected on the virions. These sera, made against uninfected HeLa cells, did not neutralize VSV(HeLa), but in an assay using *Staphylococcus aureus*, 77%-98% of the VSV(HeLa) was immunoprecipitated (Table 2). In contrast, similar assays with VSV(BHK) or VSV grown in Chinese hamster ovary (CHO) cells showed that anti-HeLa cell serum failed to interact significantly.

As an indication that the HeLa cell determinants picked up by VSV may be related to human tumor antigens, the same anti-HeLa cell serum was tested against VSV grown in primary cultures of several human tumors. A wide range (33%-97%) of immunoprecipitation was observed, indicating possible cross reactivity between the HeLa cell determinants and other human tumors. These results substantiate similar observations by Zavada.[18,19] They also suggest the feasibility of isolating the HeLa cell determinants and studying their regulation in normal and neoplastic cells.

CONCLUSION

These are just two examples of the many interesting biological mechanisms which can be studied using such a well defined virus system as VSV. Although these studies on antigenic modulation and the formation of VSV mosaics relate

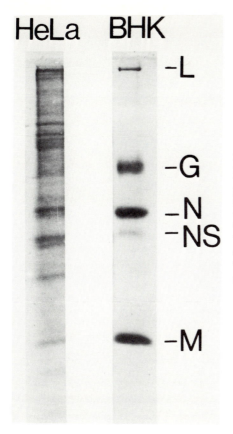

Fig. 2. Coomassie Blue Staining of polypeptides associated with preparations of VSV(HeLa) and VSV(BHK). VSV preparations were made by infecting HeLa cells or baby hamster kidney (BHK) cells or baby hamster kidney (BHK) cells and then purified by differential and rate zonal centrifugation as described previously.[17] The purified standard infectious virus was analyzed by polyacrylamide gel electrophoresis.[20]

to basic mechanisms of cellular degradation and of protein-protein recognition, respectively, they are also important to the understanding of viral pathogenesis. Virus-infected cells may escape immune surveillance by the loss of their surface neo-antigens. Also, viral progeny may escape immune surveillance by hiding in the envelope antigens of other viruses or cells.

Structurally, there are still many unanswered questions. A three-dimensional structure of the G protein is needed. Questions concerning the functional domains of G protein are only being guessed at and await the results of site specific mutagenesis. Information relating neutralizing epitopes to other functional epitopes are beginning to be gained. Given a virus which grows readily, which can be studied in cells as well as in cell-free systems, and which

66

TABLE 2

IMMUNOPRECIPITATION OF VESICULAR STOMATITIS VIRUS

PREPARATIONS BY ANTI-HeLa CELL SERA[*]

Virus	Serum	Dilution of serum	Percent of immuno-precipitated VSV
VSV(HeLa)	Anti-HeLa cell #1	1:100	77
VSV(HeLa)	Anti-HeLa cell #2	1:100	95
VSV(HeLa)	Anti-HeLa cell #3	1:100	98
VSV(HeLa)	Anti-HeLa cell #4	1:100	75
VSV(HeLa)	Anti-HeLa cell #5	1:100	95
tsO$_{45}$(HeLa)	Anti-HeLa cell #2	1:100	98
VSV(CHO)	Anti-HeLa cell #1	1:10	0
VSV(BHK)	Anti-HeLa cell #1	1:10	25
VSV(HeLa)	normal rabbit	1:10	0

[*] Virus preparations at approximately 10^7 plaque forming units, grown through the cells indicated by the parenthesis, were incubated with the indicated antiserum for 1 hr at 4° and then mixed with *Staphylococcus aureus* for another hour. The bacteria and immune complexes were removed by centrifugation and the virus remaining in the supernatants was plaque-assayed on monolayer cultures of Chinese hamster ovary cells.[16] Precipitation of VSV by bacteria alone ranged from 0-20%.

can be studied through natural mutants and genetically engineered ones, much new information concerning glycoproteins and their functional interactions at the plasma membrane is likely to be gained from this virus in the near future.

ACKNOWLEDGEMENTS

   This work is supported by research grants AI 16625 from the National Institutes of Health and PCM 8118037 from the National Science Foundation. E. J. O'Rourke and L.M. Little are awardees of the institutional National Research Service grants T32 AI 07061 and T32 CA 09031, respectively. We thank Trudy Lanman for excellent technical support and Suzanne Ress for the computer-aided preparation of this manuscript.

REFERENCES

1. Wagner, R. R. (1975) in Comprehensive Virology, Vol. 4, Frankel-Conrat, H. and Wagner, R. R., eds., New York: Plenum Press.
2. Bishop, D.H.L. (1979) *Rhabdoviruses*, Vols. I-III, Foca Raton, Fla.: CRC Press, Inc.
3. Huang, A. S. and Manders, E. K. (1972) J. Virol., 9, 909-916.
4. Clinton, G. M., et al. (1978) Cell, 15, 145-146.
5. Rose, J. K. and Gallione, C. J. (1981) J. Virol. 39, 519-528.
6. Katz, F. N., et al. (1977) Proc. Natl. Acad. Sci. USA, 74, 3278-3282.
7. Gosh, H. P. (1980) Rev. Infect. Dis., 2, 26-39.
8. Little, S. P. and Huang, A. S. (1978) J. Virol., 27, 330-339.
9. Goldberg, A. L., Strand, N. P., and Swamy, K. H. S. (1980) in Protein Degradation in Health and Disease, Ciba Foundation 75 series, Elsevier/North-Holland.
10. Pastan, I. and Willingham, M. (1981) Ann. Rev. Physiol., 43, 239-250.
11. Old, L. J., et al. (1968) J. Expt. Med., 127, 523-539.
12. Rosenthal, A. S., et al. (1973) Expt. Cell Res., 81, 317-329.
13. Yu, A. and Cohen, E. P. (1974) J. Immunol., 112, 1296-1307.
14. Perrin, L. and Oldstone, M. B. A. (1977) J. Immunol., 118, 316-322.
15. Stackpole, C. W., et al. (1980) J. Immunol., 125, 1715-1723.
16. Weiss, R. A. (1979) in Rhabdoviruses, Bishop, D. H. L., ed., Boca Raton, Fla.: CRC Press, Inc.
17. Stampfer, M., Baltimore, D., and Huang, A. S. (1969) J. Virol., 4, 154-161.
18. Zavada, J., et al. (1974) Biol., 39, 907-912.
19. Zavada, J., et al. (1974) J. Gen. Virol., 24, 327-337.
20. Laemmli, W. K. (1970) Nature (London), 227, 680-682.

REGULATION OF THE LATERAL MOBILITY OF CELL SURFACE PROTEINS BY INTERACTIONS WITH THE CYTOSKELETON

ELLIOT L. ELSON AND YOAV I. HENIS[*]
Department of Biological Chemistry, Division of Biology and Biomedical Sciences,
Washington University School of Medicine, St. Louis, Missouri

ABSTRACT

According to the simplest model cell membrane proteins should be free to move laterally limited mainly by the viscosity of the lipid bilayer in which they are embedded. There are qualitative indications that this view is too simple. These include the existence of specialized cell surface structures such as spatially confined patches of acetylcholine receptor at neuromuscular junctions, active redistribution processes such as the capping of cross-linked surface antigens on lymphocytes, and apparent interactions between cell surface proteins and the cytoskeleton revealed by immunofluorescence microscopy.

Constraints on the lateral mobility of surface proteins can be investigated quantitatively using fluorescence photobleaching methods. These reveal that proteins typically diffuse at least an order of magnitude more slowly than expected from the bilayer viscosity. The origins of the constraints on protein diffusion are still unknown but are now under active investigation. There is evidence to suggest that interactions with the cytoskeleton have an important influence on the mobility of surface proteins. This is clearest for erythro-cytes. Biochemical studies have demonstrated a linkage between spectrin in the erythrocyte cytoskeleton and the integral membrane protein, band 3, via the linker protein, ankyrin. Disruption of this linkage frees band 3 from the major constraints on its lateral mobility.

Interactions between the cytoskeleton and surface proteins on nucleated cells are less well understood both biochemically and with respect to their effects on lateral mobility. Recent studies have shown that the lateral mobility of several membrane proteins is strongly enhanced in "blebs," regions of the plasma membrane which have been detached from the underlying cytockeletal cortex. Furthermore, it has been demonstrated that the plant lectin, con-canavalin A, can reduce the mobility of other cell surface proteins in a process involving microtubules and microfilaments. Hence it appears that the cyto-skeleton in nucleated cells also influences the mobility of cell surface

---

[*]Current address: Department of Biochemistry, The George S. Wise Faculty of Life Sciences, Tel Aviv University, Tel Aviv, Israel.

proteins. Many puzzles remain, however, and the biochemical characteristics and physiological significance of the interactions involved are still unknown.

INTRODUCTION

The cell surface is a dynamic environment. Membrane proteins interact with each other and with intra- and extracellular components to carry out physiological functions and to organize structure. Lateral and rotational motion is sometimes essential for these interactions. For example, membrane receptors occupied by specific hormones[1] or immunoglobulins[2] must encounter one another on the plasma membrane to form limited aggregates or at least to come into close proximity[3] in order to trigger cellular responses. Specialized structures on the cell surface such as patches of acetylcholine receptors at the neuromuscular junction[4] or budding enveloped viruses[5,6] may assemble by trapping freely diffusing specific membrane proteins. Hence, investigation of the lateral and rotational motions of membrane proteins should provide information useful for the analysis of physiological mechanisms. Moreover, the mobility of surface components indicates interactions which constrain motion.[7,8] Therefore, measurements of mobility help to characterize physico-chemical properties of the cell surface involving its interactions with structures inside and outside the cell. This paper discusses current problems in the attempt to characterize the factors which control the lateral mobility of plasma membrane proteins.

The simplest dynamic membrane model might hold that cell membrane proteins should be free to move laterally across the cell surface limited mainly by the viscosity of the lipid bilayer in which they are embedded. There is, however, both qualitative and quantitative evidence which contradicts this view. The former includes the following:

(1) the existence of specialized structures like the spatially confined patches of acetylcholine receptors mentioned above;[9]

(2) active redistribution processes which require cellular energy and seem not to be the result of simple diffusion such as the capping of cross-linked surface antigens on lymphocytes;[10] and

(3) the apparent links between cell surface proteins and the cytoskeleton revealed by immunofluorescence microscopy[11,12] and by biochemical analysis.[13-15]

Constraints on the lateral mobility of surface proteins can be investigated quantitatively using fluorescence photobleaching methods.[16-19] These reveal that lipid-like molecules move freely through the membrane with diffusion coefficients near 10^{-8} cm^2/sec.[20] This agrees well with estimates from NMR and

EPR measurements.[21] Membrane proteins, however, diffuse much more slowly.
Indeed most membrane proteins studied, even those of defined specificity, have
been found to be divided into two mobility classes. One fraction diffuses at
a measurable rate which is, however, one to three orders of magnitude slower
than that of lipids. The balance of the protein is apparently immobile on the
time scale of the experiment, which usually means $D < 5 \times 10^{-12}$ cm^2/sec.
Lateral diffusion coefficients of mobile proteins are typically near 10^{-10}
cm^2/sec.[22-24]

Theory predicts that the translational diffusion coefficient of a cylindrical
protein molecule traversing a planar lipid bilayer should depend only weakly
(logarithmically) on the diameter of the cylinder.[25] Thus, protein molecules
would be expected to diffuse only slightly more slowly (2- to 3-fold) in
membranes than lipids. Although the theory of Saffman and Delbruck has not
been critically tested experimentally, there are data consistent with its
predictions. In the highly specialized disk membrane of the rod outer segment,
the diffusion of rhodopsin is an order of magnitude faster ($D \sim 3 \times 10^{-9}$
cm^2/sec) than for typical cell membrane proteins, as demonstrated by the
original (absorbance) photobleaching studies by Poo and Cone[26] and Liebman and
Entine.[27] Furthermore, proteins reconstituted into synthetic lipid bilayer
membranes also diffuse at least an order of faster than cell membrane
proteins[28,29] and at a rate only slightly slower than the membrane lipids.
Hence, one is led to conclude that constraints in addition to the viscosity
of the lipid bilayer limit the rate of diffusion of most membrane proteins.
The structural basis of these constraints and their physiological significance
are unknown. Currently, much attention is being focused on interactions
between membrane proteins and the cytoskeleton as a principal cause of the
inhibition of protein lateral diffusion.

Mobility of the band 3 protein in erythrocytes

The role of cytoskeletal interactions in limiting the mobility of membrane
proteins is most clearly seen in the erythrocyte. The band 3 protein is a
major intrinsic protein of the erythrocyte plasma membrane. It functions as
an anion exchange channel. Elegant biochemical studies have revealed a chain
of interactions through which the band 3 protein is linked to spectrin in the
erythrocyte cytoskeleton. Spectrin, an elongated molecule of high molecular
weight, is the principal component of this cytoskeleton. Spectrin oligomers
combined with short filaments of actin and the band 4.1 protein are supposed to
form a cortical network lying immediately beneath the plasma membrane.[30] This
cortical cytoskeleton has an important role in determining the shape and

elasticity of the erythrocyte membrane.[31]   Recently, it has been shown that
the band 3 protein is linked to spectrin via a protein of molecular weight
215,000 daltons called "ankyrin."[32]   A 72,000 dalton protolytic fragment of
ankyrin, which retains a high affinity binding site for spectrin, blocks
the binding of spectrin to band 3 protein.[33]   Hence, the band 3 and spectrin
binding sites on ankyrin are separable.

In normal erythrocytes the lateral diffusion coefficient of band 3 is low,
$D < 10^{-10}$ cm^2/sec. A role for ankyrin in restricting band 3 mobility was
first demonstrated by Fowler and Bennett.[34]   They used the 72,000 dalton fragment of ankyrin to displace spectrin from its membrane binding sites. This
resulted in an increase in the lateral mobility of membrane proteins, mainly
band 3, as measured by the rate of intermixing of fluorescence-labeled surface
proteins after fusion of labeled and unlabeled cells using Sendai virus. More
quantitative studies have measured band 3 diffusion by fluorescence photobleaching methods. Golan and Veatch have investigated the effect of changes in
temperature and ionic strength on the lateral diffusion of band 3.[35]   At low
temperature and high ionic strength (21°C, 46 mM $NaPO_4$) 90% of the labeled band
3 was apparently immobile. The remaining 10% diffused slowly with $D = 4 \times 10^{-11}$
cm^2/sec. When the temperature was increased to 37° and the ionic strength
decreased to 13 mM $NaPO_4$, the mobile fraction rose to 90% with a diffusion
coefficient, $D = 2 \times 10^{-9}$ cm^2/sec. Interestingly, the increases in mobile
fraction and diffusion rate were distinguishable both in terms of their responses
to changes in ionic strength and temperature and in their kinetics. Moreover
changes in the diffusion coefficient of the mobile band 3 molecules with
temperature were readily reversible whereas changes in the fraction of mobile
molecules were only slowly and partially reversible. More recently, Golan and
Veatch have confirmed the earlier result of Fowler and Bennett[34] by demonstrating
a 2.5-fold increase in the diffusion coefficient of band 3 protein in the
presence of the 72,000 dalton ankyrin fragment which contains the spectrin
binding site.[36]   Other studies of the effects of specific proteolysis confirm
the importance of ankyrin in regulating the lateral diffusion of band 3
protein.[36]   At 37° under extremes of both high and low ionic strength, the
diffusion coefficient of band 3 reached a value only 4-fold less than that of
lipid probes under the same conditions. Hence, under these conditions the
diffusion rate of band 3 was consistent with its being limited only by the
viscosity of membrane lipids.[35]

The existence of an hereditary blood disease in mice provided Sheetz,
Schindler, and Koppel with an elegant way to demonstrate the importance of the

cytoskeletal matrix in regulating membrane protein mobility.[37] The mouse spherocytic erythrocyte is deficient in major components of that matrix. In normal red cells the diffusion coefficient of band 3 protein was found to be 4.5×10^{-11} cm^2/sec while in spherocytes, $D = 2.5 \times 10^{-9}$. Hence, in cells with impaired cytoskeletons the band 3 protein again diffused at a rate compatible with limitation mainly by bilayer viscosity.

Although the results described above implicate the erythrocyte cytoskeleton and in particular interactions with spectrin via ankyrin in restricting the lateral diffusion of band 3, there are still unanswered question. Based on biochemical analyses there is enough ankyrin in an erythrocyte to interact with less than half of the band 3 protein on its surface.[38] Furthermore, the rotation rate of band 3 molecules, measured by the transient dichroism of bound triplet probes is relatively rapid although complex.[23] These results argue against the interpretation of the slow translational diffusion of band 3 as resulting from specific interactions of each band 3 molecule with spectrin and ankyrin. To reconcile the rapid rotational and slow translational diffusion, a model was proposed in which membrane proteins were supposed to be non-specifically entrapped in the interstices of a cytoskeletal matrix. They would therein be able to rotate freely but could translate only at a rate governed by the lability of the links in the matrix network.[23,37,39] The quantitative implications of this model have been developed and the effective surface viscosity has been related to the viscoelastic mechanical properties of red cell membranes.[40] More recently it has been observed that the fraction of band 3 protein ($\sim 40\%$) which appears to be immobile on the time scale of a conventional photobleaching recovery measurement does indeed move with an effective diffusion coefficient of $(1.5 \pm 0.8) \times 10^{-12}$ cm^2/sec, some 50-fold slower than the diffusion rate of the "mobile" fraction ($D = 8.1 \pm 3.1 \times 10^{-11}$ cm^2/sec).[41] It was suggested that the slower moving component might represent the fraction of band 3 attached by ankyrin to spectrin, while the mobility of the faster moving molecules might be limited by nonspecific steric interactions with the spectrin matrix.

Evidently the interactions which limit the lateral diffusion of band 3 are not yet understood in detail. Models based on specific interaction have been proposed to complement that of nonspecific entrapment.[7,8] Nevertheless, both biochemical studies and measurements of mobility provide convincing evidence of the central involvement of the cytoskeleton and in particular of the links between band 3, ankyrin, and spectrin. It remains for the future to interpret the mechanism by which the lateral mobility of band 3 is restricted in terms of the equilibrium and kinetic properties of its interaction with ankyrin and

of ankyrin's interaction with spectrin and possibly of the interactions of any
of these proteins with other cell surface or cytoskeletal components. This may,
perhaps, best be done either in reconstituted systems or in cells from which
specific components have been eliminated either by mutation or by selective
extraction.

Lateral diffusion of membrane proteins in nucleated cells

The molecular basis of cell surface-cytoskeletal interactions in nucleated
cells such as fibroblasts, epithelial cells, or leukocytes is both more complex
and less well characterized than in erythrocytes. In the latter cells the
cytoskeleton is composed of three systems of fibers and their associated proteins which penetrate through most of the cell interior. The microtubule and
intermediate filament systems are thought not to interact directly with the
cell surface. The microtubules seem to be responsible for cellular polarization and the orientation of directional motions.[42,43] The intermediate filaments are composed of several distinct types of protein subunits with some
degree of tissue specificity.[44] Their functions are the least understood.
They may, however, serve to anchor the nucleus and to integrate mechanical
forces through the cell.[44,45] The third cytoskeletal system, based on actin
microfilaments, is thought to interact directly with the cell surface and to
have a structural role as well as to drive active motions of the cell. In
contrast to the well characterized linkage of band 3 to spectrin via ankyrin in
erythrocytes, however, the biochemical properties of the interaction of microfilaments with the plasma membrane are still unclear. A 130,000 dalton protein
named "vinculin," has been found to be closely associated with specialized
regions of the membrane to which microfilaments are attached.[46] These include
regions of contact between the cell surface and tissue culture substratum,[47]
junctional regions of intestinal brush border cells, and dense placques in the
membranes of chicken gizzard smooth muscle cells.[48] These studies do not,
however, characterize specific molecular associations of vinculin either with
actin or with other cell surface components. Furthermore, the nature of the
association, if any, of microfilaments with the major part of the cell surface
which is not included in specialized attachment organelles is still unknown.
These latter must be important for the interpretation of measurements of
lateral diffusion, which have sampled wide areas of the cell surface. Proteins
immunologically cross-reactive with ankyrin have been found in a variety of
other cell types.[38] Curiously, their intracellular distribution as revealed by
immunofluorescence microscopy closely parallels that of the microtubules during
mitosis, leading to the suggestion that ankyrin-like proteins could link

microtubules to membranes of the nuclear envelope and endoplasmic reticulum in dividing cells.[38] There has, however, not yet been a report of a protein which functions in nucleated cells, as ankyrin does in erythrocytes, to link wide areas of the plasma membrane to the underlying cytoskeletal cortex. That such a cortex does exist and such linkages do occur, however, has recently been suggested by many studies using nonionic detergents to prepare insoluble cytoskeletal residues with membrane proteins still associated.[49] For example, an actin-containing matrix which included four additional major proteins has been isolated by detergent extraction from mouse tumor and lymphoid cells.[50] It was shown that the matrix was associated with the plasma membranes of these cells and retained the cell surface glycoprotein, 5'-nucleotidase. In another recent study, it was shown that patched or aggregated acetylcholine receptors were retained on detergent insoluble residues of embryonic muscle cells while diffusely distributed receptors were partially extracted by the detergent.[51] The biochemical basis of these interactions remains, however, to be characterized.

Recent work suggests the importance of interactions between the cell surface and interior in controlling the lateral mobility of plasma membrane proteins. One approach has been to measure diffusion on blebs, regions of the plasma membrane which have been detached from underlying structures within the cell.[52] Blebs were produced by exposing cells in culture to a solution containing low concentrations of formaldehyde (25 mM) and dithiothreitol (2 mM). This causes the plasma membrane to separate from the cytoskeleton and to form a bleb, like a blister or bubble, on the cell surface. On normal L6 rat myoblasts receptors labeled with s-ConA, a divalent derivative of the tetravalent plant lectin Concanavalin A (ConA) have a low diffusion coefficient ($D \sim 4 \times 10^{-11}$ cm^2/sec) and a substantial fraction of the receptors ($\sim 34\%$) appear immobile.[53] In blebs, however, essentially all of the receptors are mobile with a diffusion coefficient 100-fold greater ($D \sim 4 \times 10^{-9}$ cm^2/sec). This is close to the hydrodynamic limit expected if receptor diffusion is limited mainly by the viscosity of the membrane lipid bilayer. Similarly, all acetylcholine receptors at the neuromuscular endplate of isolated muscle fibers are apparently immobile. On blebbed endplate membrane, however, the receptors are released from their constraint and have a diffusion coefficient of $D \sim 2 \times 10^{-9}$ cm^2/sec, also near the hydrodynamic limit.[52] Similar results have also been found for IgE (Fc) receptors in rat basophilic leukemia cells, for low density lipoprotein receptors in human fibroblasts, and for ConA receptors on lymphocytes.[52] Although there were small increases in lipid diffusion rates in blebs relative to normal membranes, these were insufficient to account for the large increases in surface protein mobility observed in blebs.

The important result of this work is that various membrane proteins which diffuse slowly or not at all on normal cell surfaces can be shown to have diffusion coefficients near the hydrodynamic limin in blebs. Hence, the low diffusion rate observed in normal membranes is not likely to be an artifact of the measuring procedure, but rather is due to interactions with the surface which are disrupted in forming the bleb. Biochemical and microscopic data indicate that the membrane is detached from the bulk of the underlying cyto-skeleton.[52] The effective interactions cannot yet, however, be defined in molecular terms.

A different approach to study the regulation of membrane protein mobility is based on the earlier work of Edelman and coworkers.[54] They observed that the formation first of patches and then of caps of cross-linked cell surface antigen-antibody complexes was inhibited by ConA. This phenomenon, which they termed "anchorage modulation," was most thoroughly studied by observing the patching and capping of surface immunoglobulin (sIg) in B-lymphocytes.[55] A similar response was, however, observed for several different surface antigens on a variety of cell types.[54] The inhibition of patching and capping could be partly reversed by treating cells with colchicine or other agents which interfere with the assembly of microtubules.[55] Moreover, when ConA was con-fined to local regions of the cell surface by prebinding it to platelets or latex beads, patch formation was inhibited over the entire surface. Hence, the modulation propagates globally throughout the cell.[56] The inhibition of patch formation was attributed to a ConA-induced decrease in the rate of lateral diffusion of membrane antigens. This was proposed to have resulted from the action of a "surface modulating assembly" in which glycoprotein receptors trigger a reaction involving microtubules and microfilaments. This would cause anchorage of surface antigens to a cytoskeletal structure thereby inhibiting their lateral diffusion.[54]

The effect of ConA-modulation on the mobility of surface antigens was tested directly with fluorescence photobleaching measurements.[57] ConA prebound to platelets was allowed to bind to local regions of 3T3 mouse fibroblast surfaces. Then the lateral diffusion coefficients of unselected mouse cell surface antigens was measured at different locations on the surface remote from the site of ConA binding. A striking result was obtained. Below a certain threshold amount of bound ConA-platelets, there was no effect on antigen mobility. When 4% or more of the dorsal surface of the cell was covered by ConA-platelets, however, the diffusion coefficients of the mobile surface antigens decreased 6-fold. This degree of inhibition was constant at all observed extents of surface coverage above 4% threshold. The fraction of apparently immobile surface antigens was, however, unchanged by any extent of ConA-platelets bound to the cell surface. This inhibition of antigen mobility by ConA-platelets occurred with the same threshold in the presence of several drugs which inhibit microtubule assembly but with only a 3-fold rather than a 6-fold reduction in diffusion coefficient. Thus, disruption of microtubules only partially reversed the modulation of mobility. These results confirmed the hypothesis that anchorage modulation decreased the lateral mobility of membrane proteins in a microtubule-dependent process. Nevertheless, interpretation of the results was complicated by the diversity of the membrane antigens used to monitor the mobility changes.

The phenomenon was subsequently reinvestigated using a better defined system. The effect of ConA-platelets on the mobility of sIg on primary mouse B-lymphocytes was examined.[58] The results were qualitatively similar to but quantitatively even more striking than those obtained earlier. Again, a threshold was observed. On lymphocytes, however, both the diffusion coefficient and the fraction of mobile sIg molecules were strongly reduced. Thus, the interpretation of anchorage modulation as a reduction in diffusion rate was again confirmed, but this time under conditions similar to those in which most of the original studies of patching and capping were performed. When cells were treated separately with colchicine to disrupt microtubules or cytochalasin B to disrupt microfilaments, the inhibition of sIg mobility was *partially* reversed, although the threshold level of surface coverage by ConA-platelets remained unchanged. Thus, the process which initiates the modulation, characterized by the threshold, is distinguishable from that which determines the extent of the inhibition of mobility. A further striking result, however, was that colchicine and cytochalasin B added together entirely abolished the inhibition of mobility and restored the diffusion coefficient of sIg to the original level observed below the threshold level of surface coverage or in the absence of ConA-platelets. Hence, microtubules and microfilaments are both necessary but neither system is sufficient without the other for the complete expression of the modulation, and their simultaneous disruption suffices to abolish the effect on mobility. These results also demonstrate that anchorage modulation is not an all-or-none phenomenon; intermediate levels between no modulation and full modulation are observed when either

colchicine or cytochalasin D is added without the other. Finally, an important
control was the demonstration that modulation does not involve increasing the
viscosity of the membrane bilayer. This was established by verifying that the
lateral diffusion of a lipid probe was unaffected by modulation.

The physiological significance of anchorage modulation is unknown. If it is
involved in a specialized cellular function, the characteristics of the modulation response might vary among cell types depending, perhaps on differences
in cytoskeletal characteristics or cell surface-cytoskeleton interactions. This
possibility was investigated by comparing mouse lymphocytes, mouse 3T3 fibroblasts, and primary mouse macrophages and a mouse macrophage-like cell line,
P388D1.[59] The lateral mobility of membrane proteins was monitored using
rhodamine-labeled s-ConA. ConA receptors are abundant on all three types of
cells, and s-ConA cannot itself induce modulation.[55] Hence, modulation by ConA
of the mobility of its own receptors was observed. On lymphocytes both the
diffusion coefficient and mobile fraction of ConA receptors were simultaneously
decreased at a critical threshold of cell surface coverage by ConA platelets.
Hence, the measurement of s-ConA mobility faithfully reproduced the result
previously observed for the modulation of sIg mobility. On 3T3 cells the diffusion coefficient of the mobile s-ConA receptors was diminished at a critical
threshold of cell surface coverage, but the relative fractions of mobile and
immobile receptors were not changed. This reproduced the behavior previously
observed for unselected surface antigens on 3T3 cells. The consistent difference
in the responses of the mobile fractions to modulation in lymphocytes and 3T3
cells suggests some degree of cell type specificity. This suggestion was
strongly supported by the observation that ConA platelets had no effect on the
lateral mobility of ConA receptors on primary macrophages or P388D1 cells.[59]
Neither the diffusion coefficient nor the mobile fraction was diminished at all
levels of surface coverage by ConA platelets which were examined. The variation
in the degree of modulation among the three cell types studied in this work
could be due to cell-specific differences in cytoskeletal structure or regulation or in surface-cytoskeleton interactions. Because ConA binds to a diverse
collection of membrane glycoproteins (and glycolipids), it is also possible
that specific ConA receptors required to induce modulation are scarce or absent
in cell types which show little or no response. This possibility can be
addressed only by identifying the ConA receptor(s) responsible for triggering
the modulation response.

We have distinguished between the modulation observed in lymphocytes and 3T3
cells by noting a decrease of the mobile fraction in the former which is absent
in the latter. This could be due to a distinct process in lymphocytes which

does not occur in 3T3 cells. It is simpler, however, to suppose that the difference in the response of the mobile fraction results from a difference in the strength of the modulation effect. The increase in the apparently immobile fraction of ConA receptors on lymphocytes could result from a larger shift in the overall distribution of mobilities to lower values. This would increase the fraction of molecules with diffusion coefficients below the detection limit. The mobility of ConA receptors on 3T3 cells seems to be modulated less strongly. Few if any of the receptors on these cells are reduced below the detection limit, and so the mobile fraction is not detectably changed. According to this analysis we could summarize the cell-type specificity of modulation by grading the strength of the response: strongest in lymphocytes, weaker in 3T3 cells, and absent in macrophages. The molecular basis and physiological significance of these differences are a challenge for the future.

Two different demonstrations of cytoskeletal involvement in regulating the lateral mobility of cell surface proteins have been presented. Both begin at the same baseline level of the lateral diffusion coefficient observed on normal unperturbed cells. As we have observed, these diffusion coefficients indicate constraints on mobility in addition to the viscosity of the lipid bilayer. On blebs these constraints have been removed; proteins diffuse near the hydrodynamic limit. ConA-induced modulation, however, imposes additional constraints which depend on the integrity of microtubules and microfilaments. It is important to recognize that by abolishing the modulation with colchicine and cytochalasin B, diffusion coefficients are merely returned to their baseline values. The anticytoskeletal drugs do not impair the constraints which continue to inhibit mobility at that level. Hence, if cytoskeletal structures, particularly microfilaments and microtubules, are involved in determining the baseline level of mobility, as might reasonably be concluded from the measurements on blebs, these structures must be insensitive to colchicine and cytochalasin B. This apparent paradox remains to be resolved. Perhaps it is related to the two classes of microfilaments observed in studies of the membrane-associated cytoskeletal matrix: one fraction, soluble in the detergent; the other, associated with the insoluble matrix.[50]

CONCLUSION AND QUESTIONS

Lateral diffusion of cell surface proteins is much slower than expected from the viscosity of the lipid bilayer. In both erythrocytes and nucleated cells there is good evidence that interactions of membrane proteins with the cytoskeleton can play an important role in retarding mobility. In erythrocytes an inhibition of the diffusion of band 3 protein due to its linkage to ankyrin and thence to spectrin has been demonstrated. Nevertheless the extent to which this inhibition depends on specific or nonspecific steric interactions remains to be clarified. In nucleated cells measurements on blebs suggest that interactions between the surface and the interior of the cell imposes the constraints which determine the baseline rate of protein diffusion. These constraints are, however, unimpaired by the cytoskeletal disruptions caused by colchicine and cytochalasin B. In contrast, constraints above the baseline level due to ConA-induced anchorage modulation are completely abolished by the simultaneous action of colchine and cytochalasin B.

It is also worth considering whether the diffusion of surface proteins could be inhibited by interactions with the extracellular matrix. The interactions could be especially important at the specific points of contact between the matrix and the cell surface. Nevertheless, it seems unlikely that extracellular interactions have a large influence on the average diffusion behavior. There are, for example, no obvious distinctions in the observed protein mobility in lymphocytes which do not interact with a matrix and various kinds of adherent cells which do.[23,58] In both kinds of cells, proteins are divided into slowly mobile and apparently immobile classes. Furthermore, the presence of high concentrations of fibronectin, a major component of the extra-cellular matrix, does not seem to influence the lateral diffusion of membrane proteins and lipids.[60]

Evidently, much remains to be learned about the role of the cytoskeleton in regulating the mobility of membrane proteins. The chief need is for biochemical analysis of the interactions between surface and cytoskeleton. The studies of ankyrin in erythrocytes provide a model for the more complex task in nucleated cells. Once the molecules responsible for linking surface and cytoskeleton are identified and characterized, it might be possible to interpret the lateral mobilities of membrane proteins in terms of the equilibrium and kinetic properties of the relevant interactions.[7] At the same time it should be possible to investigate another question of great interest: to what extent and by what means does a cell regulate the mobility of its surface proteints to fulfill specific physiological functions?

ACKNOWLEDGEMENTS

The work from the authors' laboratory discussed in this paper was supported by NIH Grant GM 21661 and by a Chaim Weizmann postdoctoral fellowship to Yoav I. Henis. The Washington University Center for Basic Cancer Research (funded by NIH Grant 5P30 CA 16217) provided tissue culture media.

REFERENCES

1. Schlessinger, J. (1980) Trends Biochem. Sci., 5, 210-214.
2. Segal, D. M., Taurog, J. D., and Metzger, H. (1977) Proc. Natl. Acad. Sci. USA, 74, 2993-2997.
3. Balakrishnan, K., et al. (1982) J. Biol. Chem., 257, 6427-6433.
4. Poo, M-m. (1982) Nature, 295, 332-334.
5. Reidler, J. A., et al. (1981) Biochemistry, 20, 1345-1349.
6. Johnson, D. C., Schlesinger, M. J., and Elson, E. L. (1981) Cell, 23, 423-431.
7. Elson, E. L. and Reidler, J. A. (1979) J. Supramol. Structure, 12, 481-489.
8. Koppel, D. E. (1981) J. Supramol. Structure, 17, 61-67.
9. Fertuck, H. C. and Salpeter, M. M. (1974) Proc. Natl. Acad. Sci. USA, 71, 1376-1378.
10. Taylor, R. B., et al. (1971) Nature New Biol., 223, 225-229.
11. Ash, J. F. and Singer, S. J. (1976) Proc. Natl. Acad. Sci. USA, 73, 4575-4579.
12. Braun, J., et al. (1978) J. Cell Biol., 79, 409-418.
13. Koch, G. L. E. and Smith, M. J. (1978) Nature, 273, 274-278.
14. Flanagan, J. and Koch, G. L. E. (1978) Nature 273, 278-281.
15. Condeelis, J. (1979) J. Cell Biol., 80, 751-758.
16. Edidin, M., Zagyansky, Y., and Lardner, T. J. (1976) Science, 191, 466-468.
17. Jacobson, F. K., et al. (1976) J. Supramol. Structure, 5, 565-576.
18. Axelrod, D., et al. (1976) Biophys. J., 16, 1055-1069.
19. Koppel, D. E., et al. (1976) Biophys. J., 16, 1315-1329.
20. Schlessinger, J., et al. (1977) Science, 195, 307-309.
21. Edidin, M. (1974) Ann. Rev. Biophys. Bioeng., 3, 179-201.
22. Elson, E. L. and Schlessinger, J. (1979) in The Neurosciences: Fourth Study Program, Schmitt, F. O. and Worden, F. G., eds., Cambridge, Mass.: MIT Press, 691-701.
23. Cherry, R. S. (1979) Biochim. Biophys. Acta, 559, 289-327.
24. Webb, W. W., et al. (1981) Biochem. Soc. Symp., 46, 191-205.
25. Saffman, P. G. and Delbruck, M. (1975) Proc. Natl. Acad. Sci. USA, 72, 3111-3113.
26. Poo, M-m. and Cone, R. H. (1974) Nature, 247, 438-441.
27. Liebman, P. A. and Entine, G. (1974) Science, 185, 457-459.
28. Cartwright, G. S., et al. (1982) Proc. Natl. Acad. Sci. USA, 79, 1506-1510.
29. Wu, E-S., et al. (1978) Biochemistry, 17, 5543-5550.
30. Lux, S. E. (1979) 281, 426-429.
31. Evans, E. A. and Hochmoth, R. M. (1978) Curr. Top. Membr. Transp., 10, 1-64.
32. Bennett, V. and Stenbuck, P. J. (1980) J. Biol. Chem., 2540-2548.
33. Bennett, V. (1978) J. Biol. Chem., 253, 2292-2299.
34. Fowler, V. and Bennett, V. (1978) J. Supramol. Structure, 8, 215-221.
35. Golan, D. E. and Veatch, W. (1980) Proc. Natl. Acad. Sci. USA, 77, 2537-2541.
36. Golan, D. E. and Veatch, W. (1981) talk presented at the International Workshop on the Application of Fluorescence Photobleaching Techniques to Problems in Cell Biology, held at the University of North Carolina, Chapel Hill, October 25-28.
37. Sheetz, M. P., Schindler, M., and Koppel, D. E. (1980) Nature, 285, 510-512.
38. Bennett, V. (1982) J. Cell. Biochem., 18, 49-65.
39. Schindler, M., Osborn, M. J., and Koppel, D. E. (1980) Nature, 283, 346-350.
40. Koppel, D. E., Sheetz, M. P., and Schindler, M. (1981) Proc. Natl. Acad. Sci. USA, 78, 3576-3580.
41. Sheetz, M. P., Febbroriello, P., and Koppel, D. (1981) talk presented at the International Workshop on the Application of Fluorescence Photobleaching Techniques to Problems in Cell Biology, held at the University of North Carolina, Chapel Hill, October 25-28.

42. Porter, K. R. (1976) in Cell Motility, Goldman, R. D., et al., eds., Cold Spring Harbor Laboratory, New York, pp. 1-28.
43. Vasiliev, J. M. and Gelfand, I. M. (1976) in Cell Motility, Goldman, R. D., et al., eds., Cold Spring Harbor Laboratory, New York, pp. 279-304.
44. Lazarides, E. (1980) Nature, 283, 249-256.
45. Capco, D. G., Wan, K. M., and Penman, S. (1982) Cell, 29, 847-858.
46. Geiger, B. (1979) Cell, 18, 193-205
47. Singer, I. I. (1982) J. Cell Biol., 92, 398-408.
48. Geiger, B., et al. (1981) J. Cell. Biol., 91, 614-628.
49. Ben-Ze'ev, A., et al. (1979) Cell, 17, 859-865.
50. Mescher, M. F., Jose, M. J. L., and Balk, S. P. (1981) Nature, 289, 139-144.
51. Prives, J., et al. (1982) J. Cell Biol., 92, 231-236.
52. Tank, D. W., Wu, E-S., and Webb, W. W. (1982) J. Cell Biol., 92, 207-212.
53. Schlessinger, J., et al. (1976) Proc. Natl. Acad. Sci. USA, 73, 2409-2413.
54. Edelman, G. M. (1976) Science, 192, 218-226.
55. Edelman, G. M., Yahara, I., and Wang, J. L. (1973) Proc. Natl. Acad. Sci. USA, 70, 1442-1446.
56. Yahara, I. and Edelman, G. M. (1975) Proc. Natl. Acad. Sci. USA, 72, 1579-1583.
57. Schlessinger, J., et al. (1977) Proc. Natl. Acad. Sci. USA, 74, 1110-114.
58. Henis, Y. I. and Elson, E. L. (1981) Proc. Natl. Acad. Sci. USA, 78, 1072-1076.
59. Henis, Y. I. and Elson, E. L. (1981) Exp. Cell Research, 136, 189-201.
60. Schlessinger, J., et al. (1977) Proc. Natl. Acad. Sci. USA, 74, 2909-2913.

Biochemical and Biophysical Studies of Proteins and Nucleic Acids, Lo, Liu, and Li, eds.

83

DRUG CONJUGATES OF ANTITUMOR PROTEINS

JUNG-YAW LIN
Institute of Biochemistry, College of Medicine,
National Taiwan University, Taipei, Taiwan, Republic of China

ABSTRACT

Antitumor protein derivatives of methotrexate, chlorambucil and adriamycin were prepared and purified by affinity chromatography. The drug-antitumor proteins still retain the biological activities of drug and antitumor protein with a slight to moderate increase in their toxicity. A single dose of the derivative injected IP into Sarcoma 180-bearing noninbred N:NIH(S) white mice resulted in prolongation of the survival time and was more effective than an equivalent dose of free drug and antitumor protein or decreasing in solid tumor size. It was demonstrated at the molecular level that the derivatives have a stronger inhibitory effect on the DNA biosynthesis of Sarcoma 180 cells than the equivalent dose of free drug and antitumor protein.

INTRODUCTION

Though many efforts to develop chemotherapeutic agents suitable for treatments of cancer, few of antitumor agents in current use are highly effective against the growth of tumor.

Recently, the low molecular weight derivatives of antitumor drugs were prepared to increase their antitumor activity by prolonging the retention time in serum and extracellular compartments. For these purposes, alkylating agents,[1-3] MTX,[4] daunomycin[5] and enzyme[6] have been covalently linked to antibodies or bovine serum albumin with a variable improving in their antitumor activities.

The toxic lectins such as abrin or ricin consists of A and B chains linked by a disulfide bond; A-chain inhibits the protein biosynthesis while B-chain binds to cell surface receptors. The toxic lectins have a selective inhibitory effect on the growth of experimental tumor cells.[7,8] The major functions of the toxic lectin are inhibiting the protein and DNA biosynthesis of tumor cells.[9-11] Recently many attempts have been made to replace the binding subunit-B chain with other binding moieties such as protein hormones or antibodies to enhance their selective toxicity.[12] The covalent linkage of antitumor drug of low molecular weight to the chimeric toxic protein may enhance the effectiveness of their agents against tumor cells and reduce their toxicity against normal cells.

MATERIALS AND METHODS

The following chemicals were purchased from commercial sources: BSA, EDCI, and CA from Sigma Chemical Company, St. Louis, Missouri; MTX from Cyanamide Taiwan Corporation; Sepharose 4B and Sephadex G-100 from Pharmacia Fine Chemicals, Piscataway, New Jersey; (methyl-^{3}H) dThd (6.7 Ci/mmol) and L-(4,5-^{3}H) leucine (5 Ci/mmol) from New England Nuclear Corporation, Boston, Massachusetts; and RPMI-1640 medium from GIBCO, Grand Island, New York. Seeds of *Abrus precatorius*, *Ricinus communis*, and *Canavalia ensiformis* were obtained from private source in Taiwan. Adult male noninbred N:NIH(S) white mice weighing 20 $\pm$ 2 g were used for transplantation of tumor cells and study of lethality.

Isolation of lectin-AN and AG. Kernels of *A. precatorius* were extracted with PBS (pH 8.0).[16,17] The extracts were first purified by affinity chromatography on a Sepharose 4B column. The materials containing AN and AG were then separated by gel filtration on a Sephadex G-100 column.

RN and RG, like AN and AG, were first purified by affinity chromatography on a Sepharose 4B column and then by gel filtration on a Sephadex G-100 column.

Con A: Seeds of *C. ensiformis* were extracted with 0.5% acetic acid. The extracts were purified by affinity column chromatography on a Sephadex G-200 column. Con A which absorbed on the column, was eluted with 0.1 M D-glucose and then was crystallized by dialysis against distilled water.

Preparation of drug-lectin and drug-BSA conjugates. The drug-lectin conjugates were synthesized by taking advantage of affinity binding of lectin on carbohydrate. AN, AG, RN, or RG (50 mg each) was added to 10 ml of 0.01 M PBS (pH 7.0), 20 mg ECDI, and 25 mg MTX or CA dissolved in 5 ml of 0.01 M PBS (pH 7.0), 0.1 M galactose or glucose was used to protect the binding site of antitumor proteins. The reaction was terminated by gel filtration with either a Sephadex G-25 column or Biogel p-2 column (1.5 x 50 cm), the product was purified by affinity chromatography with use of Sepharose 4B or Sephadex G-100 column chromatography.

Drug-BSA conjugate was prepared under the same conditions as those described in the preparation of drug-lectin conjugate except that no monosaccharide was used in the reaction mixture. The drug-BSA conjugate produced was purified by gel filtration on a Sephadex G-25 column (1.5 x 50 cm), which was preequilibrated with 0.01 M PBS (pH 7.0), and it was eluted with the same buffer.

The degree of substitution was measured by MTX absorbance at 370 nm[4] or by alkylate activity of CA,[3] and the protein concentration was determined by the method of Lowry et al.[13]

Lethality. Male noninbred N:NIH(S) white mice that weighed 20 $\pm$ 2 g were inoculated IP with 0.14 M of normal saline containing various amounts of tested materials. After 2-5 days observation, the ID_{50} values were calculated.[14] There were 6 animals in each group.

Inoculation. Tumor inoculation was carried out by use of ascitic fluid from donor mice bearing 7-day growth of sarcoma 180 cells, and 0.4 ml of ascitic fluid containing 4×10^7 cells was injected IP into the recipient mice.

Evaluation of tumor growth. A single dose of either free drug, lectin, or drug-lectin was injected IP on the day of tumor transplantation. The highest dosage for these experiments was about one-half of the ID_{50} of lectins or drug-lectins, and the dosage of drug-lectin mixture was equivalent to the individual amount of free lectin and free drug present in the drug-lectin mixture. We evaluated the growth of the sarcoma 180 tumor by measuring the ILS, on the size of solid tumor.

Determination of DNA and protein biosynthesis inhibition. Sarcoma 180 tumor cells in suspension culture were used for the assay. The cells were washed twice with Eagle's minimum essential medium (without leucine) and were resuspended in a concentration of 1.5×10^6 cells/ml of the same medium. Various amounts of either drug-lectin, free drug, or lectin were added to each sample, and the reaction mixture was incubated at 37°C for 30 minutes. Then 1 uCi of (^{3}H) leucine was added to each sample containing 1 ml of cell culture and incubated for 1 hour. The reaction was terminated by the addition of 1 ml of 0.1 M potassium hydroxide, and the materials were precipitated by addition of trichloroacetic acid to a final concentration of 10% (wt/vol).

The acid-insoluble radioactivity was measured with a Beckman LS-250 liquid scintillation counter. To measure the DNA biosynthesis, the procedures were the same as those for protein biosynthesis except that RPMI-1460 medium was used in place of Eagle's minimum essential medium and the cell culture was incubated for 2 hours at 37°C with test reagent and further incubated with 1 uCi of (^{3}H) dThd/ml of cell culture at 37°C for 1 hour.

RESULTS AND DISCUSSION

Preparation of drug-antitumor protein conjugates

Methotrexate, chlorabucil or adriamycin was covalently linked to lectin with 1-ethyl-3-(3-dimethylaminopropyl) carbodiimide in the presence of 0.1 M D-galactose to protect the active site of B subunit of antitumor proteins such as AN, RN, AG and RG. The reaction was terminated by applying the reaction mixture on a Sephadex G-25 column (1.5 x 50 cm) (Fig. 1). Two peaks were

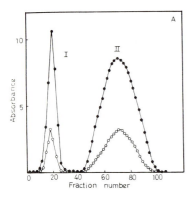

Fig. 1. Gel filtration of reaction mixture on a Sephadex G-25 column (1.5 x 50 cm). The first peak is MTX-AG and the second peak is the excess of reagents such as MTX and carbodiimide. Absorbance: ● 280 nm; o 370 nm.

obtained; the first peak contained the native and denatured drug-antitumor protein conjugates while the second peak was the excess of reagents of drug and carbodiimide. The native drug-antitumor protein conjugate was separated from the denatured one by passing the first peak of gel filtration through a Sepharose 4B column (2.0 x 15 cm). The denatured drug-antitumor protein conjugate was not absorbed on the column and eluted with 0.01 M phosphate buffered

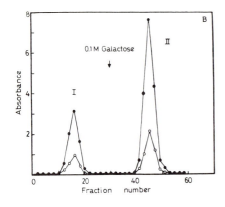

Fig. 2. Separation of native MTX-AG and denatured MTX-AG by affinity chromatography. Affinity chromatography was carried out by using a Sepharose 4B column (2.0 x 15 cm). The second peak is native MTX-AG, and the first peak is denatured MTX-AG.

saline, pH 7.0 while the native one was bound on the column and eluted with
0.1 M D-galactose in the buffer. The yield of native drug-antitumor protein
conjugate was about 75% (Fig. 2). Therefore, it was possible to obtain the
highly active preparation of drug-antitumor conjugates with a relatively higher
yield.

The degree of drug substitution for AN was 9.5 molecules of MTX per molecule
of the protein and that of AG was 21 molecules. The extent of drug substitution
correlates very well with the number of lysine residues in the antitumor proteins.

The agglutinating activity of drug-antitumor protein conjugates against the
human O type red blood cells or Sarcoma 180 cells is very similar to that of
antitumor protein. The slight to moderate increasing in the toxicity of drug-
antitumor protein conjugates was observed. For example, the LD_{50} for MTX-AN
conjugate and MTX-AG conjugate was 6 ug and 2 mg per Kg body weight while that
of AN and AG was 8 ug and 4 mg, respectively. It indicates that the covalent
linkage of drug to lysine residues of antitumor protein does not affect the
biological activities of drug-antitumor protein conjugate very much. Therefore,
none of lysine residues of antitumor proteins is essential for the activities
of antitumor proteins.[15]

Antitumor activity of drug-antitumor protein conjugates

The effects of drug-antitumor protein conjugates on the survival of sarcoma
180 cells were studied. The results demonstrated that 5 ug of MTX-AG conjugate
injected IP was more effective in reducing the number of Sarcoma 180 cells than
the mixture of an equivalent dose of free MTX and AG (Fig. 3). A single dose
of MTX-AG conjugate could reduce the number of tumor cells from 9×10^8 cells
to 1.2×10^6 cells on the third day after treatment; however, the mixture of
drug and antitumor protein was only able to reduce the number of cells to
3×10^7.

The results of drug-antitumor protein conjugates on the growth of solid tumor
of Sarcoma 180 cells are shown in Fig. 4. A single dose of 3.0 ug of MTX-RG
was able to reduce the tumor size of Sarcoma 180 much stronger than an equivalent dose of free MTX and RG. The size of tumor of the mice treated with MTX-RG
conjugate was about 30% that of control group but that of the mice treated with
the equivalent dose of free MTX and RA was about 90% that of control group.
For comparative study, MTX-BSA conjugate was used. As shown in Fig. 4, MTX-BSA
conjugate did not have any significant inhibitory effect on the growth of solid
tumor of Sarcoma 180 cells.

It is conceivable that BSA is a non-antitumor protein and its drug conjugate
does not possess the strong antitumor activity as drug-antitumor protein

88

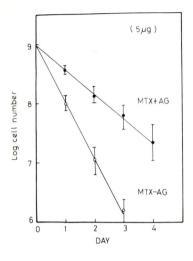

Fig. 3. The killing of Sarcoma 180 cells by MTX-AG. 9×10^8 of Sarcoma 180 cells were injected IP on the 0 day and MTX-AG was injected IP one hour after inoculation of tumor cells. The results are the average of three experiments, each with three mice.

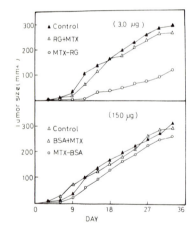

Fig. 4. The effects of MTX on RG on the growth of solid tumor of Sarcoma 180 cells. To male mice weighing 20 ± 2 g, 1×10^6 cells of Sarcoma 180 were injected i.p. on 0 day and on the 1st day various amounts of MTX-RG or MTX-BSA was injected subcutaneously. The tumor size was measured every other day. Six mice were used for each group.

conjugate does. The effects of AD-AG conjugate on the growth of solid tumor were very similar to those of MTX-RG conjugate (Fig. 5). The effects of the drug-antitumor conjugate on the life span of the Sarcoma 180 tumor bearing mice were also investigated. The results are summarized in Fig. 6. The mice treated with drug-antitumor conjugates all survived longer than those treated with the mixture of an equivalent dose of free drug and antitumor protein did. For example, the mice treated with 2.5 ug MTX-AG conjugate showed more than 300% increase in life span but the group receiving 2.25 ug AG and 0.25 ug MTX revealed 95% increase in life span. The mice treated with drug-AN conjugate or drug-Con A conjugate also exhibited a considerable increase in life span. Drug-BSA conjugate could not have such a remarkable increase in the life span of the Sarcoma 180 bearing mice as drug-antitumor protein conjugate did.

It has been shown that the polyoma transformed 3T3 cells were more sensitive to ricin than normal 3T3 cells.[16] The possible mechanism of this selectively inhibitory effect was that the antitumor protein could penetrate tumor cells and inhibit their DNA or protein biosynthesis.[10,11] MTX was demonstrated to inhibit the dihydrofolate reductase while CA could alkylate the DNA of tumor cells. The covalent linkage of MTX or CA to antitumor protein makes the two antitumor agents act independently in a synergestic manner due to their different mechanism of antitumor activity.

Effects of drug-antitumor protein conjugate in DNA and protein biosynthesis of tumor cells

The inhibition of DNA biosynthesis by drug-antitumor protein was shown in Fig. 7. It indicates that the inhibitory effect of drug-antitumor proteins on the DNA biosynthesis was more effective than the mixture of free antitumor proteins and drugs. The median inhibitory dose (ID_{50}) of MTX-AG and CA-AG was 0.67 and 1.02 ug, respectively, whereas that of free AG and MTX and CA was 3.5, 150 and 150 ug, respectively. If MTX and AG are mixed at the molar ratio equivalent to that of covalent conjugates, the ID_{50} was determined to be very similar to that of free AG. Drug-AN conjugate has similar properties as drug-AG conjugate does.

The effects of drug-antitumor protein conjugates on the protein biosynthesis are summarized in Fig. 8. Both MTX-AN and CA-AN had the strong inhibitory effect on protein biosynthesis of tumor cells, and the degree of inhibitory effect was almost the same as that of free AN. It suggests that covalent linkage of MTX or CA to abrin does not affect the inhibition of protein biosynthesis by abrin. Therefore, it was able to demonstrate at molecular level

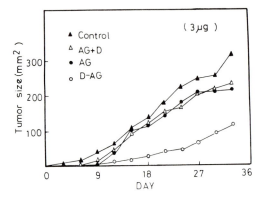

Fig. 5. The effects of AD-AG on the growth of solid tumor of Sarcoma 180 cells. The experimental methods are the same as described in Fig. 4.

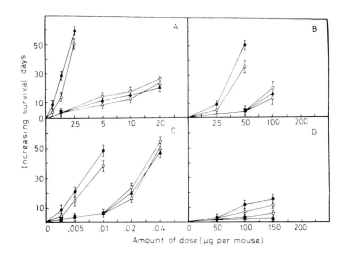

Fig. 6. The effects of drug-antitumor protein conjugates and free antitumor proteins plus free drug on the survival time of mice inoculated with Sarcoma 180 cells. (A) AG, (B) ConA, (C) AN, (D) BSA; ● , conjugated MTX; o , conjugated CA; ▼, free MTX (10 : 1 wt/wt); ▲, free CA (10 : 1 wt/wt); Δ , no drug.

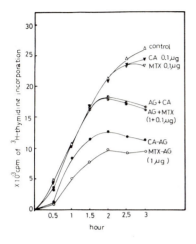

Fig. 7. The inhibitory effects of drug-antitumor proteins on the DNA biosynthesis of Sarcoma 180 cells. The details of experimental procedures are described in Materials and Methods. ● , MTX-AG; o , CA-AG; ▼ , AG + MTX; ▲, AG + CA; Δ , AG.

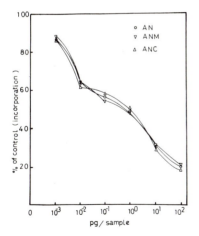

Fig. 8. The inhibitory effects of MTX-AN and CA-AN on the protein biosynthesis of Sarcoma 180 cells. The details of the experimented procedures are described in Materials and Methods. o , AN; ▽ , AMN : MTX-AN; Δ, CA-AN.

that drug-antitumor proteins were more active in inhibiting the DNA or protein
biosynthesis of tumor cells than the mixture of equivalent amount of free drug
and antitumor protein. It would be useful for preparation of more active
chimeric molecule of antibody-A chain of antitumor protein. When the drugs such
as MTX, CA or AD are covalently linked to A-chain of antitumor proteins for
preparation chimeric antitumor protein, the product would be more effective than
A chain itself linked to antibody in inhibiting the growth of tumor cells.

REFERENCES

1. Flechner, I. (1973) Eur. J. Cancer, 9, 741-745.
2. Davis, D. A. and O'Nell, G. J. (1973) Br. J. Cancer, suppl., 28, 285-298.
3. Ghose, T., Path, M. R., and Nigam, S. P. (1972) Cancer, 29, 1398-1412.
4. Chu, B. C. and Whiteley, J. M. (1979) J. Natl. Cancer. Inst., 62, 79-82.
5. Hurwitz, E., et al. (1975) Cancer Res., 35, 1175-1181.
6. Philpott, G. W., et al. (1973) J. Immunol., 111, 921-929.
7. Fodstad, O., Olsnes, S., and Pihl, A. (1977) Cancer Res., 37, 4559-4567.
8. Lin, J. Y., et al. (1970) Nature, 227, 292-293.
9. Bhattacharya, P., Simet, I., and Basu, S. (1979) Proc. Natl. Acad. Sci.
   USA, 76, 2218-2221.
10. Lin, J. Y., et al. (1970) Cancer Res., 30, 2431-2433.
11. Refsnes, K., Olsnes, S., and Pihl, A. (1974) J. Biol. Chem., 249, 3557-3562.
12. Olsnes, S. and Pihl, A. (1982) Pharmac. Ther., 15, 355-381.
13. Lowry, O. H., et al. (1951) J. Biol. Chem., 193, 265-275.
14. Reed, L. J. and Muench, H. (1938) Am. J. Hyg., 27, 493-497.
15. Sandvig, K., Olsnes, S., and Pihl, A. (1978) Eur. J. Biochem., 84, 323-331.
16. Nicolson, G. L., Lacorbiere, M., and Hinter, T. R. (1975) Cancer Res., 35,
    144-155.
17. Lin, J. Y., et al. (1981) Toxicon., 19, 41-51.

TAIWAN COBRA CARDIOTOXIN (A MEMBRANE-DISRUPTIVE POLYPEPTIDE)

YEE-HSIUNG CHEN, CHIEN-TSUNG HU, RUEY-FEN LIOU, AND CHAN-PIN LEE
Institute of Biochemical Sciences, College of Sciences, National Taiwan
University and Institute of Biological Chemistry, Academia Sinica,
Taiwan, R.O.C.

SUMMARY

Chronologically, there were three stages in the hemolysis of 0.25% (v/v)
human erythrocytes caused by 7.2 μM cardiotoxin in plasma expander. First, the
toxin bound to membrane rapidly before the cells disrupted. Binding of $[^{125}I]$-
toxin to intact erythrocytes was temperature independent and did not inhibit by
any one of the four inorganic salts including NaCl, KCl, $CaCl_2$ and $MgCl_2$ each
at 10 mM concentration. Kinetic studies revealed the presence of at least two
toxin binding sites on the erythrocyte membrane: one with high affinity
(equilibrium association constant K_a = 2.7 x 10^6 M^{-1}), low capacity (4.76 x 10^6
molecules/cell); the other with low affinity (K_a = 9.33 x 10^3 M^{-1}), high capacity
(2.12 x 10^8 molecules/cell). Antitoxin antibody could not displace the bound
toxin from the cell surface. Second, the bound toxin at 37°C reached its
action sites which might be at membrane spanning domain or/and at inner membrane
whereon somehow an interaction resulted in loosing the membrane rigidity and
the cells became less resistant to hypotonic NaCl solution. This event took
place within 25 min in the nonlytic period. Third, the hemolysis subsequent
to the second stage took place and it took around 6 h to complete the lysis of
all cells. The events in both second and third stages did not appear at lower
temperatures far apart from the physiological range and were suppressed by
either the inorganic salts at 10 mM concentration or antitoxin antibody.
Phospholipase A_2 enhanced the disruption of human erythrocytes once their membrane rigidity was loosed by the toxin.

Unlike the situation of the toxin, polylysine did not lose the membrane
rigidity. Its induction in hemolysis showed no temperature dependence and was
neither suppressed by the inorganic salts nor potentiated by phospholipase A_2.

The toxin caused the destruction of liposome vesicles in Tris-buffered
saline at pH 7.4, irrespective of whether the liposomes bore net negative
charges or not. At lower temperatures, the extent of leakage diminished. It
was found further that the release of umbelliferone phosphate entrapped within
the liposomes was enhanced in the presence of phospholipase A_2 but was suppressed
by divalent metal ions.

The toxin could induce G-actin polymerization. The polymerization was enhanced by the presence of 0.4 mM $MgCl_2$ but was suppressed if G-actin was preexposed to deoxyribonuclease. This enzyme could depolymerize also the actin polymer formed by interaction of G-actin with the toxin.

INTRODUCTION

Cardiotoxin, hereafter we refer to it as CTX, is a nonneurotoxic protein in the venom of many snake species.[1-3] This toxic polypeptide appears also in the literature by the name of cytolysin, cytotoxin, direct lytic factor, membrane-active polypeptide, etc., because it not only causes cardiac arrest, which is the main lethal effect of this toxin, but also shows other activities such as hemolysis, cytotoxicity, inhibition of accumulation of anions, amino acids and glucose in some organs, and depolarization of excitable membranes.[4] All of these biological activities are ascribed to the damage of cell membrane caused by CTX. Recently, it has been demonstrated that CTX affects the activities of several membrane enzymes.[5]

The primary structure of Taiwan cobra CTX[9] has been reported[6] (Fig. 1). This protein molecule contains 60 amino acid residues crosslinked with 4 disulfide bonds. It is a very basic polypeptide having 9 lysine and 2 arginine on the molecular surface.[7] Tyr-51 is buried and Tyr-11 and Tyr-22 are exposed.[8] Generally, CTX isolated from the venom of other snake species shares the same structural features.

This work concerns mainly three aspects. In the first part, action of the CTX on human erythrocytes will be discussed. In the second part, the capability of the CTX to destroy the liposome vesicles will be presented. In the last part, evidence for the actin polymerization induced by the CTX will be presented.

EXPERIMENTAL PROCEDURES

Materials. The venom of Taiwan Cobra (*Naja naja atra*) was supplied by Cheng Hsin Tong Chemical Co., Ltd., Taipei, Taiwan, R.O.C. CM-Sephadex G-25 and Sephadex G-25, Sepharose 4B and Dextran T 70 were purchased from Pharmacia Fine Chemicals, Sweden. DE52 was from Whatman Ltd., England. Poly-L-Lysine (Molecular weight 4,000), heparin, Tris, NaDodSo$_4$ were supplied by Sigma Co., USA. 1,3,4,6-tetrachloro-3α,6α-diphenylglycouril was available as Iodogen from Pierce Chemical Co., USA. Na^{125}I was bought from New England Nuclear Co., USA. The plasma expander whose trade name was Moriamin D-2 was supplied by China Chemical & Pharmaceutical Co., Ltd., Taipei, Taiwan, R.O.C. According to the formula, its composition (w/v%) was: L-arginine HCl, 0.135;

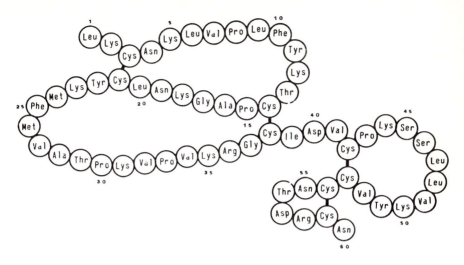

Fig. 1. The primary structure of cardiotoxin.[6]

L-histidine HCL·H$_2$O, 0.065; L-isoleucine, 0.090; L-leucine, 0.205; L-lysine
HCl·2H$_2$O, 0.370; L-methionine, 0.120; L-phenylalanine, 0.145; L-threonine,
0.090; L-tryptophan, 0.030; L-valine, 0.100; aminoacetic acid, 0.170; Dextran
T 70, 6.000; D-sorbitol, 5.000. All chemicals were reagent grade. Water was
double distilled.

Purification of CTX and Phospholipase A$_2$. These two proteins were purified
from the crude venom by CM-Sephadex C-25 column chromatography according to
Lo et al.[9] Phospholipase A$_2$ was further purified by a DE 52 column using
ammonium bicarbonate as eluent. Only preparations demonstrated to be
homogeneous by polyacrylamide disc electrophoresis were used for the study.

Preparation of anti-CTX antibody. Antisera against CTX were prepared
according to our previous procedure.[8] The antibody was purified from the antisera through affinity chromatography of Sepharose 4B attached with CTX. Coupling of CTX to the gel generally followed the procedure of Porath.[10]

Radioiodination of CTX. We followed generally the chloroglycouril
method.[11,12] Twenty mg of CTX and 5 mCi of Na^{125}I in 1.0 ml of 10 mM phosphate
buffer (pH 7.4) was incubated in a vessel plated with suitable amount of
Iodogen for 20 min at room temperature. The vessel was further rinsed twice
with the buffer. The rinses combined with the solution mixture was promptly
applied to a CM-Sephadex C-25 column (1.5 x 40 cm) preequilibrated with 0.2 M

NaCl-0.05 M Tris (pH 8.6). Washing the column with the initial buffer could
eliminate the unbound iodide. The [^{125}I]-CTX could be separated from its
parent toxin by eluting the column linearly from 0.2 to 0.4 M NaCl in 0.05 M
Tris (pH 8.6). The radioactive fraction was collected and desalted by a
Sephadex G-25 column.

Preparation of erythrocytes. RBC were collected from fresh human whole
blood by centrifugation at 1500 g for 5 min to remove the buffy coats. The
cell pellets were washed at least three times with plasma expander and were
used immediately after the treatment. Aged and young RBC were prepared by
centrifugation of RBC in a discontinuous gradient comprising 36%, 30%, 24% and
21% of Dextran T 40 solution according to Abraham et al.[13] The cells collected
at the bottom of centrifuge tube were aged RBC. Those gathered in the inter-
phase of 36-30%, 30-24% and 24-21% of dextran solution were named near-aged,
intermediate-young and young RBC, respectively. All the cells were washed at
least three times with plasma expander before use.

Cardiotoxin-pretreated erythrocytes were prepared by incubating RBC with
7.2 μM CTX for 15 min at 15° and 37°, respectively. CPE were collected by
centrifugation at 4° and washed several times with plasma expander solution
to remove free CTX. Finally, 0.25% CPE were resuspended in plasma expander
solution.

Hemolytical assay. RBC could disperse homogeneously in plasma expander and
did not sediment within a certain period. The intact and the lysed RBC gave
unequal optical density at 576 nm. Up to 0.25% of RBC in plasma expander, its
optical density was simply the linear addition of the contributions from both
intact and the lysed cells when they were coexisted. For most of the cases,
we investigated the interaction of 7.2 μM CTX with 0.25% of RBC (v/v) in plasma
expander. Hence the percentage of hemolysis (H) caused by CTX could be
estimated by the equation, $H = (A_{int} - A_{obs})/(A_{int} - A_{lys}) \times 100$. Here, A_{int}
and A_{lys} were the optical densities of the intact and the lysed RBC at 0.25%
concentration, respectively. A_{obs}, the optical density of 0.25% of RBC after
they had interacted with CTX, was measured by the method of Louw and Visser[14]
with some modification. We used recording spectrophotometer (Cary 14) attached
with jacket cell holders through which water was circulated to maintain the
temperature. The hemolysis could be monitored continuously in a cuvette containing the reaction mixture. The hemolysis could be determined also by
measuring the optical density of the incubation solution at 576 nm after the
unbroken RBC were removed. The percentage of the hemolysis evaluated by
either method agreed very well. The standard deviation for each hemolytical
assay was within 5-10%.

Determination of membrane fragility. Osmotic sensitivity of human RBC was measured according to Detraglia.[15] It was represented in terms of mean fragility and breadth of fragility distribution, which could be determined from the plot of percentage of hemolysis vs. NaCl concentration.

Binding assay. In the binding assay an aliquot of 0.25% of fresh RBC containing approximately 2.03×10^7 cells and suitable amount of the $[^{125}I]$- and unlabelled CTX in a total volume of 1.0 ml of plasma expander in a plastic tube was incubated at 15° or 37°C for 15 min. Routinely 8,400-9,800 cpm per μg of the $[^{125}I]$-CTX were used per assay. At the end of incubation, the tubes were centrifuged at 4°C to collect the cells. The RBC pellets were washed several times with plasma expander to remove the free CTX. They were then solubilized in 2 ml of 1% of $NaDodSO_4$ and transferred to a counting vial. The radioactivity was counted in a Packard A5000 gamma counter. The nonspecific binding was determined by measuring cpm of the $[^{125}I]$-CTX in the presence of 1000-fold of unlabelled CTX. It did not exceed 10-16% of the total count. Binding was measured in triplicate and the standard deviation was less than 10%. Cells were counted in a Coulter ZF-6 cell counter. Each experiment reported herein was repeated several times.

Actin polymerization. Actin of skeletal muscle of rabbit was prepared according to the method of Spudich and Watt.[16] Its polymerization was determined by the increase in either the solution viscosity or the optical density at 232 nm.[17,18]

RESULTS

Figure 2 gives the hemolysis caused by action of CTX at 7.2 μM or (Lys)n at 2.5 μM or phospholipase A_2 at 7.5 μM on human red cells at 0.25% in plasma expander at 37°C. Within 2 h, the phospholipase A_2 induced no hemolysis. On the other hand, the hemolysis caused by the CTX and the (Lys)n were comparable. In the CTX-induced hemolysis, it took around 20 min, before the cells started to rupture and the cells lasted for 6 h. This nonlytic period did not appear in the $(Lys)_n$-induced hemolysis. The inorganic salts, including NaCl, KCl, $CaCl_2$ and $MgCl_2$, each at 1.0 mM gave no influence on the CTX-induced hemolysis. However, when the salt concentration in the reaction mixture was raised to 10 mM, any one of the four salts inhibited the CTX-induced hemolysis considerably, despite that the divalent cations were more potent than the monovalent cations. One case for the $CaCl_2$ inhibition is shown in Fig. 2. At a fixed reaction time the percent hemolysis increased with increasing CTX concentration up to 7.2 μM and leveled off at higher CTX concentration (Fig. 3).

98

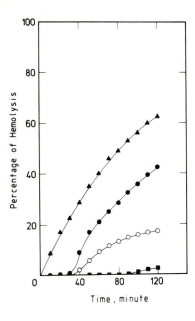

Fig. 2. The time course of hemolysis induced by various proteins. In plasma expander at 37°C, 0.25% (v/v) of RBC was interacted with polylysine at 2.50 μM (▲) or phospholipase A_2 at 7.5 μM in the presence of $CaCl_2$ at 1 mM (■) or CTX at 7.2 μM in the absence (●) or in the presence (○) of $CaCl_2$ at 10 mM.

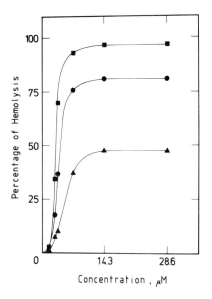

Fig. 3. The hemolysis of RBC at 0.25% (v/v) with CTX at various concentrations in plasma expander at 37°C. The incubation time is 2 h (▲) or 4 h (●) or 6 h (■).

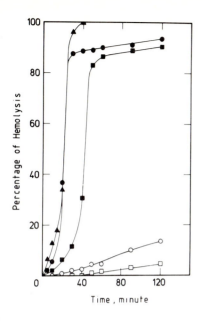

Fig. 4. The enhancement of phospho-
liapse A$_2$ on the hemolysis by CTX in
plasma expander containing 1 mM CaCl$_2$
at 37°C. RBC at 0.25% (v/v) were used
and phospholipase A$_2$ at 7.5 µM was
present for all assays. Symbols:
regular RBC in the presence of CTX at
7.2 µM ( ); CPE (37°C) in the absence
( ) and in the presence ( ) of anti-
CTX antibody at 300 µg/ml; CPE (15°C)
in the absence ( ) and in the presence
( ) of anti-CTX antibody.

In the situation of the (Lys)n-induced hemolysis, there appeared no nonlytic
period and addition of the inorganic salts gave no significant effect.

The red cells with different life spans responded to the CTX action unequally.
For instance, the percentage of the CTX-induced hemolysis at 37°C for 2 h was
20 for the young cells and 100 for the aged cells.

The CTX-induced hemolysis in plasma expander was temperature dependent. It
did not take place at temperature below 20°C. The middle point of the hemolysis
vs. temperature was at 29°C (not shown). Such a temperature-dependent property
did not appear in the (Lys)n-induced hemolysis. Even at the temperature as low
as 4°C, (Lys)n caused the hemolysis considerably.

The effect of phospholipase A$_2$ on the CTX-induced hemolysis or vice versa
is shown in Fig. 4. When the CTX at 7.2 µM coexisted, all the red cells were
broken within the nonlytic period of RBC in the absence of the enzyme. Phospholipase A$_2$ enhanced greatly also the hemolysis of CPE (37°C) and CPE (15°)
If CPE (37°C) and CPE (15°C were preexposed to anti-CTX antibody (300 µg/ml) at
15°C for 15 min, this synergistic effect of phospholipase A$_2$ was suppressed.
On the other hand, preexposure of the red cells to phospholipase A$_2$ followed
by the removal of free enzyme, showed no enhancement of the CTX-action; the
hemolysis of these cells was virtually the same as that of the cells without
pretreatment with phospholipase A$_2$. Apparently, phospholipase A$_2$ did not

facilitate the invasion of CTX into the cell membrane. Instead, the interaction of CTX with the red cell membrane made it easy for the enzyme to attack
membrane phospholipid and this accelerated the hemolysis.

In contrast, $(Lys)_n$-induced hemolysis was not enhanced by the presence of
7.5 μM phospholipase A_2. This revealed that the molecular mechanism of
hemolysis of RBC by CTX and $(Lys)_n$ was markedly different.

To assess how CTX affected the cell membrane, we compared the osmotic
sensitivities of red cells after they had interacted with CTX. The cells'
resistence to hypotonic media is shown in Fig. 5. The red cells in plasma
expander had mean fragility at 0.08 M NaCl and their breadth of fragility
from 0.06 to 0.09 M NaCl. In comparison with the red cells in phosphate
buffered saline, the red cells in plasma expander became slightly fragile.
Interaction of red cells with either phospholipase A_2 or $(Lys)_n$ did not result
in any change in membrane fragility before the rupture of cells. CPE (37°C)
had mean fragility at 0.115 M NaCl and their breadth of fragility distribution
extended a wide range. Apparently, the membrane of CPE (37°C) became more
fragile. If CPE (37°C) were preexposed to 300 μg/ml anti-CTX antibody for 15
min at 15°, their mean fragility and breadth of fragility distribution tended
to return to the values for the control cells in plasma expander solution. The
mean fragility and breadth of fragility distribution of CPE (15°C) remained
the same as the regular cells. If they were preincubated at 37°C for 15 min,
they became more fragile. When any one of the four aforementioned inorganic
salts each at 10 mM was included in the preparation of CPE (37°C), the
capability of these cells to resist hypotonic media did not deviate greatly
from the control cells. Apparently, the inorganic salts inhibited CTX to
loose the membrane structure. The breadth of fragility distribution suggested
that the inhibitory effect was more potent for divalent cations than for monovalent ones.

In the nonlytic period, the CTX could bind to the membrane of intact red
cells rapidly and specifically. A Scatchard plot[19] for the binding (Fig. 6)
revealed the presence of multiple classes of binding sites or negative
cooperativity among these binding sites on the intact cells. We analyzed the
plot according to two independent site models[20] and Table 1 lists the equilibrium constant and the capacities of binding at several situations. There
were 10^6 high affinity sites per cell with equilibrium constant of 10^6. For
the low affinity sites, there were 10^8 per cell with equilibrium constant of
10^3-10^4. Despite that the CTX-induced hemolysis did not take place at 15°C,
the binding at this temperature was virtually the same as that at 37°C. The
inorganic salts showed slight effect on the CTX-erythrocyte binding. Addition

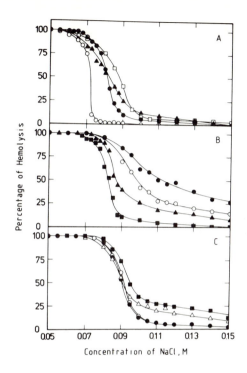

Fig. 5. Osmotic sensitivity of human erythrocytes at several conditions.
A. Membrane fragility was determined after 0.25% (v/v) of RBC had been
incubated at 37°C for 15 min in phosphate-buffered saline (○) or plasma
expander (●) or plasma expander containing 7.5 µM phospholipase A$_2$ (▲) or
plasma expander containing 2.5 µM polylysine (□). B. Membrane fragility of
CPE; (■), CPE (15°); (●) CPE (37°C); (▲), CPE (37°C) preexposed to anti-CTX
antibody at 300 µg/ml for 15 min at 37°C; (○), CPE (15°C) preincubated at 37°C
for 15 min. C. Membrane fragility of CPE (37°C) when they were prepared in
the presence of 10 mM CaCl$_2$ (●) or 10 mM MgCl$_2$ (○) or 10 mM KCl (△) or
10 mM NaCl (■). See the text for details.

of the inorganic salts to CPE (37°C) could not displace the bound CTX from the
cell membrane although they inhibited the hemolysis considerably (Table 1).

    To study further the interaction of CTX with membrane lipids, we prepared
unilamellar liposome vesicles entrapped with umbelliferone phosphate.[21] When
the liposomes were interacted with the CTX the releasing of umbelliferone
phosphate from the lipid vesicles reflected the capability of the CTX to

TABLE 1

THE EQUILIBRIUM CONSTANTS AND CAPACITIES OF
THE BINDING OF [^{125}I]-CTX TO HUMAN ERYTHROCYTES[a]

Assay Condition	Equilibrium Constant, M^{-1}		Binding Capacity, Sites/Cell	
	$K_{a1} \times 10^{-6}$	$K_{a2} \times 10^{-3}$	$R_1 \times 10^{-6}$	$R_2 \times 10^{-8}$
37°C	2.70	9.33	4.76	2.12
15°C	2.65	1.86	4.96	9.17
NaCl (10 mM)	3.03	6.70	4.03	3.01
KCl (10 mM)	2.82	4.75	4.66	2.33
CaCl$_2$ (10 mM)	4.77	5.73	2.49	5.19
MgCl$_2$ (10 mM)	4.76	5.36	2.67	3.38

[a] K_{a1} and K_{a2} represent the equilibrium constants of high and low affinity sites, respectively. R_1 and R_2 are the corresponding binding capacities.

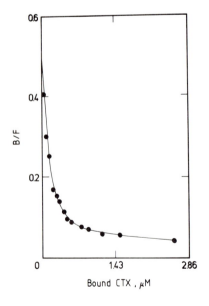

Fig. 6. Scatchard plot for the binding of [^{125}I]-CTX to RBC at 0.25% (v/v) in plasma expander at 37°C in the non-lytic period. The B/F on the ordinate represents the ratio of the bound to the free [^{125}I]-CTX.

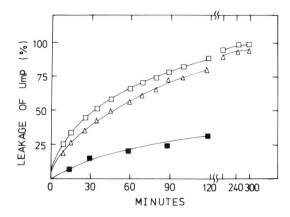

Fig. 7. CTX released umbelliferone phosphate (UmP) from the liposome bearing net charges. At 37°C, the liposome (0.014 mg of egg lecithin/ml) was incubated together with CTX: △- △-△ , 49.0 µM CTX; □-□-□, 71.5 µM CTX; ■-■-■, 10 mM Ca++ and 49.0 µM CTX.

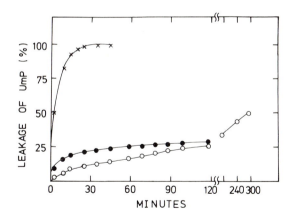

Fig. 8. The synergic effect of phospholipase A₂ on CTX action. The experimental condition was the same as described in Fig. 7: ○-○-○, 9.9 µM CTX; 9.9 µM CTX; ●-●-●, 0.86 µM phospholipase A₂; X-X-X, 9.9 µM phospholipase A₂.

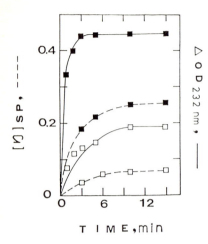

Fig. 9. Actin polymerization induced by cardiotoxin (2.3 μM) at 24°C in the presence (■) and in the absence (□) of 0.4 mM MgCl$_2$. Dashed and solid lines are the specific viscosity and the optical density at 232 nm, respectively. G-actin at 500 μg/ml in 2 mM Tris-HCl-0.2 mM CaCl$_2$-0.2 mM ATP-0.5 mM mercaptoethanol (pH 8.0) was used for the experiments.

destroy the membrane lipid structure. Figure 7 shows how the CTX caused the leakage of umbelliferone phosphate from liposomes bearing net negative charges at 37°C. The effect of 10 mM CaCl$_2$ on the leakage is shown also in the same figure. The CTX action was inhibited by 10 mM divalent cations including Ca^{++}, Mg^{++} and Ba^{++} and was diminished at lower temperature. Phospholipase A$_2$ enhanced the leakage caused by the CTX a great extent (Fig. 8). The CTX could release umbelliferone phosphate from liposomes bearing either no net charges or net positive charges although the releasing rates were slower than that from liposomes bearing net negative charges. Apparently, the negative charges in the liposome vesicles was not essential but promoted the CTX action to destroy the membrane lipid structure.

Incubating G-actin and CTX together at 27°C resulted in an increase in both the viscosity and the optical density (232 nm) of the reaction mixture (Fig. 9). Apparently, CTX induced the polymerization of G-actin, and the higher the CTX concentration in the reaction mixture, the greater extent the polymerization (Fig. 10). The CTX-induced polymerization was enhanced in the presence of 0.4 mM MgCl$_2$ (cf. all the curves in Fig. 9). Preexposure of G-actin to deoxyribonuclease I suppressed the capability of CTX to induce the polymerization (Fig. 11). On the other hand, deoxyribonuclease I could depolymerize the actin polymer formed by the interaction of CTX with G-actin (Fig. 11). Such an inhibitory effect of deoxyribonuclease I was observed also in the (Lys)$_n$-induced actin polymerization (Fig. 11).

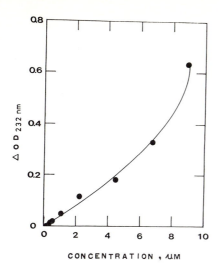

CONCENTRATION , μM

Fig. 10. Action polymerization induced by various amounts of cardiotoxin. G-actin at 100 μg/ml in the buffer of Fig. 9 was incubated together with cardiotoxin at 24°C for 15 min.

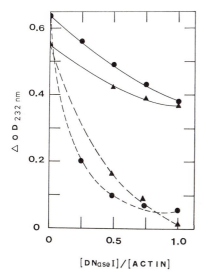

[DNase I]/[ACTIN]

Fig. 11. The effect of deoxyribo-nuclease I(DNase I) on the capability of cardiotoxin (●) and polylysine (▲) to induce actin polymerization. Depoly-merization (——) caused by DNase was measured after various amounts of the enzyme and the actin polymer which was formed by preincubation of G-actin and CTX or polylysine at 24°C for 10 min. The inhibitory effect of DNase I on the actin polymerization (----) was determined after various amounts of the enzyme had incubated with G-actin and CTX or polylysine at 24°C for 10 min. In all experiments, the final concen-trations were 8.4 μM for cardiotoxin, 2.2 μM for polylysine (MW 4,000), and 100 μg/ml for G-actin.

DISCUSSION

Evidence accumulated from pharmacological studies support that the mode of action of CTX is on the cell membranes of different origins.[22] Several biochemical studies show that CTX can affect the activity of several enzymes in the isolated cell membrane.[4,5] Nevertheless, evidence collected thus far does not seem sound enough to explain the molecular mechanism of CTX action. The present work may shed some light on how and why CTX causes the disruption of cells.

The characteristic of how CTX destroyed the liposome vesicles reveals that the destruction of lipid micelle by CTX cannot be accounted for solely by its binding to the negative charged groups of liposome. It is likely that CTX may be able to intercalate into the micelle and hence weaken the hydrophobic interaction among the hydrocarbons of lipid structure.

The length of the nonlytic period in the CTX-induced hemolysis in plasma expander solution allowed us to study how the CTX affected the cells before their rupture. We demonstrated that the CTX bound to intact cells and disintegrated the membrane structure very rapidly before the cells began to disrupt and the bound CTX molecules were mostly on the cell membrane. This kind of binding is apparently influenced neither by the status of membrane nor by the extracellular environments; note that temperature showed no effect and inorganic salts gave only slight effect on the CTX erythrocyte binding (Table 1). In line with the finding that anti-CTX antibody could not displace the bound CTX, it is likely that part or whole of these bound CTX molecules are apparently on the extracellular membrane. Which membrane component(s) can bind with CTX remains to be solved.

We assume that there are two kinds of affinity sites in the erythrocyte membrane for the CTX binding. Comparing the effects of the CTX dosages on the hemolysis (Fig. 3) and the binding (Fig. 6) reveals that increasing the binding of the CTX to the low affinity site did not result in the enhancement of hemolysis. It is apparent that action of the CTX on the high but not the low affinity sites may be associated closely with the events leading to membrane disintegration and cell rupture. Hence, searching the membrane component(s) for the high affinity sites is crucial to understand the molecular mechanism of the CTX action. On the basis of the distribution of proteins and lipids in the red cell ghost,[23,24] the major membrane components of one human red cell are about 4.5×10^8 phospholipid molecules, 1.2×10^6 band-3 molecules, 3.0×10^5 spectrin molecules and 5.0×10^5 actin molecules. If the two CTX affinity sites are not at the same kind of molecule, we suspect from the CTX binding capacity that the high affinity sites may be at the major membrane protein(s) and the low affinity sites should be at the membrane phospholipids. Having treated intact RBC with neuraminidase or chymotrypsin to cleave sialic acid residues or band-3 at the external cell surface according to the procedure of Massamiri et al.[25] or Steck et al.,[26] we found that the treated cells did not lose the affinity sites for the CTX binding (unpublished observation). Hence, the high affinity sites seem not at the external polypeptide fragment of band-3. On the other hand, the high and low affinity sites may be both at the membrane phospholipids. If it is so, the negative cooperativity among these two kinds of affinity sites may account for the characteristic of the Scatchard plot shown in Fig. 6. Hence, binding of the CTX molecules to the high affinity sites of phospholipids may decrease the affinity of the further binding of the CTX to the phospholipids. No data are available now to support this possibility.

Does the event(s) which effects membrane disintegration and cell rupture occur at outer membrane? This question can not be answered straightforwardly right now but several findings of the present work are helpful to deduce it. First, the event did not occur at lower temperature apart from the physiological range. Second, the anti-CTX antibody could prevent the event though it could not displace the bound CTX from the cell membrane. Third, the inorganic salts inhibited the event significantly but they were not able to reduce the CTX-erythrocyte binding. Obviously, the sole interaction of the bound CTX molecules with their high affinity sites seems sufficient to cause membrane disintegration and cell disruption. We suspect that the CTX action sites may be distinct from the affinity sites. If it is so, the former must be no more than the latter and the event may proceed at the membrane spanning domain and/or at inner membrane rather than at the outer membrane. Hence, transport of the bound CTX molecules to their action sites is crucial for the event. This process is not impossible, considering the compatibility of the CTX in solvents less polar than water[27] and the capability of the CTX to intercalate or penetrate into membrane lipids.[21,28] At lower temperatures the cell membranes are in more or less rigid state and it is difficult, if not impossible, to transport the CTX molecules across such membranes to their action sites. The anti-CTX may draw the bound CTX molecules to prevent them from reaching their action sites. The inorganic salts seem to inhibit the interaction of the CTX molecules with their action sites. To sum up the characteristics of the CTX-erythrocyte binding and the possible action site of CTX, the relevance of actin polymerization induced by CTX to how it caused membrane disintegration and cell disruption, though not clear at the present time, should not be overlooked.

Although the synergistic effect of phospholipase A_2 and CTX on the hemolysis was observed at early time,[29] it is not well understood whether the CTX action on the cell membrane facilitates the enzyme to hydrolyze the membrane phospholipids or the enzyme action makes the invasion of CTX easy. We demonstrated that the former is indeed the right one. Unlike CTX, polylysine showed no strong activity to disintegrate membrane structure before the cell disruption. The $(Lys)_n$-induced hemolysis was neither enhanced by the presence of phospholipase A_2 nor inhibited by inorganic salts. It showed also no temperature dependence. Perhaps it happened at the outer membrane. Clearly, the molecular mechanisms of the CTX- and the (Lys)n-induced hemolysis can not be the same.

In summary, the present work illustrated the chain events in cell membrane during the CTX-induced hemolysis. The information obtained will be beneficial to elucidate the molecular mechanism of CTX function.

ACKNOWLEDGEMENT

This work was sponsored partly by the National Science Council, Taipei, Taiwan, R.O.C.

REFERENCES

1. Lee, C. Y. (1972) Ann. Rev. Pharmacol., 12, 265-286.
2. Tu, A. T. (1977) in Chemistry and Molecular Biology, Chapter 19, New York: John Wiley and Sons, pp. 301-320.
3. Yang, C. C. (1974) Toxicon, 12, 1-43.
4. Karlsson, E. (1979) Hand. Exp. Pharmacol., 52, 159-212.
5. Condrea, E. (1974) Experientia (Basel), 30, 121-129.
6. Narita, K., et al. (1978) Int. J. Peptide Protein Res., 11, 229-237.
7. Chen, Y. H., Pan, B. T., and Lee, C. P. (1982) Biochem. Biophys. Acta, 702, 193-196.
8. Hung, M. C., et al. (1978) Biochem. Biophys. Acta, 535, 178-187.
9. Lo, T. B., Chen, Y. H., and Lee, C. Y. (1966) J. Chinese Chem. Soc., 13, 25-37.
10. Porath, J., Axen, R., and Ernback, S. (1967) Nature, 215, 1491-1492.
11. Markwell, M. A. K. and Fox, C. F. (1978) Biochemistry, 17, 4807-4817.
12. Fraker, P. J. and Speck, J. C., Jr. (1978) Biochem. Biophys. Res. Commun., 80, 849-857.
13. Abraham, E. C., et al. (1975) Biochem. Medicine, 13, 56-77.
14. Louw, A. I. and Visser, L. (1978) Biochem. Biophys. Acta 512, 163-171.
15. Detraglia, M., et al. (1974) Biochem. Biophys. Acta, 345, 213-219.
16. Spudich, J. A. and Watt, S. (1971) J. Biol. Chem., 246, 4866-4871.
17. Higashi, S. and Oosawa, F. (1965) J. Mol. Biol., 12, 843-865.
18. Spudich, J. A. and Cooke, R. (1975) J. Biol. Chem., 250, 7485-7491.
19. Scatchard, G. (1949) Ann. N.Y. Acad. Sci., 51, 660-672.
20. Thakur, A. K., Jaffe, M. L., and Rodbard, D. (1980) Anal. Biochem., 107, 279-295.
21. Chen, Y. H., Lai, M. Z., and Kao, L. S. (1981) Biochem. Internat., 3 385-390.
22. Chang, C. C. (1979) Hand. Exp. Pharmacol., 52, 309-376.

23. Guidotti, G. (1972) Arch. Int. Med., 129, 194-201.
24. Branton, D., Cohen, C. M., and Tyler, J. (1981) Cell, 24, 24-32.
25. Massamiri, Y., et al. (1979) Anal. Biochem., 97, 346-351.
26. Steck, T. L., Ramos, B., and Strapazon, E. (1976) Biochemistry, 15, 1154-1161.
27. Hung, M. C. and Chen, Y. H. (1977) Int. J. Peptide Protein Res., 10, 277-285.
28. Bougis, P., et al. (1981) Biochemistry, 20, 4915-4920.
29. Vogt, W., et al. (1970) Naunyn Schmiedebergs Arch. Pharmak., 265, 442-454.

STUDIES ON MITOCHONDRIAL ENERGY TRANSDUCTION: UTILIZATION OF

9-AMINO-ACRIDINE DYES AS FLUORESCENCE PROBES

C. P. LEE
Department of Biochemistry, Wayne State University School of Medicine,
Detroit, Michigan 48201

It is now well established that mitochondrial metabolism is chiefly concerned
with liberation of energy from substrates by respiration in a form that can be
used for various energy-consuming cell functions, and the inner mitochondrial
membrane is the site of the cellular respiratory chain and related energy
transducing and carrier facilitated ion translocating systems. The respiratory
chain-linked oxidative phosphorylation system consists of four electron transfer complexes (complexes I, II, III, and IV) and an ATP synthetase complex
(complex V). The energy-coupling sites 1, 2 and 3 are associated with complexes
I, III, and IV, respectively. Considerable evidence has accumulated which
indicates that complexes I, III, IV and V can generate an H^+ gradient across
the inner mitochondrial membrane in which they are located; and that energy
transfer between energy-transducing units can take place via an H^+ gradient.
It is, therefore, apparent that H^+ transfer is a key component of the primary
energy-coupling reaction. However, it has yet to be settled as to how H^+
function in this reaction.[1-3] Among the various models and mechanisms proposed, one can identify two categories which are fundamentally different.[4] One
category of such mechanisms has an underlying postulate that energy coupling
is a transmembrane reaction generating an electrochemical proton gradient,
$\Delta\tilde{\mu}_{H^+}$, across the inner membrane. This category includes all the variants and
derivations of the original chemiosmotic hypothesis.[5] The other group of
mechanisms has an underlying postulate that energy coupling is an intramembrane
reaction involving direct H^+ transfer between membrane proteins, or assemblies
of proteins, in localized areas within the membrane.[6] This group includes the
limited energy transfer domain model,[7,8] the conformational hypothesis of
energy coupling,[9] and the concept of energization occurring through a membrane
Bohr effect.[10]

The chemiosmotic hypothesis offers a general mechanism for transfer of
energy between various membrane-associated energy transducing units, and has
made great impact on the advancement of our knowledge in the field of bioenergetics. However, it should be pointed out that the fundamental problem,

namely, the chemical reaction mechanism, of how energy is conserved and
utilized by the energy transducing units at the molecular level remains to be
elucidated.

For the elucidation of the molecular reaction mechanisms of energy coupling,
various approaches have been applied, which can be classified into four
categories: (A) Resolution and reconstitution; (B) ion translocation;
(C) utilization of probes; and (D) molecular structure and membrane topology
of constituents of energy transducing units. In this communication, I shall
try to summarize the experimental findings of our recent studies on the
utilization of 9-aminoacridine dyes as fluorescence probes for energized
submitochondrial membranes. All the experiments are made with tightly coupled
beef heart submitochondrial membranes.[11]

9-Aminoacridines as fluorescence probes. Figure 1 shows the structural
formulas and the pKa values of the ring nitrogen of the acridine nucleus of
three derivatives of 9-aminoacridine which have been widely used as fluorescence
probes for energy transducing membranes.[3,12] A typical experiment is also
shown in a schematic diagram illustrating that when any one of the 9-amino-

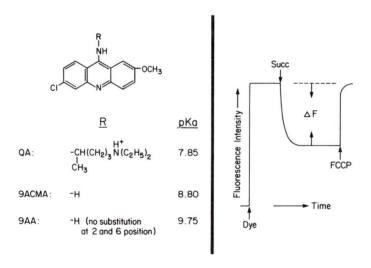

R	pKa
QA: $-CH(CH_2)_3 \overset{H^+}{N}(C_2H_5)_2$ $\quad CH_3$	7.85
9ACMA: $-H$	8.80
9AA: $-H$ (no substitution at 2 and 6 position)	9.75

Fig. 1. Structural formulas and pKa values of 3 derivatives of 9-aminoacridine
and the fluorescence responses of the dye associated with submitochondrial
membranes. QA: Quinacrine; 9ACMA: 9-amino-6-chloro-2-methoxy-acridine;
9AA: 9-aminoacridine.

$$\Delta pH = log \left[\frac{Q}{V (I - Q)} \right]$$

Q: FRACTION OF FLUORESCENCE THAT WAS QUENCHED.

V: RATIO OF PARTICLE INTERNAL VOLUME TO THE
     EXTERNAL SOLUTION VOLUME.

ASSUMPTIONS:

A. THE FLUORESCENCE INTENSITY HAS A LINEAR RELATIONSHIP
   WITH THE DYE CONCENTRATION.

B. THE FLUORESCENCE INTENSITY OF THE DYE IS COMPLETELY
   QUENCHED WHEN TAKEN INTO THE PARTICLES.

Fig. 2. Equation for the estimation of ΔpH across the membrane formulated by
Schuldiner et al.[17] with the assumptions indicated.

acridine derivatives is incubated with tightly coupled submitochondrial
membranes, a fluorescence lowering is induced upon energization of the membrane
by the addition of respiratory substrate (succinate); the fluorescence intensity
is restored upon the subsequent addition of an uncoupler, FCCP (p-trifluoro-
methoxycarbonyl cyanide phenylhydrazone). The kinetics and extent of the
energy-linked fluorescence lowering differ from dye to dye and also depend
on the experimental conditions.[12-16] Furthermore, it has also been shown[12-15]
that any agents which abolish the energy-linked proton uptake will also abolish
the energy-linked fluorescence lowering of these dyes. Therefore, the energy-
linked fluorescence lowering of 9-aminoacridines is closely related to the
energy-linked proton uptake of energy-transducing membranes. Schuldiner and
associates[17] have used the fluorescence lowering of 9-aminoacridine to calculate
the magnitude of the pH gradient across the membrane. The equation for the
calculation of ΔpH and the assumptions for its formulation are shown in Fig. 2.
This equation has been widely used by many investigators to estimate the
magnitude of ΔpH across the membrane of a variety of systems, e.g., chloroplasts,[17] chromatophores,[18] and submitochondrial membranes.[19] It is apparent

114

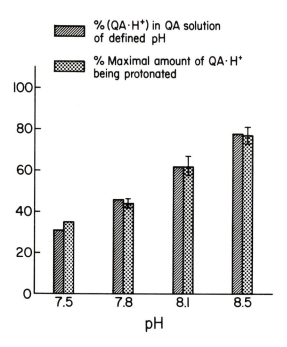

Fig. 3. Comparison of the maximal amount of QA·H$^+$ being protonated into QA·2H^{++} with the percentage of QA·H$^+$ present in a quinacrine solution of defined pH. Data from Huang et al.[20]

that the accuracy of the estimated magnitude of the pH gradient calculated using Schuldiner's equation is dependent on the validity of the assumptions.

Quinacrine as a quantitative probe for H$^+$ taken up by energized submito-chondrial membranes. Based on the studies of the absorption spectra, the fluorescence excitation and emission spectra, and the fluorescence quantum yield of quinacrine (QA) either alone or associated with energized and non-energized submitochondrial membranes, we have concluded that the energy-linked fluorescence lowering of QA is due to the protonation of the monoprotonated QA molecules into the diprotonated form.[20] This conclusion is further sub-stantiated by the quantitative correlation between the maximal amounts of QA being protonated when associated with energized membranes and the amounts of

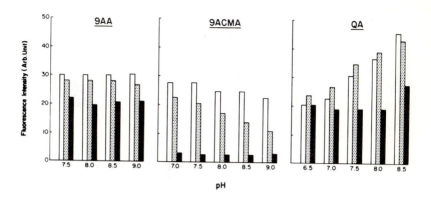

Fig. 4. The fluorescence intensity of 9AA, 9ACMA and QA alone (▯), associated with nonenergized (▨) and energized (■) membranes, as a function of pH. The reaction mixture consisted of 150 mM sucrose, 30 mM Tris-sulfate buffer, pH 7.5, 1.6 mM NaSCN, 1.7 µM dye and 0.3 mg protein of tightly coupled beef heart submitochondrial membranes. 3.3 mM succinate was employed as the energy-yielding substrate.

monoprotonated QA molecules initially in the reaction mixture as controlled by the pH of the media (Fig. 3). The virtually complete conversion of $QA \cdot H^+$ into $QA \cdot H_2^{++}$ implies that the $QA \cdot H_2^{++}$ molecules formed are associated with the energized membranes in such a fashion that they are not readily equilibrated with the external medium. Quinacrine can therefore serve as a quantitative probe for the H^+ taken up by the membrane upon energization. Two immediate questions arise: (A) Can this conclusion be applied to the other two 9-amino-acridine dyes, e.g., 9AA and 9ACMA (cf. Fig. 1)? and (B) Where are the diprotonated QA molecules located?

Energy-linked fluorescence responses of 9AA and 9ACMA. A comparative study of the energy-linked fluorescence lowering of 9AA, 9ACMA and QA as a function of the pH of the reaction mixture is presented in Fig. 4. Upon energization of the membrane even at pH 7.5, a more than 90% decrease in the fluorescence intensity of 9ACMA is seen. Similar results, though to a lesser extent, are also seen with 9AA. Since the protonated forms of 9ACMA and 9AA are more abundant than the neutral species in this pH range (cf. Fig. 1), and also exhibit greater fluorescence than the corresponding neutral species, the observed energy-linked fluorescence decreases of 9ACMA and 9AA cannot be viewed solely as resulting from protonation of the neutral species as in the case of QA. The fluorescence quantum yields of 9AA and 9ACMA are virtually constant over the pH range from 7.0 to 9.5,[21] in contrast to that of QA the quantum yield of QA at pH 9.0 is approximately 3 times that at pH 6.0.[20] All these data indicate that the energy-linked fluorescence lowering of 9AA and 9ACMA must result from mechanisms other than the protonation of the neutral species of the dye molecules, presumably from the formation of nonfluorescent complex(es) between the dye molecules and some membrane components.

Localization of diprotonated QA molecules in energized submitochondrial membranes. We have just demonstrated that QA can serve as a quantitative probe for the H^+ taken up by the energized submitochondrial membranes. One of the immediate questions concerns the location of the diprotonated QA molecules. Are they free in the inner aqueous vacuole of the vesicles, as predicted by the transmembrane (chemiosmotic) hypothesis, or are they bound to the membrane, as predicted by the intramembrane hypothesis? One way to study the microenvironment of a fluorescent dye molecule is to measure its degree of fluorescence polarization. The fluorescence polarizations of 9AA, 9ACMA, and QA either alone or associated with nonenergized and energized submitochondrial membranes are summarized in Fig. 5. In the case of QA, submitochondrial membranes even in the nonenergized state induce a significant increase in polarization. A further increase can be seen when the membranes become energized. On the other hand, in the case of 9AA and 9ACMA, the degree of fluorescence polarization of the dye associated with submitochondrial membranes in the nonenergized or energized state does not differ from that of the dye in the medium (free from membrane), and is approximately zero. These data indicate that the molecular rotation of QA is strongly hindered when the QA molecules are associated with the submitochondrial membrane even in the nonenergized state, whereas the rotation of those molecules of 9AA and 9ACMA reflected by the polarization data are unaffected by the membranes. The fact that QA possesses a long side-chain substitution at the 9-amino group of the acridine nucleus (cf. Fig. 1) which is not present in 9AA and 9ACMA suggests that the hindrance of the molecular rotation of QA may result from a specific interaction between the membrane and the QA molecules via this side-chain. A change in the microenvironment of QA upon energization of the membrane is reflected by the further increase in the degree of fluorescence polarization of QA. The close association of QA molecules with the energized membrane is further supported by the abrupt discontinuity at 15° in the slope of the plot of polarization vs. 1/absolute temperature,[21] a temperature at which a phase

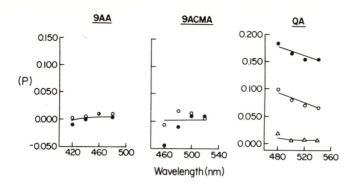

Fig. 5. The degree of fluorescence polarization (P) of 9AA, 9ACMA, and QA under various conditions. Data from Huang and Lee.[16] The dye alone in the medium (Δ——Δ), associated with nonenergized (O——O) and energized (●——●) membranes.

transition of the mitochondrial inner membrane has been shown to occur.[15,22] These data further substantiate our original contention[20] that the energy-linked fluorescence lowering of QA results from the protonation of the mono-protonated QA into the diprotonated form, and that these diprotonated QA molecules are tightly bound to the membrane.

The apparent zero polarizations of both 9AA and 9ACMA, when associated with either nonenergized or energized membranes, indicates that the polarization data reflect only those dye molecules which are free in the reaction medium, and that any dye molecules associated with the membrane are nonfluorescent. The decreases in the fluorescence intensities of 9AA and 9ACMA in the presence of submitochondrial membranes (cf. Figure 4) must be due to the decreases in their effective concentrations in the reaction medium resulting from the binding of the dye to the membrane. The binding of dyes to the membrane increases greatly upon energization (unpublished results). Unlike QA, the membrane-bound 9AA and 9ACMA molecules do not exhibit fluorescence, perhaps as a result of the formation of nonfluorescent complex(es) between the dye molecules and some charged membrane components. Naturally, these membrane-bound 9AA and 9ACMA molecules would not be reflected in the polarization data.

Effect of local anesthetics. Another line of evidence supporting the idea that there are specific binding sites for QA, 9AA and 9ACMA with the membrane comes from studies with local anesthetics.[21,23] Chlorpromazine, at low concentrations which do not exhibit any uncoupling effect, inhibits the energy-linked fluorescence lowering of these dyes, and decreases the binding of them to the membrane. If there are specific binding sites for QA, 9AA and 9ACMA with the membrane, then the ΔpH determinations derived from studies with these probes according to the equation (cf. Fig. 2) formulated by Schuldiner and associates[17] would be severely overestimated, since it has always been assumed that there are no interactions between the membrane and the probe. Indeed, such interaction may also occur between the membrane and other amines, e.g., methylamine and ammonia. Whether our conclusion is applicable to other membrane systems remains to be elucidated, since the nature of the interaction of any specific 9-aminoacridine probe may differ considerably among the various membrane systems. These differences could be due, in large part, to the exact experimental conditions employed or the composition of the membrane system under investigation, e.g., chloroplasts,[24] which are significantly different from submitochondrial membranes in their composition of both lipids and proteins.

Photoaffinity analog (QA-N$_3$) of QA. In order to identify the binding sites of QA with the membrane, we have synthesized the azido-derivative of QA as the photoaffinity label for the interaction of QA with submitochondrial membranes.[25,26] QA-N$_3$ differs from QA only in the substitution at the 6-position of the acridine nucleus. The Cl was replaced by the azido group (cf. Fig. 1). We have shown that QA-N$_3$ is similar in chemical and spectroscopic properties to QA.[26] The fluorescence response of QA-N$_3$ in relation to changes in the metabolic state of submitochondrial membranes as compared with that exhibited by QA are shown in Fig. 6. The same extent of energy-linked fluorescence lowering, approximately 30%, is seen in both cases. These data indicate that QA-N$_3$ can serve as a photoaffinity label for the interaction of QA with the membrane.

Cross-linking of QA-N$_3$ to submitochondrial membranes. The photolyzed products of QA-N$_3$ bind covalently to both energized and nonenergized membranes, with yields of 60% and 40%, respectively. Sodium dodecyl sulfate electrophoresis of the labeled membranes showed that there are primarily three peptides labeled, with corresponding molecular weights of 54K, 33K and 30K. No significant difference in the labeling pattern are seen between the membranes which were labeled while in the nonenergized or the energized state.[26]

The covalently bound dye retained capability to produce the energy-linked fluorescence response if the dye was bound to the membrane while in the energized state, but not if bound to the membrane while in the nonenergized state. Enzymatic activity profiles revealed that there are at least two distinct sites of interaction of the dye with the membrane. One site is in the vicinity of succinate dehydrogenase, while the other is along the respiratory chain on the oxygen side of NADH dehydrogenase. In addition, the latter site appears to be more susceptible to photo-induced inactivation

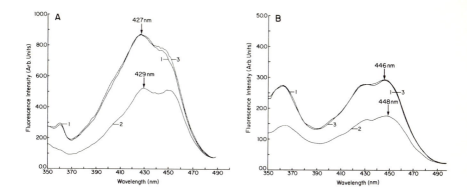

Fig. 6. Fluorescence excitation spectra of QA (A) and QA-N$_3$ (B) under various conditions. The reaction mixture consisted of 150 mM sucrose, 30 mM phosphate buffer, pH 7.5, 0.9 mg protein of tightly coupled beef heart submitochondrial membranes and 3.3 μM dye. (1) no further addition; (2) with 3.3 mM sucinate; (3) with 3.3 mM sucinate and 1 μM FCCP. From Mueller et al.[26]

when the membrane is in the nonenergized state as compared to the energized state.[26,27]

The fluorescence excitation and emission spectra of the dye (QA-N$_3$) cross-linked to the membranes under energized and nonenergized conditions are shown in Fig. 7A. A distinctive difference is seen in the 440 nm region of the excitation spectra, with a much larger maximum at 440 nm for the dye bound to the nonenergized membrane. A shift in the emission maximum from 516 to 505 nm is also seen for the dye bound to the nonenergized membranes.[26] These data suggest either that the covalently bound products differed when produced in the nonenergized and energized membranes (i.e., the photolyzed dye bound to different membrane components), or that the dye was covalently bound to the same components (sites), but that the environment surrounding the site differed in the energized and nonenergized membranes. If the former is the case, then the spectra should still differ after the membranes are dissolved. Fractionation of the labeled membranes into lipid and protein products have therefore been made. As shown in Fig. 7, the fluorescence spectral difference seen for the energized and nonenergized labeled membranes were very similar to those observed for the extracted lipid products (Fig. 7B), but distinct from those obtained for the extracted protein products (Fig. 7C). Furthermore, no

120

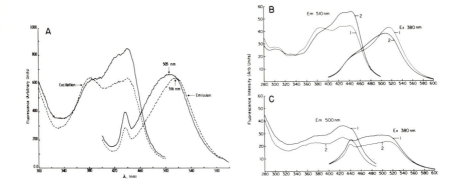

Fig. 7. Fluorescence excitation and emission spectra of the dye cross-linked
to the membrane. Intact membrane (A) labeled while in the energized (----) and
nonenergized (——) state. Fluorescence spectra of the lipid (B) and the
protein (C) components derived from the membranes labeled with QA-N$_3$ while in
the energized (1) and nonenergized (2) state. From Mueller et al.[26]

significant difference was seen in the excitation and emission spectra of the
labeled protein products while the membranes were labeled in the energized or
nonenergized states. It is apparent that the selectivity of the insertion
reaction appears to be associated with the hydrophobic region of the membrane.[26]
Identification and quantitation of the lipid components being labeled are now
under investigation.

ACKNOWLEDGEMENTS

I would like to acknowledge the experimental collaboration of Dr. C. S.
Huang, Dr. D. M. Mueller and Mr. S. J. Kopacz. Thanks are also due to Dr. M. E.
Martens for valuable discussions and editorial assistance in preparing this
manuscript. Supported by grants from NSF, PCM 7808549 and NIH, GM 22751.

REFERENCES

1. Boyer, P. D., et al. (1977) Ann. Rev. Biochem., 46, 955-1025.
2. Wikström, M. and Krab, K. (1980) in Current Topics in Bioenergetics,
   Sanadi, R., ed., Vol. 10, New York: Academic Press, pp. 51-101.
3. Lee, C. P. and Storey, B. T. (1982) in Mitochondria and Microsomes, Lee,
   C. P., et al., eds., Boston: Addison-Wesley, pp. 121-154.
4. Storey, B. T. and Lee, C. P. (1981) Trends in Biological Sciences, 6, 166-
   170.

5. Mitchell, P. (1966) Biol. Rev., 41, 455-502.
6. Williams, R. J. P. (1979) Biochim. Biophys. Acta, 505, 1-44.
7. Ernster, L. (1977) Ann. Rev. Biochem., 46, 981-995.
8. Williams, R. J. P. (1978) FEBS Lett., 85, 9-19.
9. Boyer, P. D. (1975) FEBS Lett., 58, 1-6.
10. Chance, B. (1972) FEBS Lett., 23, 3-20.
11. Lee, C. P. (1979) Methods Enzymol., 55, 105-112.
12. Kraayenhof, R., Brockelhurst, J. R. and Lee, C. P. (1976) in Concepts in Biochemical Fluorescence, Chan, R. F. and Edelhoch, H., eds., New York: Marcel Dekker, pp. 767-809.
13. Lee, C. P. (1971) Biochemistry, 10, 4375-4381.
14. Lee, C. P. (1973) in International Symposium on Mechanisms of Bioenergetics, Azzone, G. F., et al., eds., New York: Academic Press, pp. 115-126.
15. Lee, C. P. (1974) BBA Library 13, 337-353.
16. Huang, C. S. and Lee, C. P. (1978) in Frontiers of Biological Energetics From Electrons to Tissues, Dutton, L. P., Leigh, J. S., and Scarpa, A., eds., New York: Academic Press, pp. 1285-1292.
17. Schuldiner, S., Rottenberg, H., and Avron, M. (1972) Eur. J. Biochem., 25, 64-70.
18. Baccarini-Melandri, C. A. and Melandri, B. A. (1974) Eur. J. Biochem., 47, 121-131.
19. Rottenberg, H. and Lee, C. P. (1972) Biochemistry, 14, 2675-2680.
20. Huang, C. S., Kopacz, S. J., and Lee, C. P. (1977) Biochim. Biophys. Acta, 459, 241-249.
21. Huang, C. S., Kopacz, S. J., and Lee, C. P. (1983) Biochim Biophys. Acta, 722, 107-115.
22. Vanderkooi, J., et al. (1974) Biochemistry, 13, 1589-1595.
23. Mueller, D. M. and Lee, C. P. (1982) FEBS Lett., 137, 45-48.
24. Haraux, F. and DeKonchkovsky, K. (1980) Biochim. Biophys. Acta, 592, 153-169.
25. Mueller, D. M., Hudson, R. A., and Lee, C. P. (1981) J. Amer. Chem. Soc. 103, 1860-1862.
26. Mueller, D. M., Hudson, R. A., and Lee, C. P. (1982) Biochemistry, 21, 1445-1453.
27. Mueller, D. M., et al. (1979) in Membrane Bioenergetics, Lee, C. P., et al., eds., Boston: Addison-Wesley, pp. 507-520.

STEROID RECEPTOR MECHANISM: ANDROGEN ACTION IN THE RAT VENTRAL PROSTATE

SHUTSUNG LIAO, RICHARD A. HIIPAKKA, SHEILA M. JUDGE,

ALAN G. SALTZMAN, KAREN SCHILLING, AND DAVID WITTE
Ben May Laboratory for Cancer Research and Department of Biochemistry,
University of Chicago, Chicago, Illinois 60637

ABSTRACT

Testosterone, the major testicular androgen, is converted by a NADPH-dependent 5α-steroid reductase to 5α-dihydrotestosterone (DHT) in the rat ventral prostate. DHT is the active intracellular androgen which binds to a high affinity steroid-specific receptor. 5α-Reductase deficiency can limit normal male sexual differentiation while the use of a reductase inhibitor can suppress the growth of the prostate. Certain antiandrogens also act by competing with DHT for binding to the receptor.

DHT-receptor complex interacts with nuclear chromatin and presumably enhances the synthesis of certain RNAs that are necessary for the production of specific proteins. The receptor-chromatin interaction can be controlled by factors such as pyridoxal phosphate and certain phosphatase inhibitors. Also a specific protein factor may be involved in feed-back control of receptor activity by inhibiting the interaction of the DHT-receptor complex with DNA.

The receptor complex may be released from chromatin by binding to specific sequences of RNA. Such an interaction may be involved in the processing and stabilization of specific mRNA for hormone-induced proteins.

Androgen-receptor complex that is released from nuclear chromatin appears to go through a dynamic process of inactivation and reactivation by an energy-dependent mechanism. This process may involve dephosphorylation and phosphorylation of the receptor or a closely related regulatory factor.

INTRODUCTION

The biochemical mechanisms involved in androgen action have been thoroughly studied in the rat ventral prostate.[1,2] In this organ, testosterone, the major blood androgen produced by testis, is converted to 5α-dihydrotestosterone (DHT) which binds to a specific receptor protein and promotes the interaction of receptor with nuclear chromatin. This process is believed to play a key role in the modulation of the biosynthesis of certain RNAs that are necessary for the growth and function of the prostate. The details of the molecular mechanisms involved in this process, however, are not clear. In addition, very

124

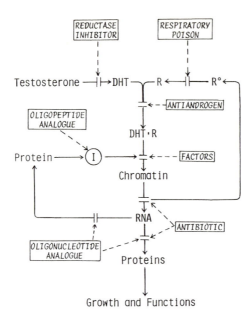

Fig. 1. A working model for the steps involved in intracellular cycling of the androgen receptor in target cells. In this hypothetical model, the receptor protein (R°) is activated by an energy-dependent process that is sensitive to respiratory poisons such as CN, azide, and DNP. The activated receptor (R) then binds an active androgen such as DHT that may be formed from a precursor such as testosterone. The androgen-receptor complex (DHT·R) is transformed in a temperature-dependent step to the form that can be retained tightly by chromatin. This receptor-chromatin interaction may modulate the synthesis of mRNA and certain proteins necessary for the growth and functions of the prostate. Some protein factors may play a feed-back control in regulating the interaction of the receptor complex with chromatin.

little is known about the dynamic status and regulation of the receptor activity in prostate target cells. This article summarizes our own work on some of these aspects. The individual areas to be discussed are shown in Fig. 1.

CONVERSION OF TESTOSTERONE TO DHT

Our studies in the early 1960s showed that administration of androgens to rats *in vivo* very rapidly enhances the RNA synthesizing activity of prostate nuclear chromatin[3] and increases the mRNA content of polysomes.[4] These results prompted us to investigate whether testosterone can directly affect the RNA polymerase activity of isolated nuclei. Because of the failure of these attempts, we suspected that testosterone may act through a metabolite in prostate cell nuclei, and that this metabolite may bind to a specific receptor protein.

To find this nuclear androgen, we injected [^{3}H] testosterone into castrated rats, purified prostate cell nuclei and identified radioactive androgens that were present in the nuclei. Our work[5] and the independent study of Bruchovsky and Wilson[6] demonstrated that the major androgen retained by the nuclei was [^{3}H] DHT. This finding was also supported by an autoradiographic study.[7] Since we could show the nuclear retention of DHT in prostate chromatin by *in vitro* incubation of minced prostate and radioactive testosterone, testosterone is apparently taken up by prostate cells, reduced to DHT and retained by the chromatin.[5,8] The enzyme that catalyzes the conversion of testosterone to DHT is a NADPH-dependent Δ^4-3-ketosteroid-5α-oxidoreductase (5α-reductase) that appears to be associated with a cytoplasmic membrane fraction of the prostate.

The importance of the nuclear retention of DHT is emphasized by the observation that the selective nuclear retention of DHT was not detected in organs (lung, spleen, thymus, and diaphragm) that are not very sensitive to androgen and that cyproterone, an antiandrogen, strongly reduced the nuclear retention of DHT *in vivo* and *in vitro*.[9] Since other androgens, such as androstenedione and androstenediols,[10] can be converted to DHT, the actions of these androgens may also depend on the nuclear retention of DHT.

More recently, inhibitors of 5α-reductase, such as diethylcarbamoyl-methyl azaandrostanone (Fig. 2) have been shown to reduce nuclear retention of DHT[11] and also suppress the growth of rat prostates.[12] Earlier, an inherited form of male pseudohermaphroditism in man had been related to 5α-reductase deficiency.[13] In these patients, the decrease in DHT is believed to cause incomplete masculinization of the external genitalia, a reduction in prostate growth, alteration of hair growth patterns and a reduced prevalence of acne. The apparently normal differentiation of the epididymis and seminal vesicles in these individuals appears to be due to the stimulatory effects of testosterone, which is present in sera at normal concentrations.

ANDROGEN-RECEPTOR INTERACTION

Since DHT was suspected to be the active intracellular androgen, the search for a specific receptor for androgen was initiated by using [^{3}H] DHT as the ligand. Proteins that can bind radioactive DHT *in vivo* and *in vitro* were soon found in the cytosol and nuclear preparations of the ventral prostate of adult male rats.

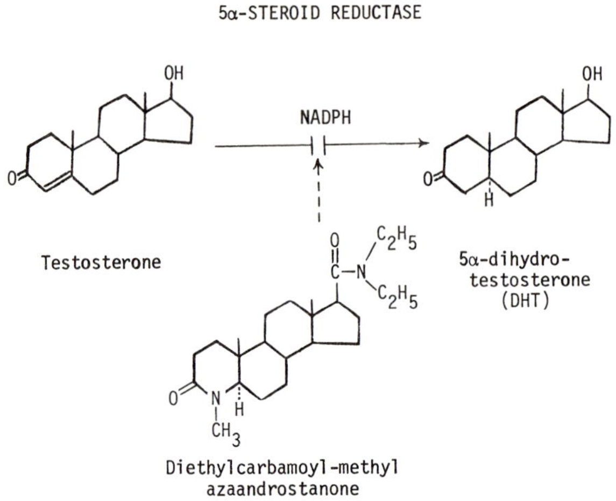

5α-STEROID REDUCTASE

Fig. 2. Structures of testosterone, 5α-dihydrotestosterone and a synthetic inhibitor of an NADPH-dependent Δ^4-3-keto-5α-oxidoreductase (5α-reductase).

Since it has not been possible to assay steroid receptors by analyzing their biochemical activity, steroid receptors have been characterized by measurements of their physical properties, such as sedimentation behavior and steroid binding specificity and affinity. Different forms of radioactive steroid-receptor complexes have often been identified due to variations in their rates of sedimentation in sucrose gradients. DHT-receptor complexes from rat ventral prostate have been found to sediment in a gradient of low ionic strength as two forms which are usually designated as 8 ± 1S and 4 ± 1S. The 8 S form of the receptor that is found in cytosol preparations dissociates into a 4 S form in gradients containing 0.4 M or higher concentrations of KCl, whereas the steroid-receptor complexes that are extracted from nuclei by a medium with high ionic strength (0.4 M KCl) sediment as a 3-5 S component in the high ionic strength medium.[14]

The biochemical relationship between different forms of the receptor is not clear. The 8 S form may represent the association of either identical or dissimilar macromolecules. Since the quantity of the receptor protein in the cytosol fractions appears to be very small (about 0.002% of total cytosol proteins), fortuitous interactions of the steroid-receptor complex with other cellular components in the tissue extracts may occur. Thus, androgen-receptor complexes have been seen to sediment differently (3 to 10 S) depending on the pH of the medium.[14] Since proteases in the tissue extracts may hydrolyze the receptor proteins without damaging the steroid binding capabilities of the receptors, including the receptor for androgens,[15] it is not possible to determine whether any of the forms identified so far are the native form of the receptor that functions in intact target cells.

Besides sucrose gradient centrifugation, other methods have been used for the separation of unbound hormones from receptor-bound hormones. These include removal of free or nonreceptor bound steroids by dextran-coated charcoal, Sephadex gel filtration, agar gel electrophoresis, isoelectric focusing, and immobilized antisteroid antibodies as well as selective precipitation or adsorption of the receptor-steroid complexes by protamine sulfate, ammonium sulfate, hydroxylapatite, and glass powders or beads.

Purification of androgen receptor has not been easy because concentrations of receptor in tissue extracts is only on the order of 2 to 100 fmoles per mg of protein (about 0.1 µg to 5 µg receptor protein per 1 kg fresh tissue). In addition, the purification process is usually followed by the quantitation of bound radioactive ligands (steroids) rather than the receptors themselves; this makes it difficult to determine whether the receptor proteins have been adversely altered during the isolation. One technique that is suitable for the purification of steroid receptors is affinity chromatography. A receptor-specific steroid covalently attached to an insoluble support is used to trap the receptor and separate it from other proteins in the crude extract. One of the affinity adsorbents we have made for the purification of the androgen receptor is DHT linked to Sepharose by a thioether bridge (Fig. 3).

The androgenicity and the receptor-binding affinity are closely correlated for many steroids. For receptor binding, the bulkiness and flatness of the steroid molecule, especially in the ring A area, appear to play a more important role than the detailed electronic structure of the steroid nucleus. The prostatic androgen receptor does not bind steroids with an A/B cis structure, such as the inactive 5β isomer of DHT. In addition, the receptor binding affinity of relatively flat steroids with rings A/B in the trans structure is also sensitive to the bulkiness in the ring A/B area.[16] These findings and our other studies suggest that androgens are "enveloped" in the hydrophobic cavity of the receptor.[16] The location of steroid-binding sites well inside the receptor may be responsible for the very slow rate of association or dissociation of steroids from the receptor at low temperatures, the very high steroid binding affinity, and the acceleration of the rate of exchange of unbound steroid with bound steroid by freezing and thawing.

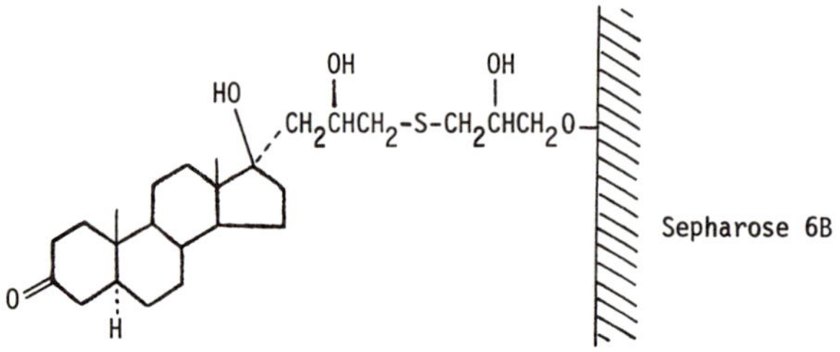

Fig. 3. A DHT-Sepharose adsorbent for affinity chromatography of androgen receptor.

Additional support for this view came from our study on the ability of anti-steroid antibodies to interact with steroids bound to various proteins.[17] Antibodies against DHT or testosterone were effective in removing steroids bound to nonreceptor proteins of the blood and prostate, since these proteins (steroid-metabolizing enzymes or blood steroid-binding globulins) generally recognize only a portion of the steroid molecule, and since the steroids dissociate much more rapidly from these nonreceptor proteins than from the receptor. The steroids bound well inside the receptor have very low rates of dissociation and, therefore, the steroid is not readily removed by the antibody. In addition, the antibody does not form a ternary complex with the androgen-receptor complex.

TRANSFORMATION AND NUCLEAR RETENTION OF ANDROGEN-RECEPTOR COMPLEX

The cytosol androgen receptor is not bound by prostate nuclei unless it first interacts with DHT.[8] Thus, the first recognizable cellular action of androgen is to alter the cytoplasmic receptor to a form that is retained by nuclei. A similar observation has been made in rat uterus, where estrogen is believed to cause a temperature-dependent transformation of the steroid-receptor complex into a form that is retained firmly by cell nuclei.[18] Also the retention of DHT-receptor complex by nuclear chromatin of rat ventral prostate occurs more effectively at 20°-40°C than at 0°C as demonstrated by autoradiography[7] and by reconstitution of cellular components.[19,20] It is not clear whether the transformation of the receptor involves a proteolytic action, a bimolecular reaction, or a simple change in receptor conformation. It is very likely, however, that binding of steroid to a receptor is accompanied by a conformational alteration of the receptor.

Pyridoxal phosphate effectively prevents the androgen-receptor complex from binding to chromatin or DNA-cellulose *in vitro*. This inhibition is apparently due to the formation of a Schiff base between pyridoxal phosphate and an amino group of the receptor.[21] Molybdate also inhibits receptor retention by chromatin. The exact mechanism involved in the molybdate effect is not clear. It has been suggested that dephosphorylation of the steroid-receptor complex may be necessary before the receptor complex can be retained by chromatin and that molybdate acts by inhibiting this dephosphorylation.

The first indication that there might be a specific nuclear acceptor which interacts with and retains steroid receptor complexes came from studies in which the retention of DHT-receptor complexes by prostate nuclei was demonstrated.[19,22] Soon afterwards a similar observation was made on the retention of progestin-receptor complex by chick oviduct nuclei.[23]

These acceptor molecules appear to be more abundant or more active in nuclei of androgen-sensitive tissues than in less responsive tissues, such as liver.[19] Solubilization and isolation of nuclear protein fractions that had acceptor-like activity were first achieved by Tymoczko and Liao.[24] Prostate nuclear proteins were extracted from nuclei with salt solutions and were mixed with DNA to form a nucleoprotein aggregate. The reconstructed chromatin was then tested for its acceptor activity by determining its capacity to bind the prostate [3H]DHT-receptor complex. Hiremath et al.[25] also purified a salt-extractable acceptor-like protein from rat ventral prostate. Klyzejko-Stefanowicz and co-workers[26] studied the chromatin acceptor by sequentially removing urea-soluble chromosomal nonhistone proteins, histones, and DNA-associated nonhistone proteins from the chromatin of the rat ventral prostate and testis. The prostate [3H]DHT-receptor complex was found to bind much more readily to partially deproteinized chromatin, which still contains the DNA-binding nonhistone proteins, than to purified DNA. The receptor-binding activity of rat DNA was enhanced significantly by the addition of DNA-associated nonhistone protein of the prostate or testis, but not that of the liver. Using covalent linkage of nuclear components to Sepharose, Mainwaring et al.[27] found that certain nonhistone basic proteins from prostate nuclei could retain the prostate cytosol [3H]DHT-receptor complex.

INACTIVATION, REACTIVATION, AND RECYCLING OF ANDROGEN RECEPTORS

Very little is known about the fate of the steroid-receptor complex after its interaction with chromatin. The nuclear receptor or the steroid may be modified in such a way that they no longer bind to each other tightly, causing the release of the receptor from the genomic structure. Another possibility

we have considered is that the receptor complex and other proteins may bind to
nuclear RNA thereby facilitating the release of the complex from nuclei (see
below). By pulse-chase techniques with labeled steroid we have estimated that
the steroid-receptor complex has a half-life of about 70 min in the nucleus.
This long half-life may reflect the involvement of the steroid-receptor complex in a time-consuming mechanism that is essential for hormone responses.

Although the events that follow the nuclear retention process are not known,
our studies in 1969[28] showed that, in the prostate cells, the androgen receptor
is rapidly deactivated (half-life : 2 min) in the presence of respiratory
poisons, such as 2,4-dinitrophenol or cyanide. This inactivation is reversible
if the poison is removed even in the presence of cycloheximide which should
inhibit *de novo* synthesis of receptor.[29] Purified prostate nuclei contain
enzyme(s) that inactivate the steroid binding ability of the receptor protein.
Whether this inactivating enzyme is a phosphatase, as has been suggested for
other steroid receptor systems[30,31] is not known. Reactivation of L-cell or
thymocyte glucocorticoid receptor[32] and uterine estrogen receptor,[33] in a cell-
free system, requires ATP. No direct evidence is available, however, to show
whether it is the receptor itself or another closely related molecule that is
phosphorylated or dephosphorylated during the activation-inactivation processes.

Since the androgen-receptor complex is maintained at a constant level for
at least two hours even in the absence of new protein synthesis, the early
effects of androgen on RNA synthesis (within one hour of androgen injection
into castrated rats) probably do not require a depletion of a major portion
of cellular receptors. Since the inactivation, reactivation, and nuclear
retention processes are more rapid than receptor degradation, the androgen
receptor appears to go through a dynamic process of recycling.[29]

RECEPTOR EFFECT ON RNA SYNTHESIS AND UTILIZATION

Interaction of steroid-receptor complex with nuclei is presumably the key
step by which RNA synthesis is modulated in the target cells. Although
certain steroid-receptor preparations have been shown to enhance RNA synthesis
and nuclear RNA polymerase activity, these studies have not been able to
provide much information on the actual role of the steroid receptors. Some of
the drawbacks of these studies are that they were carried out with crude
receptor and RNA polymerase preparations, and also our lack of knowledge concerning the regulatory mechanisms of eukaryotic RNA synthesis.

The study of the induction of specific proteins in certain steroid hormone
target cells has provided considerable information on the structure of the
genes and mRNA for specific steroid hormone sensitive proteins, such as chick

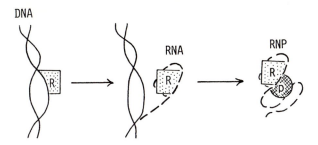

R:steroid-receptor complex
p:other protein components

Fig. 4. A hypothetical model of hormone action on gene transcription and receptor binding to RNA. Simplified from Liao and Fang.[28]

oviduct ovalbumin,[34] rat liver proteins,[35,36] and prostate secretory proteins,[37,38] However, there is some doubt as to whether the genes for these proteins are the primary sites of steroid receptor action.[39,41] Recently presented evidence indicates that steroid-receptor may interact with specific locations on certain genes[42,42a] but the significance of such an observation is still not clear.

Elucidation of a common mechanism of action for all steroid hormones may be difficult since some of the effects of steroids appear to be independent of the genomic response (RNA synthesis). For example, oocyte maturation can be induced by progesterone and other steroid hormones in an enucleated oocyte[43] or by a steroid hormone linked to a polymer that can not enter the cell.[44] Thus, the possibility that steroid hormones may act at multiple cellular sites should be pursued further.

In 1969, we suggested that nuclear RNA may bind to the receptor complex and facilitate the release of the receptor from nuclei (Fig. 4). We proposed that the receptor complex, in turn, may play an important role in processing, stabilization, and/or utilization of RNA.[28,45] Although direct evidence supporting this idea is still lacking, our studies have shown that both the estrogen- and androgen-receptor complexes can bind to certain ribonucleoprotein (RNP) particles from the uterus and prostate.[46,47]

Since this scheme suggests that certain RNA molecules can promote the release of DHT-receptor complex from DNA or chromatin, we have tested this possibility by DNA-cellulose column chromatography, gradient centrifugation, and nuclear incubation techniques.[48] The results of these investigations have indicated that certain polyribonucleotides can release the DHT-receptor complex from DNA or nuclei. The ability of polyribonucleotides to facilitate this release is dependent on the type of nucleotide base (presence of an oxygen at C-6 of purines or C-4 of pyrimidines), nucleotide sequences (U and G content), and size (10 to 20 nucleotide length) of the polymer. These observations have been confirmed by other investigators.[49]

We have carried out similar experiments with estrogen-receptor complex from rat uterus and with glucocorticoid-receptor complex from rat liver and have found that these steroid-receptor complexes behave like the DHT-receptor complex of rat ventral prostate. Although we have not been able to detect differential receptor specificity for RNA in these experiments, it is conceivable that there are certain natural RNA molecules with the proper nucleotide sequence that bind to the steroid-receptor complex more effectively and with greater specificity than the polymers that we have tested. In addition, different RNA molecules may contain identical or similar nucleotide sequences, so that more than one RNA species can be selected by the same steroid-receptor complex. Such a scheme may explain the selectivity and multiplicity observed in the induction of different proteins by steroid hormones.

The receptor may lose its capacity to bind to RNA at various stages of RNA processing and utilization, especially when the steroid hormone concentration in the cell is low. Both receptor proteins and acceptor proteins may reassociate with RNP particles when the steroid hormone content of the cell is replenished. Thus, binding of receptor to RNA may be reinitiated by the steroid hormone at different points, either in the nucleus or in the cytoplasm. This emphasizes the role of gene transcription (RNA synthesis) in relation to gene translation (protein synthesis) for the overall functioning of a steroid hormone in target cells. If the target cells contain sufficient amounts of RNA and the protein constituents of RNP, the early effects of the hormone may simply be dependent on the processing and utilization of RNP, and not upon RNA synthesis.

A HYPOTHETICAL FEED-BACK CONTROL OF RECEPTOR ACTIVITY

The major protein induced by androgen stimulation in the rat ventral prostate is α-protein (> 30% of total cytosol proteins). This protein was identified by us[19] during studies on the cytosol androgen-receptor complex. Several investigators have studied this low affinity, high capacity steroid-binding protein as a marker for androgen action in the prostate.[37,38]

133

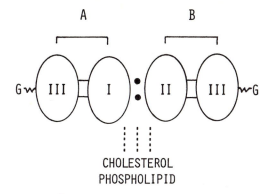

Fig. 5. α-Protein subunits and polypeptide components. α-Protein is a cholesterol and phospholipid binding protein and has carbohydrates attached to Component III.

Amino acid sequence

NH$_2$- Ser-Gln-Ile-Cys-Glu-Leu-Val-Ala-His-Glu-Thr-Ile-Ser-Phe-Leu-Met-Lys-Ser-Glu-Glu-Glu-Leu- (10...20)
Lys-Lys-Glu-Leu-Glu-Met-Tyr-Asn-Ala-Pro-Pro-Ala-Ala-Val-Glu-Ala-Lys-Leu-Glu-Val-Lys-Arg- (30...40)
Cys-Val-Asp-Gln-Met-Ser-Asp-Gly-Asp-Arg-Leu-Val-Val-Ala-Glu-Thr-Leu-Val-Tyr-Ile-Phe-Leu- (50...60)
Glu-Cys-Gly-Val-Lys-Gln-Trp-Val-Glu-Thr-Tyr-Tyr-Pro-Glu-Ile-Asp-Phe-Tyr-Tyr-Asp-Met-Asn-OH (70...80...88)

Amino acid composition: Molecular weight:

Glu$_{13}$, Gln$_3$, Asp$_5$, Asn$_2$, Lys$_6$, Arg$_2$, His$_1$,   10,191

Val$_9$, Leu$_8$, Ala$_6$, Ser$_4$, Ile$_4$, Thr$_3$, Gly$_2$,

Tyr$_6$, Phe$_3$, Pro$_3$, Trp$_1$, Met$_4$, Cys$_3$

Fig. 6. Amino acid sequence of component I of α-protein isolated from rat ventral prostate. This component inhibits the chromatin retention of the androgen-receptor complex in cell free systems.

134

In cell-free systems, this protein inhibits binding of DHT-receptor complex to DNA and prostate nuclei[19] and promotes the release of the androgen-receptor complex already attached to chromatin (or DNA) in a temperature-dependent process.[50] These effects of α-protein are not due to irreversible destruction of the receptor complex or to damage to the nuclear (or DNA) binding sites.

α-Protein has two subunits (A and B), each composed of two polypeptide chains[51,52] as shown in Fig. 5. Of the three polypeptide components, only Component I possesses the inhibitory activity described above. The amino acid sequence of Component I (Fig. 6) indicates that most of the glutamic acid and lysine residues are localized in the NH_2-terminal half, whereas all aspartic acid residues and nearly all of the aromatic acids are in the COOH-terminal half of the protein.[53] Whether these features are important in the inhibition of the chromatin retention of the receptor complex is not clear. In fact, the inhibitory activity may be due to a small oligopeptide stretch in Component I and may not require a complex structure.

There is no direct evidence that α-protein or its components act *in vivo* as regulators, of receptor action in the prostate. However, our observations provide the first model for the study of protein-mediated control of the interaction of steroid-receptor complexes with nuclear chromatin.

REFERENCES

1. Liao, S. (1975) Int. Rev. Cyt., 41, 817-872.
2. Liao, S. (1977) Biochem. Action of Hormones, 4, 351-406.
3. Liao, S., et al. (1965) Endocrinology, 77, 763-765.
4. Liao, S. and Williams-Ashman, H. G. (1962) Proc. Nat. Acad. Sci. USA, 48, 1956-1964.
5. Anderson, K. M. and Liao, S. (1968) Nature, 219, 277-279.
6. Bruchovsky, N. and Wilson, J. D. (1968) J. Biol. Chem., 243, 2012-2021.
7. Sar, M., Liao, S., and Stumpf, W. E. (1970) Endocrinology, 86, 1008-1011.
8. Fang, S., Anderson, K. M. and Liao, S. (1969) J. Biol. Chem., 244, 6584-6595.
9. Fang, S. and Liao, S. (1969) Mol. Pharmacol., 5, 428-431.
10. Bruchovsky, N., et al. (1975) Vitam. Horm., 33, 61-102.
11. Liang, T. and Heiss, E. (1981) J. Biol. Chem., 256, 7988-8005.
12. Brooks, J. R., et al. (1981) Endocrinology, 109, 830-836.
13. Imperato-McGinley, J., et al. (1974) Science, 186, 1213-1215.
14. Liao, S., et al. (1975) Vitam., Horm., 33, 297-317.
15. Wilson, E. M. and French, F. S. (1979) J. Biol. Chem., 254, 6310-6319.
16. Liao, S., et al. (1973) J. Biol. Chem., 248, 6154-6162.
17. Castaneda, E. and Liao, S. (1975) J. Biol. Chem., 250, 883-889.
18. Jensen, E. V., et al. (1974) Vitam. Horm., 32, 89-127.
19. Fang, S. and Liao, S. (1971) J. Biol. Chem., 246, 16-24.
20. Mainwaring, W. I. P. and Peterken, B. M. (1971) Biochem. J., 125, 285-295.
21. Hiipakka, R. A. and Liao, S. (1980) J. Steroid Biochem., 13, 841-846.
22. Liao, S. and Fang, S. (1970) in Some Aspects of Aetiology and Biochemistry of Prostate Cancer, Griffiths, K., and Pierrepoint, C. G., eds., Cardiff: Alpha Omega Alpha Publ., pp. 105-108.

23. O'Malley, B. W., et al. (1972) Nature (London), 235, 141-144.
24. Tymoczko, J. L. and Liao, S. (1971) Biochim. Biophys. Acta, 252, 607-611.
25. Hiremath, S. T., Loor, R. M., and Wang, T. Y. (1980) Biochem. Biophys. Res. Commun., 97, 981-986.
26. Klyzsejko-Stefanowicz, L., et al. (1976) Proc. Natl. Acad. Sci. USA, 73, 1954-1958.
27. Mainwaring, W. I. P., Symes, E. K., and Higgins, S. J. (1976) Biochem. J., 156, 129-141.
28. Liao, S. and Fang, S. (1969) Vitam. Horm., 27, 17-90.
29. Rossini, G. P. and Liao, S. (1982) Biochem. J., 208, 383-392.
30. Nielsen, C. J., et al. (1977) J. Biol. Chem., 252, 7568-7578.
31. Auricchio, F., Migliaccio, A., and Castoria, G. (1981) Biochem. J., 194, 569-574.
32. Wheeler, R., et al. (1981) J. Biol. Chem., 256, 434-441.
33. Auricchio, F., Migliaccio, A., and Castoria, G. (1981) Biochem. J., 198, 699-702
34. O'Malley, B. W., et al. (1979) Rec. Progr. Horm. Res., 35, 1-46.
35. Kurtz, D. T. and Feigelson, P. (1978) in Biochemical Actions of Hormones, Litwack, G., ed., Vol. 5, New York: Academic Press, pp. 433-455.
36. Roy, A. K. (1979) in Biochemical Actions of Hormones, Litwack, G., ed., Vol. 6, New York: Academic Press, pp. 481-517.
37. Parker, M. G. and Scrace, G. T. (1979) Proc. Natl. Acad. Sci. USA, 76, 1580-1584.
38. Peeters, B. L., et al. (1980) J. Biol. Chem., 255, 7017-7023.
39. Delap, L. and Feigelson, P. (1978) Biochem. Biophys. Res. Comm., 82, 142-149.
40. Chen, C.-L.C. and Feigelson, P. (1979) Proc. Natl. Acad. Sci. USA, 76, 2669-2673.
41. McKnight, G. S. (1978) Cell, 14, 403-413.
42. Payvar, F., et al. (1981) Proc. Natl. Acad. Sci. USA, 78, 6628-6632.
42a Mulvihill, E. R., LePennec, J-P. and Chambon, P. (1982) Cell, 24, 621-632.
43. Smith, L. D. and Ecker, R. E. (1971) Develop. Biol., 25, 232-247.
44. Godeau, F., et al. (1978) Proc. Natl. Acad. Sci. USA, 75, 2353-2357.
45. Liao, S., et al. (1972) Proceedings of the 4th International Congress on Endocrinology, Excerpta Medica Int. Congr. Ser. No. 273, 404-407.
46. Liao, S., Liang, T. and Tymoczko, J. L. (1973) Nature, New Biol., 241, 211-213.
47. Liang, T. and Liao, S. (1974) J. Biol. Chem., 249, 4671-4678.
48. Liao, S., et al. (1980) J. Biol. Chem., 255, 5545-5551.
49. Feldman, M., Kallos, J., and Hollender, V. P. (1981) J. Biol. Chem., 256, 1145-1148.
50. Shyr, C. and Liao, S. (1978) Proc. Natl. Acad. Sci. USA, 75, 5969-5973.
51. Heyns, W., et al. (1978) Eur. J. Biochem., 89, 181-186.
52. Chen, C., et al. (1982) J. Biol. Chem., 257, 116-121.
53. Liao, S., Chen, C., and Huang, I-Y. (1982) J. Biol. Chem., 257, 122-125.

STRUCTURE AND MECHANISM OF ACTION OF CHOLERA ENTEROTOXIN

CHUN-YEN LAI
Roche Institute of Molecular Biology, Nutley, New Jersey

ABSTRACT

Cholera toxin is composed of two functionally distinct subunits, A and B;
the former is responsible for the ability of the toxin to stimulate adenylate
cyclase and the latter, for binding of the toxin to cell surfaces. The subunit
A consists of two polypeptides, A_1 (M_r = 23000) and A_2 (M_r = 7500) linked to each
other by a single disulfide bond. The polypeptide A_1 has been shown to contain the active site of the subunit; it is solely responsible for the stimulation of adenylate cyclase in a broken cell system. Subunit B has a strong
affinity to ganglioside G_{M1}. It is a polypeptide of 103 amino acid residues
whose sequence has been determined. Using chemical modification studies, the
region in the molecule involved in the ganglioside binding has been identified.
The holotoxin contains one A subunit and five B subunits. When exposed to
cells, the toxin binds to the cell surfaces through the B subunits. Polypeptide A_1 then enters the cell and exerts its action. Recent evidence
indicating the formation of polypeptides A_1 and A_2 from a single precursor
protein suggests the role of A_2 in the holotoxin as that of an anchor. The
mechanism of the adenylate cyclase activation by cholera toxin appears to
involve ADP-ribosylation of a regulatory protein. The separated A_1 polypeptide
also exhibited the ADP-ribose transferase activity. This property has been
used to study the structure activity relationship. Considerable progress
has been made in the elucidation of the primary structure of polypeptide A_1
and the COOH-terminal half of the molecule has been found to contain the active
site. These studies provide insights to the mode of action of cholera toxin
at the molecular level.

INTRODUCTION

Cholera toxin is a diarrheagenic protein first isolated from the culture
filtrate of *Vibrio cholerae* in 1969 by Finkelstein and LoSpalluto.[1] It has
attracted considerable attention among biochemists and pharmacologists in the
subsequent years because of its unexpected, hormone-like activity; the toxin
has been found to stimulate adenylate cyclase in virtually all types of
mammalian cells, causing various cyclic-AMP mediated changes in cell function.[2]
The massive outpouring of fluid observed in cholera is now understood to be a

consequence of the increased production of cyclic AMP in the epithelial cells of the small intestine in response to the toxin released by the bacteria *in situ*.

For the past several years, we have been studying the structure-function relationship of cholera toxin, in order to understand the molecular mechanism involved in the membrane-mediated control of cell function. In this paper, we present experimental results which have led to elucidation of the subunit structure, the subunit responsible for the adenylate cyclase activation, the primary structure and the site of membrane interaction in the binding subunit, and location of the active center in the active subunit of cholera toxin.

MATERIALS AND METHODS

Cholera toxin was purified from the culture filtrate of *Vibrio cholera* by Na-metaphosphate precipitation[3] followed by gel filtration on Sephadex G-75.[1,2] Subunits were isolated from the purified toxin as described.[4] [^{3}H]- or [^{32}P]-labeled NAD was synthesized from [2,8-^{3}H]ATP or [α-^{32}P]ATP and an excess of nicotinamide mononucleotide in the presence of NAD-pyrophosphorylase.[5] The procedures for the adenylate cyclase stimulation assay[6] and the ADP-ribosylation assay[7] of cholera toxin and its active subunits, and methods for the protein structural analyses were described previously.[6-9]

RESULTS AND DISCUSSION

Subunit structure and identification of the active component

On dissolution of cholera toxin (M_r = 86,000) in 6M Guanidine-HCl containing 5% formic acid and gel filtration on a Sephadex G-75F column in 5% formic acid, two proteins designated A (M_r = 30,000) and B (M_r = 11,000) were obtained in 85%-95% yield (Fig. 1a). The high recovery of protein indicated that these were the only constituents of cholera toxin. Polyacrylamide gel electrophoresis in 0.1% Na-dodecylsulfate and 8M Urea indicated each was homogeneous as to its size, and corresponded to the heavy (subunit A) and light (subunit B) component of the holotoxin (Fig. 1b). Amino acid analyses of the subunits indicated that each contained 2 cysteine residues per mole.

On complete reduction and S-carboxymethylation, subunit A was dissociated further into two components that could be separated on Sephadex G-75 in 5% formic acid (Fig. 1c). The apparent molecular weights estimated by the SDS-Urea electrophoresis were 23,000 and 7,500 respectively, for the polypeptide A_1 and A_2. Each was found to contain a single cysteine residue per mole. The experiments clearly indicated that subunit A was composed of two peptides

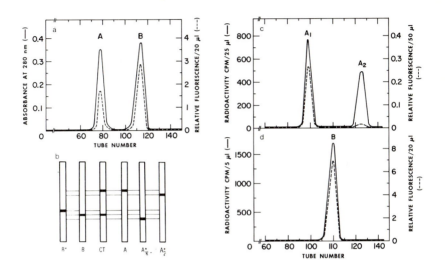

Fig. 1. Separation of cholera toxin subunits.[4] (a) Gel filtration of cholera toxin (5 mg) dissolved in 5% HCOOH containing 6M guanidine HCl (2 ml) on a column (1.5 x 180 cm) of Sephadex G-75 in 5% HCOOH. (B) A sketch of gel-electrohoretic pattern of cholera toxin and separated subunits in 0.1% SDS and 8M urea. B^*, A_1^*, and A_2^* denote S-carboxymethylated subunits B, A_1 and A_2, respectively. S-carboxymethyl B subunit (B^*) showed higher apparent molecular weight than the unmodified B subunit. (c) Gel-filtration of reduced and S-carboxymethylated subunit A as in (a). (d) Same treatment on subunit B.

linked to each other by a disulfide bond. Subunit B_1 on the other hand, remained a single component after the same treatment (Fig. 1d). Incorporation of radioactivity by S-carboxymethylation using [^{14}C]iodoacetic acid indicated that the two cysteine residues in this subunit formed an intrachain disulfide bond.

Subunits A and B as isolated in 5% formic acid are not soluble at neutral pH. When these were dissolved in 0.05 M Tris-Cl buffer, pH 7.5 containing 8M urea and dialyzed against a gradient of decreasing concentration of urea in the same buffer containing 0.1 M NaCl, the subunits remained in solution, apparently having assumed their natural conformations. Sedimentation

140

TABLE 1

MOLECULAR WEIGHTS OF CHOLERA TOXIN AND ITS SUBUNITS:

FROM SEDIMENTATION EQUILIBRIUM EXPERIMENTS

Proteins	Solution	Molecular weight[a]
Cholera Toxin	Tris-HCl buffer pH 7.5, 0.2 M NaCl	84,000
Subunit A	Tris-HCl buffer pH 7.5, 0.2 M NaCl	Aggregate
Subunit B	Tris-HCl buffer pH 7.5, 0.2 M NaCl	55,200
Subunit A	6 M Guanidine HCl	30,000
Subunit B	0.1 M NH_4-formate 5% HCOOH	9,400
Subunit B	6 M Guanidine HCl	10,300

[a]Partial specific volume of all samples were assumed to be 0.74 for the calculation of molecular weights.

equilibrium experiments revealed that subunit B existed as a pentamer in its natural conformation in a physiological solution (Table 1); the molecular weight of the renatured B-subunit was approximately 5 times that of the denatured form in 5% HCOOH or 6 M Guanidine HCl. The molecular weights of subunit A (30,000 M_r) and the pentameric form of subunits B add up to the molecular weight of the holotoxin (84,000) as determined by sedimentation equilibrium, suggesting that cholera toxin is composed of 1 subunit A and 5 subunits B. The isolated A subunit solubilized in a physiological solution was apparently in a metastable state, for it precipitated out of solution quickly on ultracentrifugation (Table 1). The subunit composition of AB_5 for cholera toxin was also determined by quantitative analysis of the COOH-terminal dipeptide, Ala-Asn, of subunit B, released upon treatment of cholera toxin with cyanogen bromide.[4] In addition, reduction and S-carboxymethylation of the holotoxin yielded 5.4 moles of B subunits per mole each of A_1 and A_2.[4] With the use of a bifunctional reagent to effect inter-subunit crosslinking, Gill demonstrated the existence of molecular species corresponding to A, AB, AB_2, AB_3, AB_4, and AB_5 by SDS-gel electrophoresis, confirming the above result.[10]

Availability of separated subunits in a renatured state has now enabled us to examine their biological activities. In 1975, Gill and King reported that the activity of cholera toxin could be observed with pigeon erythrocyte lysates.[11] Activation of adenylate cyclase occurred immediately upon addition of the toxin, indicating that the binding of the toxin to cell surfaces was no

TABLE 2

EFFECT OF CHOLERA TOXIN ON THE ACTIVITY OF ADENYLATE CYCLASE
IN THE WASHED PIGEON ERYTHROCYTE MEMBRANE[a]

	Adenylate cyclase activity
	pmoles of cAMP/min/mg membrane protein
No additions	1.47
+ cholera toxin	1.79
+ cholera toxin + cytosol	2.54
+ cholera toxin + cytosol + NAD	15.42
+ cholera toxin + DTT + NAD	13.22

[a] Pigeon erythrocyte "membrane" (8.3 mg/ml) was incubated with cholera toxin (145 μg/ml), cytosol (29.1 mg/ml), NAD (1 mM) and DTT (1 mM) as indicated, for 30 min at 37°. Aliquots were then assayed for the adenylate cyclase activity as described in the Methods.

longer involved in the process. We refined this assay system and used the washed erythrocyte membrane to study the activity of isolated subunits.

With washed pigeon erythrocyte membranes cholera toxin did not show activity unless NAD and cytosol were added (Table 2). The requirement for NAD in the broken cell system had previously been noted by Gill.[12] Dithiothreitol was found to substitute for cytosol in the activation of adenylate cyclase. When subunits of cholera toxin were tested in this assay system in the presence of 1 mM each of DTT and NAD, subunits A and polypeptide A_1 were found to stimulate adenylate cyclase more efficiently than the holotoxin (Fig. 2). Subunit B and polypeptide A_2 were completely inactive. The amount required for the half-maximal activation of adenylate cyclase was approximately 9 μg (0.4 nmol)/ml for peptide A_1 (Fig. 2). Not only was the amount required less, but poly-

TABLE 3

EFFECT OF THIOL ON ACTIVATION OF ADENYLATE CYCLASE IN
PIGEON ERYTHROCYTE MEMBRANE BY CHOLERAGEN AND ITS SUBUNITS

Addition	Adenylate cyclase activity pmol of cAMP produced/min/mg		
	Choleragen	A	A_1
None	2.7	8.84	23.07
Dithiothreitol (1 mM)	10.68	12.71	18.81

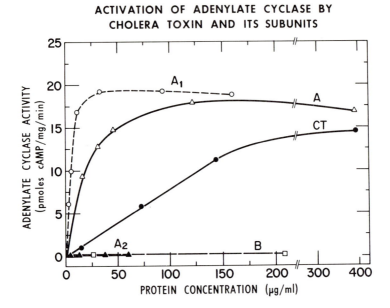

ACTIVATION OF ADENYLATE CYCLASE BY
CHOLERA TOXIN AND ITS SUBUNITS

Fig. 2. Adenylate cyclase activity of washed pigeon erythrocyte membrane in the presence of increased amount of cholera toxin and isolated subunits. The reaction mixture contained 1 mM each of DTT and NAD.

peptide A_1 stimulated adenylate cyclase activity in erythrocyte membrane instantaneously, as compared to the holotoxin (Fig. 3). These results clearly demonstrate that polypeptide A_1 is solely responsible for the action of cholera toxin in the stimulation of adenylate cyclase.

Since A_1 is linked to A_2 by a single disulfide bond in the native cholera toxin, the requirement of DTT in the activation of adenylate cyclase by the holotoxin (Table 2) was thought to be related to the release of A_1 polypeptide from the rest of the molecule. Indeed, with isolated A_1, DTT was no longer required for the adenylate cyclase activation (Table 3). The effect of DTT was less pronounced with subunit A than with the holotoxin, suggesting that the active site was partially exposed in the isolated subunit A. The free sulfhydryl group in polypeptide A_1 was, however, not involved in the peptide's action on adenylate cyclase. S-alkylation of the single cysteine residue in polypeptide A_1 did not affect its activity (Table 4). The result indicates that the active site of cholera toxin is located at the vicinity of the cysteine residue in the A_1 polypeptide, and shielded in the holotoxin by the

143

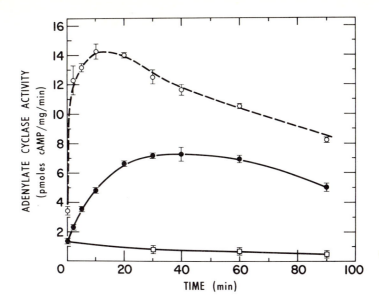

Fig. 3. Time course study of adenylate cyclase activation by cholera toxin and subunit A_1.

disulfide linkage to polypeptide A_2 or it could be a conformational change of A_1 after release from A_2.

Mode of action of cholera toxin

Although DTT was not required for the action of subunit A_1, addition of cytosol was found to further enhance its activity (Table 5) suggesting the presence of an effector in the cytosol of pigeon erythrocytes. A kinetic study indicated that cytosol acted nonenzymatically in the enhancement of A_1 activity (data not shown). Isolated polypeptide A_1 is unstable in solution at ambient temperatures and loses its activity within 30 min at 37°.[7] The cytosol may act to stabilize A_1 in its active conformation. Attempts to isolate the cytosol factor were not successful, but resulted in a heat-stable component (UF) and a heat-labile protein fraction (RI). The latter was about 30-fold purified over the cytosol (Table 6).

Experimental results described above have provided some insight to the mode of action of cholera toxin. With broken cells or crude membrane preparation, cholera toxin is not active unless conditions to dissociate polypeptide A_1, such as addition of DTT or the presence of cytosol (Table 2), are provided.

144

TABLE 4

STIMULATION OF ADENYLATE CYCLASE IN PIGEON ERYTHROCYTES MEMBRANE

BY A_1 POLYPEPTIDE MODIFIED WITH IODOACETIC ACID AND N-ETHYLMALEIMIDE

Activator added	Adenylate cyclase activity pmol of cAMP produced/min/mg
None	0.88
A_1 unmodified	8.09
A_1-carboxymethylated	9.91
A_1 treated with NEM	8.46

TABLE 5

ACTIVATION OF MEMBRANE ADENYLATE CYCLASE

BY A_1 - SUBUNIT AND THE EFFECT OF CYTOSOL

	activity[*]
"Membrane	1.5
"Membrane" + A_1	11.3
"Membrane" + A_1 + cytosol	25.7

[*] pmol cAMP/min/mg membrane protein

TABLE 6

EFFECT OF CYTOPLASMATIC FRACTIONS ON THE ACTIVATION

OF ADENYLATE CYCLASE BY CHOLERA TOXIN

Additions	Activity[*]
-	1.5
+ Cytosol (3.6 mg)	2.3
+ A_1	10.0
+ A_1 + Cytosol (3.6 mg)	18.6
+ A_1 + R_I (0.12 mg) + UF	19.0
+ A_1 + R_I (0.12 mg)	15.9
+ A_1 + R_{II} (1.5 mg)	11.7
+ A_1 + R_{II} (1.5 mg) + UF	13.8
- (assayed with 8 mM NaF)	100.0

[*] pmol cAMP/min/mg membrane protein

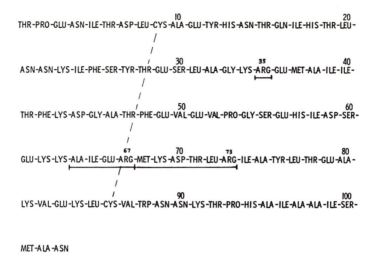

Fig. 4.   The primary structure of cholera toxin subunit B.

The freed A_1 still required NAD to activate the membrane adenylate cyclase, and the process is greatly augmented by the presence of cytosolic effectors.   This is in agreement with the notion that A_1 acts from within the cell.   In the intoxication process, cholera toxin first binds to the cell surfaces through b-subunits (see below).   Dissociation of A_1 from the rest of the molecule then occurs and A_1 penetrates into the cell.   The free polypeptide A_1 subsequently interacts with the adenylate cyclase system in the presence of NAD and cytosol, and causes stimulation of its activity.

Structure and function of subunit B

   Availability of subunit B in relatively large quantity prompted us to investigate its primary structure.[8]   It contains 103 amino acid residues, with only 3 arginines at positions 35, 67, and 73 (Fig. 4).   Cysteine-9 and Cys-86 form an intrachain disulfide as shown previously.[8]

In the last step of purification of cholera toxin from the culture filtrate of *Vibrio cholerae*, Finkelstein and LoSpalluto obtained a protein which cross-reacted with the antibody against cholera toxin, but did not show diarrheagenic activity in infant rabbits.[1,13] The protein was named choleragenoid to indicate its antigenic similarity to choleragen. It was found to block the action of cholera toxin on intact cells, and considered to bind the same sites on cell surfaces as choleragen.[13] In 1971 van Heyningen et al. demonstrated that the brain ganglioside G_{M1} inhibited the action of choleragen toward various tissues.[14] At a relatively high concentration of 2.5 mg/ml ganglioside and 5 mg/ml cholera toxin, precipitation was obversed, and the supernate was found to contain no toxin activity.[14] The results indicated that ganglioside G_{M1} and choleragen formed a complex. Similar complex formation was observed between ganglioside G_{M1} and choleragenoid.

As stated before, when subunit B was separated and renatured, it formed a pentamer with $M_r = 55,000$ (Table 1). The molecular weight was similar to that of choleragenoid (56,000 M_4) previously reported.[15] The renatured subunit B was found to form precipitate with anticholeragen antibody in a double-diffusion plate, as well as with ganglioside G_{M1} (Fig. 5). These data indicate that subunit B is identical to choleragenoid, and that it is responsible for the binding of cholera toxin to the cell surfaces.

With the use of cyclohexanedione to modify the arginine side chain group, we have recently been able to identify the site of ganglioside-binding in subunit B.[9] By selecting a proper reagent concentration and reaction pH, it was possible to modify 1 or 2 arginine-guanido groups in the renatured subunit B. The ability to form a precipitate with ganglioside G_{M1} or anticholera toxin antibody was intact when 1 arginine residue was modified, but was completely abolished when 2 arginines had reacted (Fig. 5). This indicated that the second arginine was or at the region involved in the ganglioside and the antibody binding. The residues reacted with cyclohexanedione were analyzed by comparison of tryptic peptide patterns between native and two modified subunit B. Arg-35 is located next a lysine residue and is released free on tryptic digestion, as can be determined directly on an amino acid analyzer. This residue was lost only when 2 arginine residues were modified [(CHD)$_2$B]. The peptides containing Arg-67 and Arg-73 are both in the T2 fraction from the Sephadex G25 gel-filtration of the tryptic digests (Fig. 6). On Aminex-A5 chromatography of T2 fractions from (CHD)$_1$-B and (CHD)$_2$B (Fig. 6), peptide 64-67 was recovered in high yield whereas no peptide 68-73 was found in either tryptic digest. Arg-73 was therefore modified in both (CHD)$_1$-B and (CHD)$_2$-B,

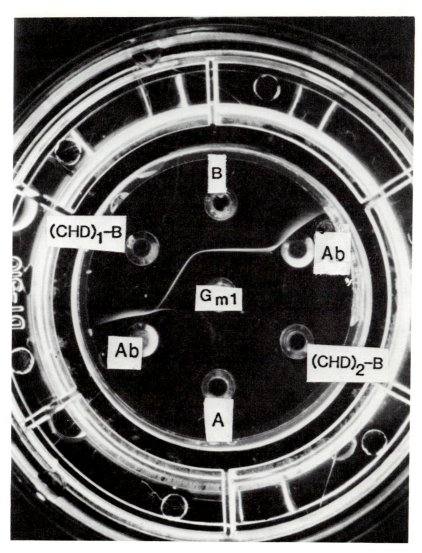

Fig. 5. Double diffusion test of the native and chemically modified subunit B against ganglioside G_{M1} and anticholera toxin antibody from rabbit. $(CHD)_1$-B and $(CHD)_2$-B are subunit B containing 1 and 2 cyclohexanedione modified arginines, respectively.

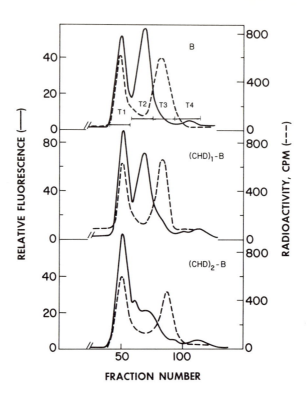

Fig. 6. Separation of tryptic digests of the native and modified subunit B on Sephadex G-25 gel filtration. The T2 fraction which contained peptide 64-67 and peptide 68-73 decreased in amount in the digest of $(CHD)_1$-B, and further in that of $(CHD)_2$B (see legend to Fig. 5).

but Arg-67 was not reacted at all. These results (Table 7) indicated that Arg-35 is at the site of ganglioside/antibody binding, and is thus responsible for the binding of cholera toxin to cell surfaces. The secondary structure of B-subunit as deduced by the method of Chou and Fasman is shown in Fig. 7. The diagram was constructed to take into account the disulfide bridge in the molecule.[9]

The mechanism of action and structural studies of polypeptide A_1

The experiments described above have established that polypeptide A_1 is solely responsible for the toxins' ability to stimulate adenylate cyclase. The peptide must be dissociated from the holotoxin in order to manifest its activity, in which NAD is a required cofactor.

TABLE 7

DETERMINATION OF UNMODIFIED ARGININE AFTER TRYPTIC DIGESTION

	% Yield on Trypsinolysis		
	Arg-35[a]	Peptide 64-67[b] (Arg-67)	Peptide 68-73[b] (Arg-73)
Subunit B	70 (1)[c]	90 (1)	46 (1)
CHD$_1$-B	72 (1)	68 (0.8)	0 (0)
CHD$_2$-B	25 (0.3)	61 (0.7)	0 (0)

[a] On exhaustive digestion with trypsin of the subunit B (see text), Lys-63 Arg-35 released from the -Gly-Lys-Arg-Glu- sequence. The digests were analyzed directly on an amino acid analyzer.
[b] The peptides containing Arg-67 and Arg-73 were isolated from fraction T2 (Fig. 3) by ion-exchange chromatography and quantitated by amino acid analysis. The overall yields are shown.
[c] The ratios to the values to that from the unmodified subunit B are shown in parenthesis.

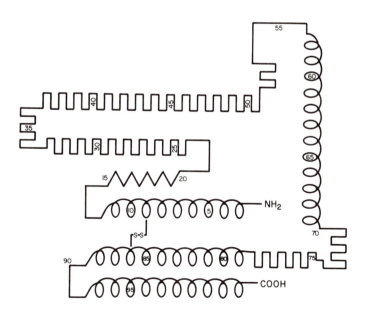

Fig. 7. The secondary structure of subunit B as deduced from the primary structure (Fig. 4) Arg-35 is indicated to be involved in the ganglioside and antibody binding.

TABLE 8

ADP-RIBOSYLATION OF POLYARGININE BY CHOLERA TOXIN AND ITS ACTIVE SUBUNITS

Enzyme	Incorporation[a] pmol	
	−DTT	+DTT
A_1 (1.85 µg)	536	611
S-cm A_1 (1.75 µg)	520	555
Subunit A (2.77 µg)	263	530
Cholera toxin (6.57 µg)	72	272

[a] Incorporation of [3H]ADP-ribose moiety to 0.1 mg polyarginine in 30 min at room temperature. To 0.1 ml 15 mM Na-PO4 buffer, pH 6.6, 10 µl enzyme in 0.1% BSA and 10 µl 2 mM NAD (0.3 µCi) were added to start the reaction. Tubes containing 1 mM DTT were preincubated with the enzyme for 15 min at room temperature.

The requirement of NAD in the intoxication process was previously demonstrated with diptheria toxin. In the late 1960s, it was discovered that diphtheria toxin catalyzed the transfer of the ADP-ribose moiety of NAD onto the elongation factor II, causing its inactivation.[17] Gill suggested a similar mechanism for the action of cholera toxin when he discovered the NAD requirement for the toxin's activity in a cell-free system.[11] Evidence for such a mechanism has been obtained since 1978 by several investigators.[18,19] Cholera toxin has been found to catalyze the NAD-dependent ADP-ribosylation of a membrane GTP-binding protein which is involved in the regulation of adenylate cyclase activity. In all of these studies, cholera toxin was first "activated" by preincubation with or without Na-dodecylsulfate. The necessity of this pretreatment indicated that dissociation of A_1 was required in the ADP-ribosyl transferase activity of cholera toxin. We have recently obtained direct evidence that isolated polypeptide A_1 catalyzed ADP-ribosylation of a component of pigeon erythrocyte membrane, in the absence of DTT. The enzymatic activity could be assayed using polyarginine as the acceptor of ADP-ribose moiety (Table 8). S-carboxymethylated A_1 subunit was found to be equally active, indicating that free SH-group in the polypeptide was not directly involved in the catalysis, as in the case of adenylate cyclase activation (see above). Cholera toxin and subunit A require the presence of 1 mM DTT for maximal incorporation of ADP-ribosyl groups onto polyarginine, confirming the previous observations. The results have provided an indication that the same active

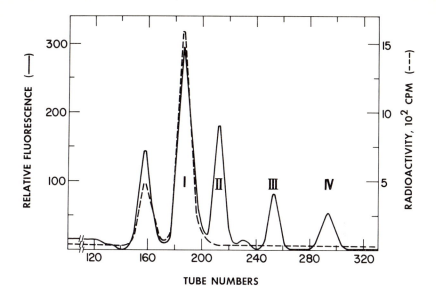

Fig. 8. Separation of the BrCN fragments of subunit A$_1$ on Sephadex G-50F. The largest fragment, Cn 1 is derived from the COOH-terminus and contains the only cysteine residue in the molecule.

site in polypeptide A$_1$ is involved in the ADP-ribosyl transferase activity and in the stimulation of adenylate cyclase. It thus provides a convenient assay for the biological activity of the A$_1$ subunit.

Polypeptide A$_1$ is a chain of approximately 200 amino acid residues containing 3 methionine residues. On cleavage of methionyl bonds with BrCN and gel filtration on Sephadex G-50 we have obtained four peptides in approximately equimilar ratio (Fig. 8). The NH$_2$- and COOH-terminal analyses revealed that peptide Cn III and Cn I were respectively derived from the NH$_2$- and COOH-termini of A$_1$ molecule. In separate experiments, the tryptic peptides containing Met residues were isolated and characterized, from which the information for the arrangement of 4 BrCN peptides in the A$_1$- subunit has been obtained (Fig. 9). Some interesting features of the primary structure of A$_1$ became apparent from the gross structure. The only cysteine residue which provides the link between A$_1$ and A$_2$ is located in peptide Cn I, within the COOH-terminal third of the molecule. Since exposure of the region around the cysteine residue is required for the manifestation of activity, the segment near the

152

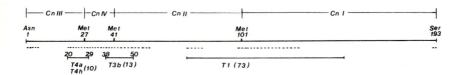

Fig. 9. The gross primary structure of polypeptide A₁.

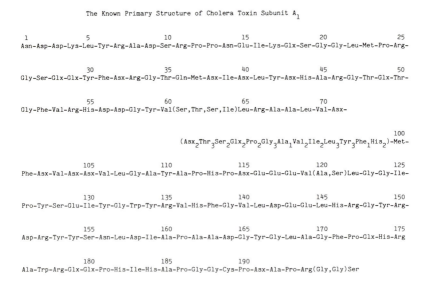

Fig. 10. Known amino acid sequence of polypeptide A₁.

(a)

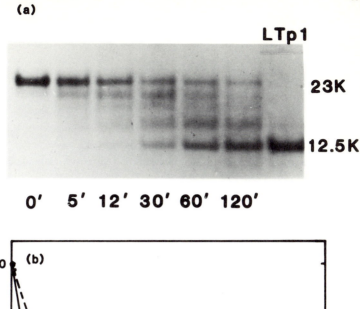

(b)

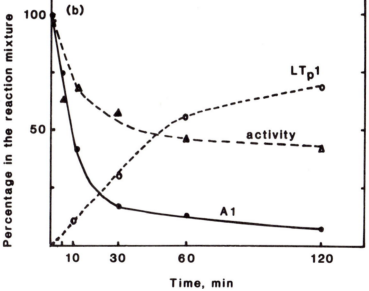

Fig. 11. Limited trypsinolysis of polypeptide A_1. The renatured S-carboxy-methyl A_1 was incubated with 0.2% (w/w) of trypsin at 0° and pH 7.5. At the indicated times, aliquots were removed and analyzed by gel-electrophoresis in 0.1% SDS and 8M urea, and for ADP-ribose transferase activity. (a) Electro-phoretic pattern. (b) Percentage of activity and A_1 remaining and of 12.5K band formed (T_p1, estimated by densitometry) at various time intervals.

COOH-terminus must also contain the active site (see above). Polypeptide A_1 is characterized by the high content of arginine (14/mole) and only 2 lysine residues which are located within 10% of the peptide chain from the NH_2-terminus. The sequence analysis of Cn I and Cn II peptides are still in progress, but over 80% of the sequence have now been determined (Fig. 10).

Chemical modification studies to locate the active center in polypeptide A_1 has been hampered by the instability of A_1-subunit at an ambient temperature. In either adenylate cyclase stimulation assay using pigeon erythrocyte membrane or the ADP-ribosylation assay, no activity was detected after incubation of a solution A_1 or S-carboxymethyl (S-cm) A_1 for 30 min at 37°. When kept at 3°, the renatured S-cm A_1 remained active for at least three months. Addition of bovine serum albumin was found effective in stabilizing the active conformation, but chemical modification studies could not be performed.

In an attempt to gain insight into the active conformation of polypeptide A_1, the subunit was subjected to limited hydrolysis with 0.2% (w/w) of trypsin at pH 7.5, at 0°. The reaction was monitored by activity measurements and gel electrophoretic analyses of the degradation products. After two hrs. of digestion, 90% of A_1 had been degraded and a major fragment with M = 12,500 appeared (Fig. 11a), when approximately a half of the original activity remained (Fig. 11b). The experiments suggest that an active conformation exists which is relatively resistant to trypsinolysis at 0°. Isolation of the 12,500 M_4 core fragment has, however, resulted in the total loss of the activity, suggesting that the removed fragment(s) may also be the component of the active site. The core peptide was found to derive from the center portion of the molecule, corresponding to residues 47 to 157 of A_1 polypeptide chain. The active conformation must be constituted from this region and the COOH-terminal region containing the cysteine residue. Preliminary evidence indicated that the substrate (NAD) binding site was located around the first arginine from the COOH-terminus (data not shown).

CONCLUDING REMARKS

Over the past ten years, through the efforts of many investigators throughout the world, the pathogenesis of cholera and the mechanism of action of cholera toxin have become clear. The molecular mechanism by which cholera toxin activates adenylate cyclase in the cell can now be summarized as in Fig. 12. After binding to the cell surfaces, A_1 polypeptide is released into the cytoplasm where it catalyzes the transfer of ADP-ribose moiety onto the GTP-binding protein on the inner surface of the membrane. Upon ADP-

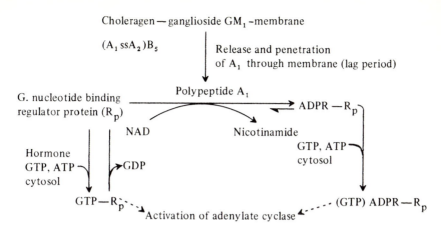

Fig. 12. The mechanism of action of cholera toxin.

ribosylation, the GTP-binding protein loses its ability to hydrolyze GTP in the system, with the result that adenylate cyclase is maintained in an activated state.

We have obtained considerable information as to the structure and active center in both the binding and active subunits of cholera toxin. The nature of the binding substance on the membrane, the subunit interaction in the cholera toxin molecule, and the structure involved in the ADP-ribosyl transfer in A_1 subunit remain subjects for our future investigations.

REFERENCES

1. Finkelstein, R. A. and LoSpalluto, J. J. (1969) J. Exp. Med., 130, 185-220.
2. Lai, C. Y. (1980) Crit. Rev. Biochem., 9, 171-206.
3. Rappaport, R., Rubin, B. A., and Tint, H. (1973) Infect. Immun., 9, 294-303.
4. Lai, C. Y., Mendez, E., and Chang, D. (1976) J. Inf. Dis., 133, S23-S30.
5. Cassel, D. and Pfeuffer, T. (1978) Proc. Natl. Acad. Sci. USA, 75, 2669-2673.
6. Wodnar-Filipowicz, A. and Lai, C. Y. (1976) Arch. Biochem. Biophys., 176, 465-471.
7. Lai, C. Y., Cancedda, F. and Duffy, L. K. (1981) Biochem. Biophys. Res. Commun., 102, 1021-1027.
8. Lai, C. Y. (1977) J. Biol. Chem., 252, 7249-7256.
9. Duffy, L. K. and Lai, C. Y. (1979) Biochem. Biophys. Res. Commun., 91, 1005-1010.
10. Gill, M. (1976) Biochemistry, 15, 1242-1248.
11. Gill, D. M. and King, C. A. (1975) J. Biol. Chem., 250, 6424-6432.

156

12. Gill, D. M. (1975) Proc. Nat. Acad. Sci. USA, 72, 2064-2068.
13. Finkelstein, R. A. (1973) Cholera. CRC Crit. Rev. Microbiol., 2, 553.
14. van Heyningen, et al. (1971) J. Infect. Dis., 124, S415-S426.
15. LoSpalluto, J. J. and Finkelstein, R. A. (1972) Biochem. Biophys. Acta., 257, 158-166.
16. Honjo, T., et al. (1968) J. Biol. Chem., 243, 3553-3555.
17. Gill, D. M. (1978) Proc. Natl. Acad. Sci. USA, 75, 2669-2673.
18. Cassel, D. and Pfeuffer, T. (1978) Proc. Natl. Acad. Sci. USA, 75, 3050-3054.

BIOCHEMICAL AND BIOPHYSICAL STUDIES OF PROTEIN-LIPID INTERACTIONS IN MEMBRANES

CHIEN HO, E. ANN PRATT, SUSAN R. DOWD,

GORDON S. RULE, VIRGIL SIMPLACEANU, AND PATRICIA F. COTTAM
Department of Biological Sciences, Carnegie-Mellon University,
Pittsburgh, Pennsylvania 15213

ABSTRACT

D-Lactate dehydrogenase of *Escherichia coli* provides a promising system for
biophysical studies of protein-lipid interactions in biological membranes. The
gene has been amplified to produce over 100 times the normal amount of protein
so that adequate amounts are available for study. Enzyme containing fluoro-
tryptophan can be made which is able to reconstitute D-lactate dehydrogenase-
deficient membrane vesicles. Fluorine-19-labeled membrane vesicles have been
prepared which can be reconstituted by unlabeled enzyme. Thus, protein and
lipid can be labeled independently; the labeled components are functional and
can be used to study protein-lipid interactions by means of nuclear magnetic
resonance. Preliminary studies show that fluorine-19-labeled phospholipids, in
the form of macroscopically oriented bilayers, exhibit Pake doublets from which
order parameters can be obtained. The spectrum of non-oriented liposomes can be
understood as resulting from a random superposition of the doublets, broadened
by heteronuclear magnetic dipole-dipole interactions. A computer simulation
gives excellent agreement with the experimental spectrum.

INTRODUCTION

In this presentation, we give a brief summary of several aspects of our
research on membranes. Recent research clearly indicates that protein-lipid
interactions are important for the activity, organization, and regulation of
enzymes in membranes. This suggests that studies of protein-lipid interactions
at a molecular level are necessary for understanding the reaction mechanisms
involved in membrane function.[1-4] During the past several years, we have been
developing a system designed to allow investigation into the nature of protein-
lipid interactions in membranes. The main component of the system is D-lactate
dehydrogenase (D-LDH) of *Escherichia coli*. D-LDH is an intrinsic membrane
protein, requiring detergents for solubilization and lipid-like molecules for
maximum activity.[5,6] In electron transfer reactions, D-LDH catalyzes the
oxidation of D-lactate coupled to the active transport of various amino acids
and sugars.[7,8] By adding purified enzyme to D-LDH-deficient membrane vesicles

prepared from E. coli ML 308-225 dld-3, the oxidase and transport activities can be reconstituted.[9,10] Thus, this system provides the possibility of studying the interaction of lipid and protein by labeling independently the membrane lipids and the exogenous D-LDH. A major constraint on this system has been the small amounts of D-LDH made by E. coli, relative to the amounts needed for biophysical studies. By means of recombinant DNA techniques, we have constructed several plasmids which result in the production of excess D-LDH.

Nuclear magnetic resonance (NMR) spectroscopy has been used extensively to investigate structural changes and dynamic processes that take place in the phospholipid bilayer during the gel-to-liquid-crystalline phase transition.[11,12] The nuclei suitable for these studies are ^{1}H, ^{2}H, ^{13}C, ^{19}F, and ^{31}P. Earlier work from this laboratory has shown that ^{19}F NMR of a difluoromethylene group incorporated into the hydrocarbon chain of a fatty acid can give information about the local motional states present in the hydrophobic region of the acyl chains in both model and biological membranes.[13-15] In addition, ^{19}F-labeled amino acids have been incorporated into proteins of bacteria.[16-19] These preliminary results indicate that the ^{19}F nucleus is of considerable interest as an NMR probe in the study of membrane dynamics and organization. There are a number of advantages in using ^{19}F as a membrane probe. First, ^{19}F has both a nuclear spin of 1/2 and a large magnetic moment, thus providing good sensitivity (83% that of ^{1}H). The natural abundance of ^{19}F is 100% so that an inherently large signal-to-noise ratio is obtained. Second, since the ^{19}F nucleus has no quadrupole moment, the analysis of the line shape by a second moment formalism is simplified by the presence of only magnetic dipole–dipole interactions. Third, the relatively large chemical shifts encountered in ^{19}F NMR enable one to probe the motional state of the molecule not only through the averaging of its magnetic dipole-dipole interaction, but also through the effect of motion upon the anisotropy of the ^{19}F chemical shift in high magnetic fields. Fourth, the ^{19}F resonance shows a complex relaxation behavior which could be very informative with regard to the dynamics of ^{19}F-labeled fatty acyl chains and amino acids in cell membranes.

We have recently extended our earlier studies on the application of ^{19}F NMR to the problem of motions present in the hydrophobic region of phospholipid dispersions and in biological membranes. 1-Myristoyl-2-(8,8-^{19}F$_2$)-myristoyl-sn-glycerol-3-phosphocholine, 2-[8,8-^{19}F$_2$]DMPC, and the corresponding analogs with the CF$_2$ group at the 4 and 12 positions in the fatty acyl chains have been investigated by ^{19}F NMR at 282.4 MHz. ^{19}F NMR spectra obtained from macroscopically oriented bilayers exhibit Pake doublets from which an order parameter

(S_{FF}) can be obtained. The ^{19}F spectra obtained from non-oriented liposomes of phospholipids can be explained in a satisfactory manner as a random super-position of the Pake doublets broadened by heteronuclear magnetic dipole-dipole interactions. From an analysis of the data, information concerning the motional state of the hydrocarbon chains in the liquid-crystalline phase can be obtained. For details, refer to Engelsberg et al.[20]

MATERIALS AND METHODS

Materials

The fluorinated fatty acids, 4,4-, 8,8-, and 12,12-difluoromyristic acids, were synthesized as described previously.[13] Palmitoleic acid was obtained from Nu-Chek Prep, ^3H-labeled proline from New England Nuclear, and 4-, 5-, and 6-fluorotryptophans, Brij 58, phenazine methosulfate (PMS), and 3-(4-,5-dimethylthiazolyl-2)2,5-diphenyltetrazolium bromide (MTT) from Sigma. Triton X-100 was obtained from Rohm & Haas. Fluorinated phospholipids, $2-[8,8-^{19}F_2]$ DMPC and the corresponding phospholipids with a CF_2 group at the 4 and 12 positions in the fatty acyl chains, were synthesized as described previously.[21,20] Other chemicals were reagent grade and were used without further purification.

Bacterial strains and plasmids

E. coli ML 308-225[22] and ML 308-225 *dld*-3[23] were gifts of Dr. H. R. Kaback. The unsaturated fatty acid-requiring mutants, ML 308-225 *ufa*-8 and ML 308-225 *dld*-3 *ufa*-2 were isolated as described in Gent et al.[14] W3110*trp*A33, which is a tryptophan auxotroph,[24] was the gift of Dr. C. Yanofsky.

Plasmid pIY2[25] contains the gene coding for D-LDH on a 7Kb HindIII fragment. This plasmid was a gift of Dr. H. R. Kaback and was used as a source of the D-LDH gene in all subsequent plasmid constructions for our work.

Preparation of membrane vesicles and enzymes, and reconstitution

E. coli membrane vesicles were prepared from all strains essentially as described by Kaback.[26]

D-LDH was prepared as described by Pratt et al.[5] through the DE52 column step. Pooled DE52 fractions were precipitated with an equal volume of acetone to remove Triton X-100 used in the purification process.

Membrane vesicles were reconstituted essentially as described by Short et al.[10] For details of all these procedures, see Pratt et al.[18,19]

Assays

D-LDH was assayed as described previously[5] using PMS and MTT in the presence of Triton X-100 at pH 8 and 23°C.

Oxidase activity was determined with a Clark-type oxygen electrode (YSI 5331) inserted into a jacketed 3 ml reaction chamber maintained at 37°C.

Transport of ^{3}H-proline into membrane vesicles was measured at 23°C essentially as described by Kaback.[26] For details of the oxidase and transport assays, see Pratt et al.[18,19]

Protein concentration was determined by the method of Lowry et al.[27] with bovine serum albumin as standard.

^{19}F NMR measurements

The ^{19}F NMR spectra were obtained at 282.4 MHz on a modified Bruker WH-300 spectrometer interfaced with an Aspect 2000A computer and operated in the Fourier-transform mode with home-built 5- and 13-mm probes. For details, see Engelsberg et al.[20]

RESULTS

Binding and reconstitution with unlabeled membrane components

Figure 1A shows the binding of D-LDH prepared from strain ML 308-225 to membrane vesicles of strains ML 308-225 and ML 308-225 dld-3. The data indicate that there is no difference in binding between the two strains.

The results given in Fig. 1B indicate that strains ML 308-225 and ML 308-225 dld-3 show equal increases in oxidase activity for units of D-LDH bound, and the oxidase activity rises to levels 7-8 times that in membrane vesicles of ML 308-225 without added D-LDH.

Figure 1C illustrates the uptake of ^{3}H-proline in the presence of increasing amounts of added D-LDH. The effect of added D-LDH on L-proline transport activity is very different from the effect on oxidase activity as given in Fig. 1B. There is little increase in accumulation of ^{3}H-proline by ML 308-225 membrane vesicles when extra D-LDH is added, and transport is inhibited by increasing levels of enzyme.

Binding and reconstitution with ^{19}F-labeled membrane components

When ML 308-225 ufa-8 or ML 308-225 dld-3 ufa-2 are grown in the presence of 8,8-difluoromyristic acid for one generation, the fluorinated fatty acid is found to comprise 10-20% of the total fatty acid content. We have found that membrane vesicles containing 8,8-difluoromyristic acid bind D-LDH. The oxidase

161

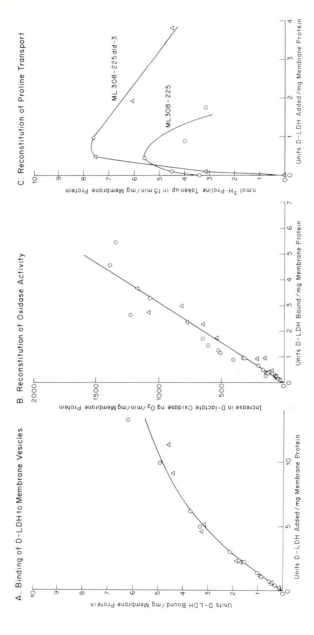

Fig. 1. Reconstitution of D-lactate dehydrogenase-deficient membrane vesicles by adding D-LDH;
A. Binding of D-LDH to membrane vesicles; B, oxidase activity of reconstituted membrane vesicles;
and C, transport of L-proline in reconstituted membrane vesicles. o—o: membrane vesicles prepared
from E. *coli* ML 308-225 and Δ-Δ: membrane vesicles prepared from E. *coli* ML 308-225 *dld*-3. Taken
from Fig. 1 of Pratt et al.,[19] with permission.

activities in ^{19}F-labeled membrane vesicles are not much different from those of unlabeled membrane vesicles prepared from the same fatty acid auxotrophs. Again, membrane vesicles prepared from fatty acid auxotrophs grown in the presence of 8,8-difluoromyristic acid can transport L-proline as well as (and in some cases, better than) unlabeled membrane vesicles from the same fatty acid auxotrophs (results not shown).

We have found that D-LDH containing 4-, 5-, or 6-fluorotryptophan binds to *dld* membrane vesicles to the same extent as unlabeled enzyme (Fig. 2A). For a given amount of enzyme bound, 5- or 6-fluorotryptophan-labeled enzyme appears to give a somewhat higher oxidase activity (Fig. 2B). Also, for a given amount of D-LDH added, 5- or 6-fluorotryptophan-labeled enzyme appears to give a higher transport activity for L-proline (Fig. 2C).

For detailed information on biochemical studies of binding and reconstitution with labeled and unlabeled membrane components, refer to Pratt et al.[18,19]

Construction of plasmids which produce higher levels of D-LDH

The D-LDH gene product in *E. coli* is normally present at approximately 1,000 copies per cell and contributes approximately 0.05% of the total cellular protein. Recombinant DNA techniques can be used to increase the production of specific proteins in *E. coli*. Some extra-chromosomal plasmids are present in many copies within the same cell. Thus, by cloning the gene of interest into such plasmids, the levels of the gene product are often elevated. Gene product amplification can also occur if the transcription of mRNA from the gene is increased. In order to increase mRNA levels, it is necessary to direct transcription of the cloned gene from a strong promoter. Two such promoters are the *lac* promoter (p_{Lac}) of the *E. coli* lactose operon and the left promoter (p_L) of the bacteriophage λ. We have been able to construct several plasmids with a high copy number whose expression of D-LDH is under the control of p_{Lac} or p_L. These plasmids can produce over 100-fold amplification of D-LDH, i.e., 5-7% of the total cellular protein is D-LDH. Furthermore, this amplification is inducible, and this characteristic should prove useful in obtaining ^{19}F-labeled D-LDH. Details of this aspect of our research will be published elsewhere (G. S. Rule, E. A. Pratt, and C. Ho, results to be published).

^{19}F NMR investigation of ^{19}F-labeled phospholipids

Figure 3A shows the ^{19}F NMR line shapes obtained for oriented multilayers of 2-[8,8-[19]F$_2$]DMPC at 27°C as a function of the orientation with respect to the magnetic field of the normal to the bilayer. The water content in this sample

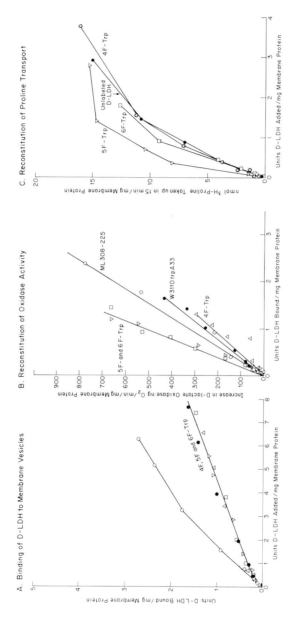

Fig. 2. Reconstitution of D-lactate dehydrogenase-deficient membrane vesicles by adding ^{19}F-labeled D-LDH: A, binding of ^{19}F-labeled D-LDH to membrane vesicles; B, oxidase activity in reconstituted membrane vesicles containing ^{19}F-labeled D-LDH; and C, transport of L-proline in reconstituted membrane vesicles containing ^{19}F-labeled D-LDH. ▲, D-LDH from W3110*trp*A33 containing 4F-Trp; ▼, D-LDH from W3110*trp*A33 containing 5F-Trp; ▢, D-LDH from W3110*trp*A33 containing 6F-Trp; ⊙, unlabeled D-LDH from ML 308-225; and ●, unlabeled D-LDH from W3110*trp*A33. Taken in part from Fig. 3 of Pratt et al.,19 with permission.

A. Oriented Sample

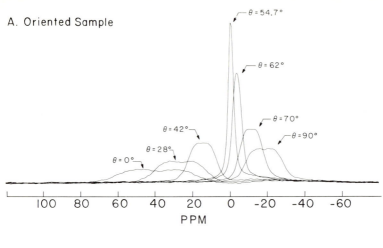

B. Non-Oriented Sample

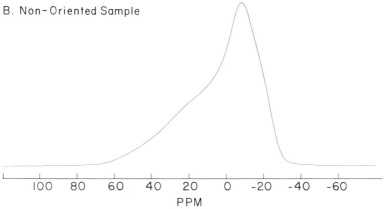

Fig. 3.   282.4 MHz ^{19}F NMR spectra of 2-[8,8-^{19}F$_2$]DMPC in H$_2$O at 27°C.
A, oriented sample with ∼ 25% water content; and B, non-oriented sample with
∼ 70% water content.

is ∼ 25%. Except for θ = 54.7° (the so-called "magic angle") where the
chemical shift anisotropy and dipolar interactions are averaged out, the line
shapes clearly exhibit a resolved "Pake" doublet pattern. The ^{19}F chemical
shift anisotropy, which can be measured directly from the separation between
the center of each of the doublets at 90° and 0°, is 64 ppm for 2-[8,8-^{19}F$_2$]
DMPC at 27°C. Figure 3B shows the ^{19}F NMR line shape of 2-[8,8-^{19}F$_2$]DMPC
liposomes in excess H$_2$O (≥ 70% in H$_2$O content) at 27°C. The ^{19}F NMR line

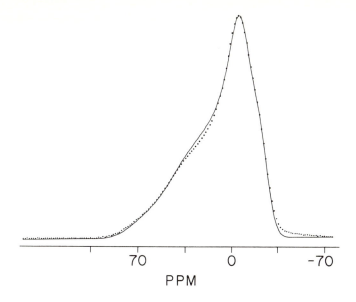

Fig. 4. 282.4 MHz ^{19}F NMR spectra of non-oriented 2-[8,8-^{19}F$_2$]DMPC in excess H$_2$O at 27°C; ····, experimental spectrum; and ———, simulated spectrum.

shapes in non-oriented liposomes (Fig. 3B) can be explained as resulting from a random superposition of the doublets broadened by heteronuclear magnetic dipole-dipole interactions as shown in Fig. 3A. By taking into account the ^{19}F chemical shift anisotropy and both the homonuclear and heteronuclear (^{19}F-^{19}F and ^{1}H-^{19}F) dipolar interactions, we have been able to simulate the ^{19}F NMR line shape of non-oriented liposomes as shown in Fig. 4. As seen in this figure, the computer-simulated ^{19}F NMR spectrum of 2-[8,8-^{19}F$_2$]DMPC in excess water at 27°C is in excellent agreement with the experimentally obtained spectrum.

In the liquid-crystalline phase, the motional state of the hydrocarbon chain at the ^{19}F site can be adequately described by an order parameter (S_{FF}). This order parameter is a measure of the degree of anisotropy of the motion. S_{FF} can be determined directly from the splitting in the Pake doublet structure (as shown in Fig. 3A) by the following equation:

$$\Delta\nu_{FF} = \frac{3}{2} \; \frac{\gamma_F^2 \, \hbar}{2\pi r^3} \; (3\cos^2\theta - 1) S_{FF}$$

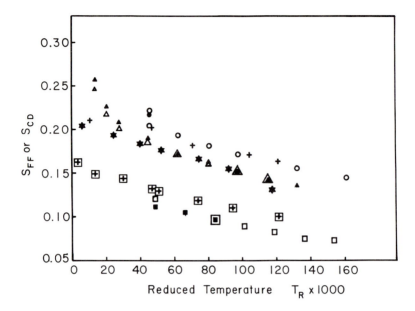

Fig. 5. Order parameters (S_{FF} or S_{CD}) as a function of reduced temperature, T_R: ✦, S_{FF} from 2-[4,4-^{19}F$_2$]DMPC non-oriented sample; ●, 2-[8,8-^{19}F$_2$]DMPC oriented sample; o, S_{FF} from 2-[2,2,7,7,9,9-^{2}H$_6$-8,8,^{19}F$_2$]DMPC oriented sample; ▲, S_{FF} from 2-[8,8-^{19}F$_2$]DMPC non-oriented sample; △, S_{FF} from 2-(2,2,7,7,9,9-^{2}H$_6$-8,8-^{19}F$_2$)DMPC non-oriented sample; ■, S_{FF} from 2-(12,12-^{19}F$_2$)DMPC non-oriented sample; □, S_{FF} from 2-(2,2,7,7,9,9-^{2}H$_6$-12,12-^{19}F$_2$)DMPC non-oriented sample; +, S_{CD} from 2-(8,8-^{2}H$_2$)DMPC non-oriented sample; and ⊞, S_{CD} from 2-[12,12-^{2}H$_2$]DMPC non-oriented sample. The data for + and ⊞ are taken from Oldfield et al.[28] Reduced temperature, T_R, is defined as $(T - T_c)/T_c$ where T is the measured temperature and T_c is the phase transition temperature of a given phospholipid.

where $\Delta\nu_{FF}$ is the frequency separation between the peaks of the Pake doublets, γ_F is the gyromagnetic ratio of the fluorine nucleus, $\hbar$ is the Planck constant divided by 2π, r is the distance between the two fluorine atoms, and θ is the angle between the rotation axis and the external magnetic field. S_{FF} measures directly the motion of the fluorine-fluorine internuclear vector. For details on our ^{19}F NMR investigation of phospholipid dispersions, see Engelsberg et al.[20]

Figure 5 gives a comparison between S_{FF} of ^{19}F-labeled phospholipids and S_{CD} derived from the corresponding [2]H-labeled phospholipids[28] as a function of reduced temperature. It is interesting to note that the values of S_{FF} and S_{CD} agree to within 10%. The implication seems to be that the CF_2 group has little additional perturbing effect in the lipid bilayer.

DISCUSSION

In order to carry out biochemical and biophysical studies on the interaction between D-LDH and phospholipids as a model for protein-lipid interactions in membranes, we need to fulfill the following requirements: (i) to select a suitable biophysical or spectroscopic tool so as to derive molecular informa-tion about the protein-lipid interactions in membranes; (ii) to label D-LDH and membrane lipids with appropriate spectroscopic probes; (iii) to reconsti-tute the system with spectroscopically labeled components; and (iv) to obtain sufficient amounts of D-LDH for our studies.

During the past two years, we have achieved the above-mentioned requirements. We shall briefly summarize our results here. ^{19}F-Labeled compounds have been incorporated into lipids and proteins of E. coli. ^{19}F-Labeled membrane vesicles, prepared by growing a fatty acid auxotroph of a D-LDH-deficient strain (E. coli ML 308-225 dld-3 ufa-2) on 8,8-difluoromyristic acid, can be reconstituted for oxidase and transport activities by binding exogenous D-LDH. ^{19}F-Labeled D-LDH prepared by adding 4-, 5-, or 6-fluorotryptophan to a tryptophan-requiring strain (E. coli W3110trpA33) is able to reconstitute D-LDH-deficient membrane vesicles. Thus, protein and lipid can be labeled independently and used to investigate protein-lipid interactions in membranes. In addition, by placing the gene coding for D-LDH downstream from the promoter, p_L or p_{Lac} in a high copy number plasmid, we have achieved over 100-fold amplification of the gene product. Furthermore, this amplification is inducible, and this property is very useful in obtaining ^{19}F-labeled D-LDH. Thus, we have fulfilled the biochemical aspects of this project.

We have also made excellent progress in the biophysical side of the project. Our early ^{19}F NMR investigation of membranes has been extended to a DMPC model membrane containing a difluoromethylene group only in the 2-acyl chain. The order parameters for the difluoromethylene group in the 4-, 8-, and 12-position as a function of temperature have been determined. The results are consistent with work[28] using deuterium atoms in the same positions. Line-shape simulations have enabled us to determine S_{FF} and the chemical shift anisotropy from powder-pattern type line shapes.

168

In conclusion, we have developed all the basic tools needed to investigate the nature of protein-lipid interactions for the D-lactate dehydrogenase system in E. coli membrane vesicles. During the next few years, we hope to gain some new insights into the molecular basis of protein-lipid interactions in membranes from our study of D-lactate dehydrogenase. It appears that ^{19}F NMR can be a powerful technique to investigate the structures and dynamics of membrane systems.

This work is supported by research grants from the National Institutes of Health (GM-26874 and HL-24525) and the National Science Foundation (PCM 82-08829.

REFERENCES

1. Cronan, J. E., Jr. and Gelmann, E. P. (1975) Bacteriol. Rev., 39, 232-256.
2. Gennis, R. B. and Jonas, A. (1977) Ann. Rev. Biophys. Bioeng., 6, 195-238.
3. Sandermann, H., Jr. (1978) Biochim. Biophys. Acta, 515, 209-237.
4. Chapman, D., Gómez-Fernández, J. C., and Goñi, F. M. (1979) FEBS Lett., 98, 211-223.
5. Pratt, E. A., et al. (1979) Biochemistry, 18, 312-316.
6. Fung, L. W.-M., Pratt, E. A., and Ho, C. (1979) Biochemistry, 18, 317-324.
7. Barnes, E. M., Jr. and Kaback, H. R. (1971) J. Biol. Chem., 246, 5518-5522.
8. Kaback, H. R. (1974) Methods. Enzymol., 31, 698-709.
9. Futai, M. (1974) Biochemistry, 13, 2327-2333.
10. Short, S. A., Kaback, H. R., and Kohn, L. D. (1974) Proc. Natl. Acad. Sci. USA, 71, 1461-1465.
11. Griffin, R. G. (1981) Methods. Enzymol., 72, 108-174.
12. Jacobs, R. E. and Oldfield, E. (1981) Proc. Nucl. Magn. Reson. Spectros., 14, 113-136.
13. Gent, M. P. N., Cottam, P. F., and Ho, C. (1978) Proc. Natl. Acad. Sci. USA, 75, 630-634.
14. Gent, M. P. N., Cottam, P. F., and Ho, C. (1981) Biophys. J., 33, 211-224.
15. Gent, M. P. N. and Ho, C. (1978) Biochemistry, 17, 3023-3038.
16. Pratt, E. A. and Ho, C. (1975) Biochemistry, 14, 3035-3040.
17. Robertson, D. E., Kroon, P. A., and Ho, C. (1977) Biochemistry, 16, 1443-1451.
18. Pratt, E. A., et al. (1982) Biophys. J., 37, 101-103.
19. Pratt, E. A., et al. (1983) Biochim. Biophys. Acta, 729, 167-175.
20. Engelsberg, M., et al. (1982) Biochemistry, 21, 6985-6989.
21. Oldfield, E., et al. (1980) J. Biol. Chem., 255, 11652-11655.
22. Winkler, H. H. and Wilson, T. H. (1966) J. Biol. Chem., 241, 2200-2211.
23. Hong, J. S. and Kaback, H. R. (1972) Proc. Natl. Acad. Sci. USA, 69, 3336-3340.
24. Drapeau, G. R., Brammar, W. J., and Yanofsky, C. (1968) J. Mol. Biol., 35, 357-367.
25. Young, I. G., Jaworowski, A., and Poulis, M. (1982) Biochemistry, 21, 2092-2095.
26. Kaback, H. R. (1971) Methods. Enzymol., 22, 99-120.
27. Lowry, O. H., et al. (1951) J. Biol. Chem., 193, 265-275.
28. Oldfield, E., et al. (1978) Biochemistry, 17, 2727-2740.

SEQUENCING OF PROTEINS BY DNA SEQUENCING

LI-HE GUO AND RAY WU
Section of Biochemistry, Molecular and Cell Biology
Cornell University, Ithaca, New York 14853

INTRODUCTION

Ten years ago, it would have taken about two years to sequence 50 nucleotides in a segment of DNA which codes for 16 amino acids, while determining the corresponding amino acids would have been faster, requiring only a few months. Today, the situation is the reverse. It would take only a few months to sequence a segment of DNA 2000 nucleotides in length, which codes for a protein containing 666 amino acids. However, it would take about two years to sequence a protein of this size.

Ten years ago, it was not practical to sequence a segment of DNA 1000 nucleotides in length. Today it is not practical to sequence certain proteins which are very difficult to isolate and purify. A good example is interferon, because only minute amounts are present in blood cells. For instance, even with the super-sensitive protein sequencing method, only a dozen amino acids were determined from the amino terminus of α-interferon.[1] In contrast, within a relatively short time the gene coding for the α-interferon was cloned and the complete DNA sequence was determined, which allows for the deduction of the complete amino acid sequence of the 180 long α-interferon.[2]

In this article we will briefly review the major developments known to us before June 1982, which made it possible to sequence almost any DNA at rapid speed. We will next describe two useful plasmids constructed in our laboratory and a new strategy for cloning and sequencing long fragments of DNA. The various methods available for cloning almost any gene can be found in the following books[3-6] and will not be discussed here.

RAPID METHODS FOR SEQUENCING DNA

The first rapid method for sequencing DNA was developed by Sanger and Coulson in 1975.[7] They used a single-stranded DNA from φX174 as the template and a synthetic decanucleotide or a restriction fragment as the primer. E. coli DNA polymerase I is used to extend the primer and copy the template sequence (in two stages) under controlled conditions. Four separate samples are used, each with one of the four dNTP missing (in the "minus" system). The incorporation stops wherever the missing dNTP is required. The DNA samples from

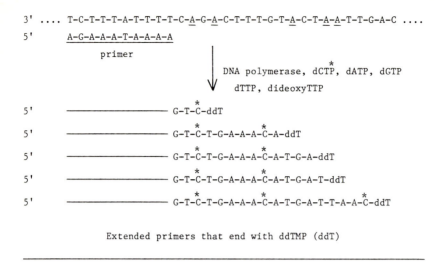

3' .... T-C-T-T-T-A-T-T-T-T-C-A-G-A-C-T-T-T-G-T-A-C-T-A-A-T-T-G-A-C ....

5'   A-G-A-A-A-T-A-A-A-A

              primer

              DNA polymerase, dCTP*, dATP, dGTP
                    dTTP, dideoxyTTP

5'   ——————————— G-T-C-ddT*

5'   ——————————— G-T-C-T-G-A-A-A-C*-A-ddT*

5'   ——————————— G-T-C-T-G-A-A-A-C*-A-T-G-A-ddT*

5'   ——————————— G-T-C-T-G-A-A-A-C*-A-T-G-A-T-ddT*

5'   ——————————— G-T-C-T-G-A-A-A-C*-A-T-G-A-T-T-A-A-C*-ddT*

Extended primers that end with ddTMP (ddT)

Fig. 1. The principle of the dideoxynucleotide chain-termination method for
sequencing DNA.[8] The upper line gives a hypothetical sequence of a single-
stranded DNA. The short sequence immediately below represents a primer, which
is complementary to a part of the upper sequence. Using E. coli DNA polymerase
and the deoxynucleotides:dideoxynucleotide mixture, only the extended primers
of five different lengths are shown in the lower part of the figure. In this
example, ddTTP was used, therefore all the primers were terminated with a ddT.
In the complete sequencing method, four different reaction mixtures were used,
each with a different ddNTP.

the four incubation mixtures are then denatured and subjected to electrophoresis
in polyacrylamide gel, which can separate DNA molecules differing in length by
a single nucleotide. The sequence of the DNA can be read off from the autoradiogram of the gel.[7]

The above method was improved by Sanger et al. in 1977,[8] who introduced
dideoxynucleotides as specific chain-terminating inhibitors of DNA polymerase,
in place of leaving one dNTP out as in the "minus" system. Wherever a dideoxynucleotide (ddN) is incorporated, the chain cannot be further extended because
of the lack of the 3' hydroxyl group at the terminal nucleotide. As shown in
Fig. 1, a primer (decanucleotide) is incubated with a template DNA (top line)
in the presence of DNA polymerase, $[\alpha\text{-}^{32}P]$dCTP, dATP, dGTP, dTTP and dideoxyTTP.
The primer is extended in copying the template sequence. At any position where
a dTTP is needed, either a dTMP or a ddTMP (ddT) is incorporated. If a ddT is
incorporated, that chain can no longer be extended. If a dT is inserted, that

chain continues to grow until the next ddT is incorporated. Thus, a mixture
of DNA fragments all having the same 5' end and with ddT residues at the 3'
ends is obtained. When this mixture is fractionated by electrophoresis in
denaturing polyacrylamide gels, the bands in the T lane show the distribution
of dTs in the newly synthesized DNA. By using a different ddNTP as terminators
in four separate samples, and running the samples in parallel in the gel, a
pattern of bands is obtained from which the sequence can be read off as in the
other rapid technique mentioned above.

Maxam and Gilbert in 1977 have developed a different rapid method for
sequencing DNA,[9] based on chemical reactions that break a terminally-labeled
DNA molecule at specific nucleotides. Reaction condition is adjusted so that
only one nucleotide, on the average, is reacted per DNA fragment. Subsequent
cleavage of the DNA at these positions will produce a family of radioactive
DNA fragments, extending from the same labeled end to each of the positions
of that nucleotide. Four chemical reactions are used that cleave DNA preferentially at each of the four nucleotides. Finally, the products of the four
specific reactions are fractionated by size in polyacrylamide gel, and the DNA
sequence can be read off from the gel pattern. This chemical method can be
applied to both double-stranded and single-stranded DNAs. However, the method
is still relatively time consuming, since it requires at least one additional
preparative gel electrophoresis step to isolate the single-end labeled DNA.
Recently, Rüther et al.[10] constructed a poly-linker plasmid, pUR222, which can
be used to clone foreign DNA and to sequence the DNA according to the method
of Maxam and Gilbert[9] without the need of an additional gel step. However, the
chemical method is basically more time consuming than the enzymatic method of
repair labeling using the dideoxynucleotides.[8] It is fair to conclude that for
relatively short DNA fragments (e.g., 100-500 base pairs long) the chemical
method is probably more convenient, but for long DNA fragment (e.g., greater
than 2000 base pairs long) the dideoxynucleotide chain-termination method is
faster.

The limitation of Sanger's method is that it requires single-stranded DNA
as a template for the DNA polymerase-catalyzed primer extension reaction. It
is suitable for certain bacteriophages, such as φX174 and M13, that occur
naturally in the single-stranded form, or for double-stranded DNA in which the
strands can be separated. It is also possible to clone a segment of double-
stranded DNA into the replicative form of M13 phage, and after transfection and
replication in the host, to isolate the single-stranded DNA from the recombinant
phage.[11-14] It is relatively easy to clone short segments of DNA, and about
250 nucleotides can be sequenced by using a primer.[13,14] However, longer DNA

172

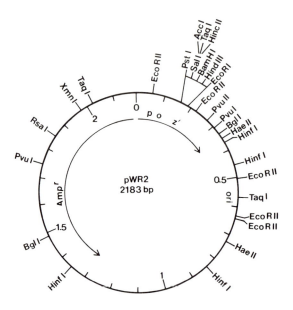

Fig. 2. A restriction map of pWR2. In pWR2 plasmid, seven unique restriction sites (*PstI*, *SalI*, *AccI*, *HincII*, *BamHI*, *HindIII* and *EcoRI*) are located between codon number 4 (thr) and number 6 (ser) of the *lac* z' gene,[10,11] and an eighth, *PvuII*, is in codon number 35 (ser).

cannot be sequenced directly (unless cut to smaller pieces first). Another problem is that certain long DNA molecules tend to be unstable when cloned into M13 phages. Furthermore, a lot of overlapping work is needed to piece together many 250-nucleotide long fragments to obtain a sequence 5000-long. We will describe an alternative sequencing procedure which overcomes these limitations.

Construction and properties of plasmid pWR2. Rüther et al.[10] constructed pUR222 which includes the coding sequence for the 59 amino acid residues of the α-peptide of β-galactosidase as well as the operator and promoter regions of the *lac* operon. Bacteria harboring this plasmid should make blue colonies on indicator plates containing isopropyl-thiogalactoside (IPTG) and 5-bromo-4-chloro-indocyl-β-D-galactoside (X-gal). If an exogenous DNA fragment is inserted in between the *lac* promoter and the *lac* z' gene, bacteria harboring this plasmid generally give rise to white colonies. This makes the selection easy for bacteria colonies carrying inserted DNA. pUR222 contains six unique cloning sites (*PstI*, *SalI*, *AccI*, *HindII*, *BamHI* and *EcoRI*) in a small region of its *lac* z'-gene. This poly-linker plasmid allows rapid cloning and chemical

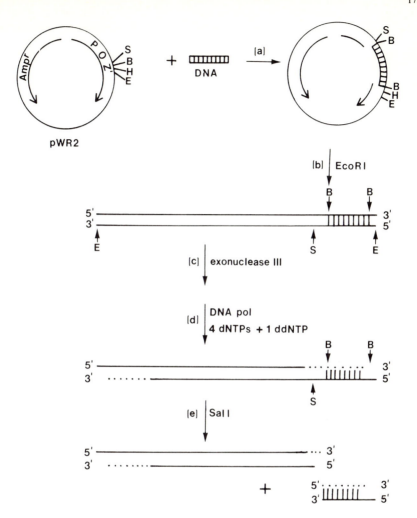

Fig. 3. The principle of the exonuclease III-repair synthesis method for sequencing DNA.[15] Plasmid pWR2 is given as an example of a cloning vehicle for facilitating direct sequencing analysis. The positions of the *lac* region (P for the *lac* promoter, O for the *lac* operator, z' for the coding sequence of the amino-terminal end of the β-galactosidase), and the Amp[r] region (coding for the protein that renders the *E. coli* ampicillin resistant) are indicated. Four important unique restriction sites are given (S for *Sal*I, B for *Bam*HI, H for *Hin*dIII and E for *Eco*RI).

After step (d), the dots at the 3' ends represent the incorporation of labeled nucleotides.

sequencing of DNA fragments, resulting from digestion by: HpaII, AccI, SalI, PstI and EcoRI, but not by BamHI, BglII and Sau3A.[10]

We have constructed an improved plasmid, pWR2, which includes two additional features. First, we have added on a HindIII site between the BamHI site and EcoRI site (Fig. 2). Second, we have deleted about 530 base pairs from pUR222, which includes the region coding for the ColE1 relaxation site (bom). Deletion of this region resulted in an increased copy number of pWR2, which is about four times higher than that of pUR222 and pBR322. The high yield of pWR2 plasmid and its derivatives allows the application of a rapid small scale (5 ml liquid culture or one-eighth of a Petri plate) DNA isolation procedure,[17] yielding plasmid of sufficient purity and quantity for DNA sequencing. In fact, any procedure for DNA isolation would result in a higher ratio of pWR2 plasmid to chromosomal DNA, which makes the sequencing gel pattern cleaner.

We have observed that bacteria carrying pWR2 are capable of growing in either M9 minimal medium or YT broth containing up to 1000 µg/ml of ampicillin. However, bacteria harboring recombinant plasmids with inserts in the poly-linker region grow slower and can tolerate only up to 20 µg/ml of ampicillin.

pWR2 can be used for rapid cloning and chemical sequencing of DNA fragments resulting from digestion by any one of the following 11 common enzymes: HpaII, AccI, SalI, PstI, EcoRI, HindIII, HincII, XhoI, BamHI, BglII and Sau3A. It is important that pWR2 contains unique restriction sites which include enzymes that recognize six base pairs, such as SalI, BamHI, HindIII and EcoRI, because long DNA fragments (average in size of 1000-4000 base pairs) can be cloned into this plasmid, and can be sequenced by the exonuclease III method (Fig. 3).

Fig. 4. Schematic representation of plasmid pWR33 construction. In step (a) pUR222 plasmid was digested with BamHI (B) and XmnI (X), and the short XmnI-BamHI fragment isolated. Also, pWR2 was digested with B and X, and the long XmnI-BamHI fragment isolated. In step (b), the appropriate fragments from step (a) were ligated with T4 DNA ligase to produce pWR31. In step (c), pWR31 was partially digested with PvuII, and a SmaI linker, d(^{32}pCCCGGG) duplex, was blunt-end ligated to the plasmid at the PvuII site near the lac promoter. After SmaI digestion to remove any multiple SmaI linker, the plasmid was ligated again and used to transform E. coli. In step (d), pWR32 was digested with TthI(T), followed by exonuclease III, and S1 nuclease digestion, to remove about 150 nucleotides from each end. The plasmid was blunt-end ligated to produce pWR33. Abbreviations for restriction enzyme sites are: B for BamHI, E for EcoRI, H for HindIII, Ps for PstI, Pu for PvuII, R for RsaI, S for SalI, T for TthI, X for XmnI.

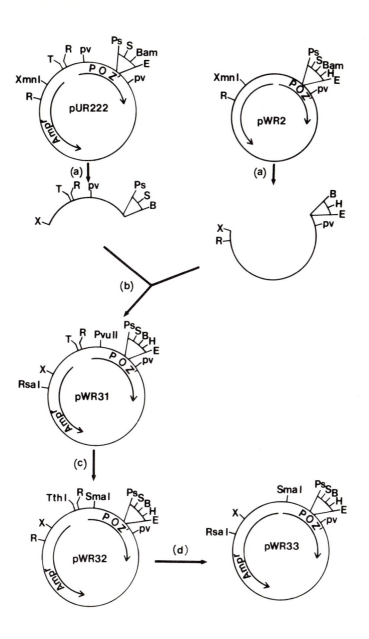

<u>Construction and properties of plasmids pWR33.</u> We have constructed pWR33
to further increase the usefulness of pWR2 plasmid for cloning and sequencing
DNA. The advantage of pWR33 is that a new unique *Sma*I site has been introduced,
which will facilitate sequencing of cloned DNA. The *Sma*I site can be used to
linearize the DNA for the primer method of sequencing, or it can serve as
either the first or second restriction enzyme to cut the DNA in order to
determine the sequence near the cloning site by the exonuclease III method.

The construction of pWR33 is shown in Fig. 4. By combining the short *Bam*HI
*Xmn*I fragment from pUR222 to the long *Bam*HI-*Xmn*I fragment of pWR2, plasmid
pWR31 was produced. The plasmid was opened up at the *Pvu*II site on the left-
hand side of the *lac* promoter and a *Sma*I linker was inserted to give pWR32.
The latter plasmid was cleaved at the *Tth*I site, and shortened by exonuclease
III and S1 nuclease digestion. After religation, plasmid pWR33 was produced,
which had its copy number control region (*bom* site) deleted so that a larger
number of plasmids could be produced per cell. The extent of deletion in
pWR33 is less than that in pWR2. As a result, pWR33 gives the same response to
a certain concentration of ampicillin (e.g., 100 µg/ml) whether or not a segment
of DNA is inserted into one of its restriction sites in the poly-linker region.

EXONUCLEASE III METHOD FOR SEQUENCING DNA

We have described an enzymatic method for sequencing double-stranded DNA.
It is based on partial digestion of any duplex DNA with exonuclease III to produce DNA molecules with shortened 3' ends, followed by repair synthesis to
extend and label the 3' ends. After asymmetrical cleavage of the DNA with a
restriction enzyme, the labeled DNA fragments are separated by gel electrophoresis and the sequence read of the autoradiogram.[15] The principle of the
method is shown in Fig. 3. (a) A fragment of DNA (hatched region) is cloned
into a plasmid, such as pWR2; (b) this recombinant plasmid is digested with a
restriction enzyme (such as *Eco*RI) to give a linear DNA; (c) the linearized
DNA is digested with exonuclease III to remove a number of nucleotides beyond
the length to be sequenced (e.g., 500 nucleotides); (d) the partially digested
DNA molecules are labeled at the 3' end by *E. coli* DNA polymerase-catalyzed
incorporation of four dNTP (one of them is labeled with ^{32}P) and one ddNTP in
each of the four tubes; (e) the labeled DNA is digested asymmetrically with
another restriction enzyme (such as *Sal*I) to produce two families of labeled
fragments (long and short). In the last step, these fragments are denatured
and fractionated by electrophoresis on a polyacrylamide gel to separate the
DNA fragments, differing in length by as few as one nucleotide. The short

fragments (labeled upper strand only) constitute a family of fragments (ranging from as short as 25 nucleotides to as long as 500 nucleotides), which can be separated on an 80 cm gel.

For DNA fragments (hatched region, Fig. 3) of different lengths, the following strategy for sequencing is used. For a DNA fragment up to 500 base pairs in length, its sequence can be completely determined after cutting the DNA with EcoRI in step (b), and cutting the labeled DNA with SalI in step (e). For a DNA fragment up to 1000 base pairs in length, the first 500 base pairs at the left-hand half is sequenced as in the above example. The sequence of the 500 base pairs at the right-hand half can be determined by cutting an aliquot of the DNA with SalI in step (b), and cutting the labeled DNA with EcoRI in step (e). For DNA fragments up to 1500 base pairs in length, it is likely that after determining 500 base pairs from each end using the method just outlined (or using the primer method to be described in the following section), additional restriction enzyme sites can be found. These sites can be used for the second digestion (step e) to give sequence information of the central portion of the DNA fragment.[17] For DNA fragments larger than 1500 base pairs, one can use the new strategy presented in the following section.

A NEW STRATEGY FOR SEQUENCING LONG FRAGMENTS OF DNA

We recently developed a new rapid method for cloning and sequencing a long fragment of DNA. The method (shown in Fig. 5) is based on the digestion of duplex DNA (e.g., 3000 base pairs in length) by exonuclease III for different lengths of time followed by nuclease S1 digestion[16,17] to produce a family of progressively shortened duplex DNA (step a). These fragments are fractionated on an agarose gel, and DNA molecules shortened by removing about 600, 1100, 1500 base pairs can be obtained (Fig. 6). DNA molecules shortened to each desired length are cut out from the gel and blunt-end ligated to HincII-cut pWR2 DNA (Fig. 5, step b). Each sample that represents a specific size class is used separately for transformation and white colonies are picked from each class. The length of the inserted DNA is confirmed after digesting a portion of each sample with PstI and EcoRI, followed by agarose gel electrophoresis. Once the insert size is confirmed, sequence analysis is carried out on another portion of each sample by applying the dideoxynucleotide chain-termination method.[8,17,19] Each sample is divided into two aliquots, and the sequence is determined by using primer I or primer II, respectively. As shown in Fig. 5 (c), about 400 nucleotides from each end can be determined. Since each size class differs by about 600 base pairs from both ends, or 300 base pairs from

Fig. 5. A new strategy for sequencing DNA. As an illustration, a DNA of
3000 base pairs in length is sequenced by the new strategy. In step (a), the
DNA is digested with econuclease III[15] for 30 and 60 minutes etc., which
removes approximately 300 and 600 nucleotides from each end (or 600 and 1200
nucleotides from both ends). The incubation mixture is adjusted to pH 4.5 and
the single-stranded ends of each DNA molecule is removed by digestion with
nuclease S1[16,17] to produce blunt-ended DNA. The DNA molecules are fractionated and sized on a 1% low-melting agarose gel. In step (b), a *Hinc*II-cut
pWR2 (or pWR33) plasmid is used as the cloning vehicle, which is blunt-end
ligated to the gel fractionated DNA of approximately 2400, 1800, 1200 and 600
base pairs in length. The recombinant plasmid containing DNA of each of these
sizes (classes) is used to transform *E. coli* cells and the transformants which
give white colonies are selected.[15] Four white colonies (isolates) are picked
from each class and streaked on a Petri dish to produce more cells and recombinant plasmids.[17] The recombinant plasmids are purified from the cells and
an aliquot (1 μg) of the plasmid DNA is digested with *Pst*I (P) and *Eco*RI (E)
restriction enzymes to release the inserted DNA. The presence and the size
of the insert in each isolate is confirmed after agarose gel electrophoresis.
Sixteen to 24 isolates from 4 to 6 size classes can be purified, digested and
electrophoresed within a day. These isolates with the correct insert sizes
are sequenced by using another aliquot of the recombinant DNA samples. Sequence
analysis is accomplished by linearizing the DNA with *Sma*I or *Pvu*II (a unique
site), and then by denaturing the DNA and hybridizing it[17,18] with either
primer I or primer II.[19] The primer is extended by using the dideoxynucleotide
chain-termination method and the sequence analyzed after gel electrophoresis.[8]
Approximately 400 nucleotides can be sequenced from each primer site using an
80 cm polyacrylamide gel (8%, 0.04 cm thick). As shown in step (c), from the
recombinant plasmid with an insert of 2400 base pairs, 400 nucleotides (300-
700) can be sequenced by using primer I, and 400 nucleotides (2300-2700) by
using primer II. By analyzing four recombinant isolates, the sequence between
300 and 2700 can be determined within a few days.

    The sequence of primer I is d(A-C-C-A-T-G-A-T-T-A-C-G-A-A-T-T), and that of
primer II is d(C-A-C-G-A-C-G-T-T-G-T-A-A-A-A-C). Both hexadecanucleotide
primers were synthesized by Fawzy Georges and Saran Narang.

    Abbreviations for restriction enzyme sites are: E for *Eco*EI, H for *Hinc*II,
P for *Pst*I.

179

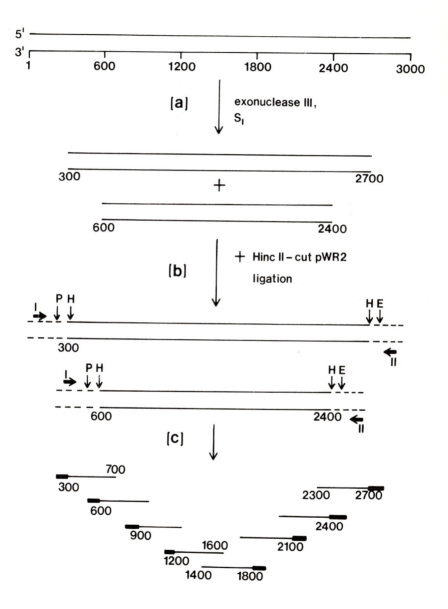

180

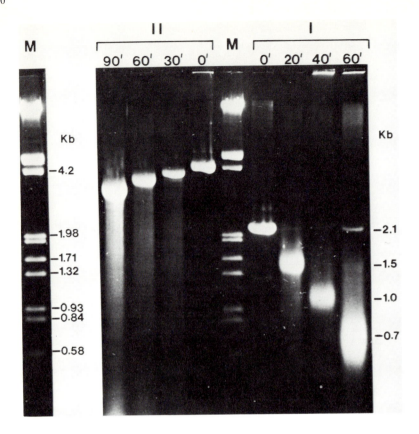

Fig. 6. Digestion of DNA with exonuclease III and nuclease S1 followed by gel electrophoresis.[16,17] pWR2 DNA (36 μg) or pBR322 DNA (36 μg) was linearized with *Hind*III in 120 μl of restriction enzyme buffer.[15] After completion of digestion 48 μl of 10X exonuclease III buffer,[16] 312 μl water and 660 units of exonuclease III (BRL Company) were added, and incubation was carried out at 23°. Aliquots were pipetted out at different times: 40 μl at 20 min. (or 30 min.), 80 μl at 40 min. (or 60 min.), and 160 μl at 60 min. (or 90 min.).

To each aliquot, 10X S1 buffer[16] was added (4.5 μl, 9 μl or 18 μl, respectively) to terminate the reaction. Nuclease S1 (BRL Company) was added (60, 120, or 240 units, respectively) and incubation was carried out at 23° for 15 min. A stop solution of 0.1 M EDTA-1.5 M NaOAc was added to each reaction mixture, and the DNA was ethanol precipitated and electrophoresed on a 1% agarose gel. Size marker (M) was *Hind*III - *Eco*RI digested λ DNA. pWR2 DNA samples (20', 40' and 60') were in the three right-hand lanes, and pBR322 DNA sample (30', 60' and 90') were in the three left-hand lanes.

each end, the sequence from each class overlaps about 100 nucleotides from that of the next shorter class. Using this strategy, the entire sequence can be obtained within a short time.

In step (b) of Fig. 5, the shortened duplex DNA fragment can be ligated to the plasmid cloning vector in either of the two orientations. On the average, 50% of the fragments are joined in each orientation. Because the sequence is analyzed from both termini with different primers, about 400 nucleotides from each terminus can be obtained regardless of the orientation. The end result of sequencing the different classes gives nucleotide sequences from 300 to 1600 of the upper strand, and from 1400 to 2700 of the lower strand. The sequence of 1 to 400 and 2600 to 3000 is determined by cloning the original DNA fragment in pWR2 (or pWR33), followed by sequence analysis using primer I and II, respectively. Since the two primers hybridize to the plasmid vector outside the cloned DNA, the same primers can be used as universal primers to sequence any DNA cloned into pWR2 or its derivatives.

This method is primarily designed for sequencing any double-stranded DNA in a plasmid such as pWR33, pWR2[17] or pUR222[10] which is easy to grow and manipulate. The principle can be applied to cloning of DNA in a M13 phage, and sequencing the single-stranded DNA by the primer method.[14]

The major advantage of this new method is that it requires only a few very simple steps for cloning and sequencing, and it does not require any prior knowledge of the restriction sites within the DNA fragment. In contrast, in the M13 "shotgun" method, a 3000 long DNA needs to be digested with one (or more) restriction enzymes to give about 15 fragments. Each fragment then needs to be cloned into M13 mp8 or M13 mp9 in both orientations, giving a total of 30 clones. After clone selection and sequencing, additional effort is needed to line up the 15 fragments. Furthermore, repeated analysis of largely duplicated regions is often required before the entire sequence can be completed.

In summary, we have constructed plasmids pWR2 and pWR33 for rapid cloning and sequencing of long segments of DNA. The DNA is first digested with exonuclease III followed by nuclease S1 to produce a family of progressively shortened duplex DNA. After agarose gel fractionation, DNA fragments differing in lengths by about 600 base pairs (300 from each end) are joined by blunt-end ligation to the plasmid and cloned. The sequence of about 400 nucleotides can be determined from each terminus of each cloned fragment. A sequence of several thousand nucleotides can be readily obtained by simply overlapping a family of progressively shortened DNA fragments.

ACKNOWLEDGEMENTS

We thank Fawzy Georges and Saran Narang for the synthetic primers I and II. This work was supported by research grant GM 29179 from the National Institutes of Health, U. S. Public Health Service.

REFERENCES

1. Zoon, K. C., et al. (1980) Science, 207, 527.
2. Nagata, S., Mantei, N., and Weissmann, C. (1980) Nature, 287, 401.
3. Methods in Enzymology, (1979) Vol. 68, Wu, R., ed., New York: Academic Press.
4. Genetic Engineering, Principles and Methods, (1979-1981) Vols. 1-3, Setlow, J. K., and Hollaender, A., eds., New York: Plenum Press.
5. Genetic Engineering Techniques (1982) Huang, P. C., Kuo, T. T., and Wu, R., eds., New York: Academic Press.
6. Methods in Enzymology, (1983) Vols. 100, 101, Wu, R., Grossman, L., and Moldave, K., eds., New York: Academic Press.
7. Sanger, F. and Coulson, A. R. (1975) J. Mol. Biol., 94, 441.
8. Sanger, F., Nicklen, S., and Coulson, A. R. (1977) Proc. Natl. Acad. Sci. USA, 74, 5463.
9. Maxam, A. M. and Gilbert, W. (1977) Proc. Natl. Acad. Sci. USA, 74, 560.
10. Rüther, U., et al. (1981) Nucleic Acids Res., 9, 4087.
11. Gronenborn, B. and Messing, J. (1978) Nature (London), 272, 375.
12. Barnes, W. M. (1979) Gene, 5, 127.
13. Sanger, F., et al. (1980) J. Mol. Biol., 143, 161.
14. Messing, J., Crea, R., and Seeburg, P. H. (1981) Nucleic Acids, 9, 309.
15. Guo, L. H. and Wu, R. (1982) Nucleic Acids Res., 10, 2065.
16. Wu, R., et al. (1976) Biochem., 15, 734.
17. Guo, L. H. and Wu, R. (1983) in Methods in Enzymology, Vol. 100, Wu, R., Grossman, L., and Moldave, K., eds., New York: Academic Press, p. 60.
18. Wallace, R. B., et al. (1981) Gene, 16, 21.
19. Bahl, C. P., Narang, S. A., and Wu, R. (1983) this volume.

Biochemical and Biophysical Studies of Proteins and Nucleic Acids,
Lo, Liu, and Li, eds.

OLIGONUCLEOTIDE SYNTHESIS AND ITS APPLICATION TO RECOMBINANT DNA TECHNOLOGY

CHANDER P. BAHL,[+] SARAN A. NARANG,[++] AND RAY WU[*]
[+]Cetus Corporation, 600 Bancroft Way, Berkeley, California 94710; [++]Division
of Biological Sciences, National Research Council of Canada, Ottawa, Ontario,
Canada; [*]Section of Biochemistry, Molecular and Cell Biology, Cornell
University, Ithaca, New York 14850

INTRODUCTION

The isolation, structure elucidation and synthesis of complex natural
products has been a challenge undertaken by organic chemists for a long time.
DNA molecule, a component of all life forms, has unique structural and biological properties. Ever since Watson and Crick[1] reported the structure of this
interesting molecule, it became a challenge for organic chemists to synthesize
this unique natural product. Chemical synthesis of DNA, which started as an
intellectual exercise, took applied dimensions with the advent of recombinant
DNA technology. The ability of chemists to assemble well-defined sequences of
oligonucleotide have provided important tools to molecular biologists for
studying function and structure of genetic elements. Synthetic DNA fragments
have been instrumental in elucidating the genetic code,[2] investigating regulatory signals,[3-5] studying gene structure,[6] and in synthesizing genes for producing important peptides.[7] Synthetic oligonucleotides have been used as
linkers and adaptors in molecular cloning,[8-10] as hybridization probes,[11] as
primers for sequencing DNA,[12-15] and as site-specific mutagens.[16]

In this article we will describe some of the current methods of DNA synthesis
and also some examples of the use of these synthetic molecules in recombinant
DNA technology.

METHODS FOR DNA SYNTHESIS

Polynucleotides of about 40 nucleotides in length have been synthesized
chemically, yet it is more practical to synthesize oligonucleotides in the size
range of 10-20. Oligonucleotides of this size range can be used for various
applications in recombinant DNA research. Larger DNA molecules can be assembled
from these short ones by ligating them using T4 polynucleotide ligase in the
presence of staggered complementary oligonucleotide fragments.[17,6,7]

Four different chemical methods have been utilized for the synthesis of
oligonucleotides. In the following section, we will discuss briefly these four
methods for building oligonucleotide chains. For a detailed review of the
chemistry of various steps involved, refer to a recent review.[18]

Phosphodiester method. This method of oligonucleotide synthesis, as outlined in Fig. 1, involves the coupling of a suitably protected deoxynucleotide
by a coupling reagent like the dicyclohexylcarbodiimide or sulfonylchlorides.
The product of coupling is a phosphodiester, that is why the method is commonly
known as the phosphodiester method. This method was developed by Khorana and
his coworkers and was the most commonly used method for the synthesis of oligonucleotides in the sixties and early seventies. Using this method Khorana et
al. were able to synthesize two transfer RNA genes.[17]

Phosphotriester method. This method of oligonucleotide synthesis is the
most commonly used method today. The basic coupling reaction is outlined in
Fig. 2. In this method the internucleotide phosphate esters are kept esterified as phenyl esters during the course of the synthesis, thereby making the
products less polar and rendering them soluble in organic solvents like
chloroform. The solubility of these compounds in organic solvents make it
possible to use standard organic chemistry purification procedures like the
silica-gel chromatography. The method is much faster than the diester method
and gives better yields. The phosphotriesters can be converted into desired
phosphodiesters, using mild deblocking conditions.[19] Figure 3 illustrates
the synthesis of a pentadecamer using this method.

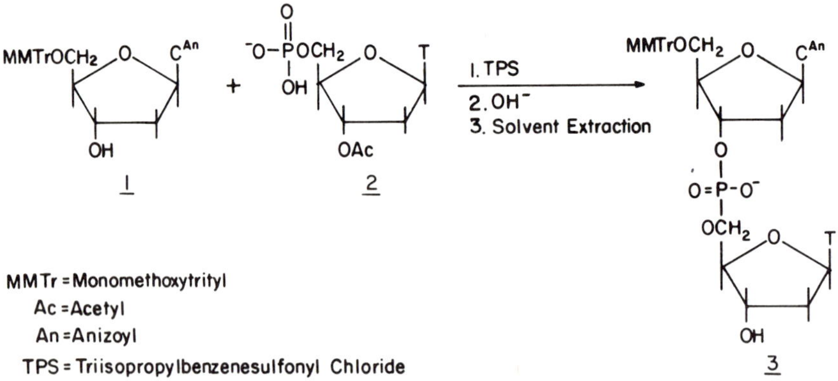

MMTr = Monomethoxytrityl
Ac = Acetyl
An = Anizoyl
TPS = Triisopropylbenzenesulfonyl Chloride

Fig. 1. Phosphodiester approach to oligonucleotide synthesis.

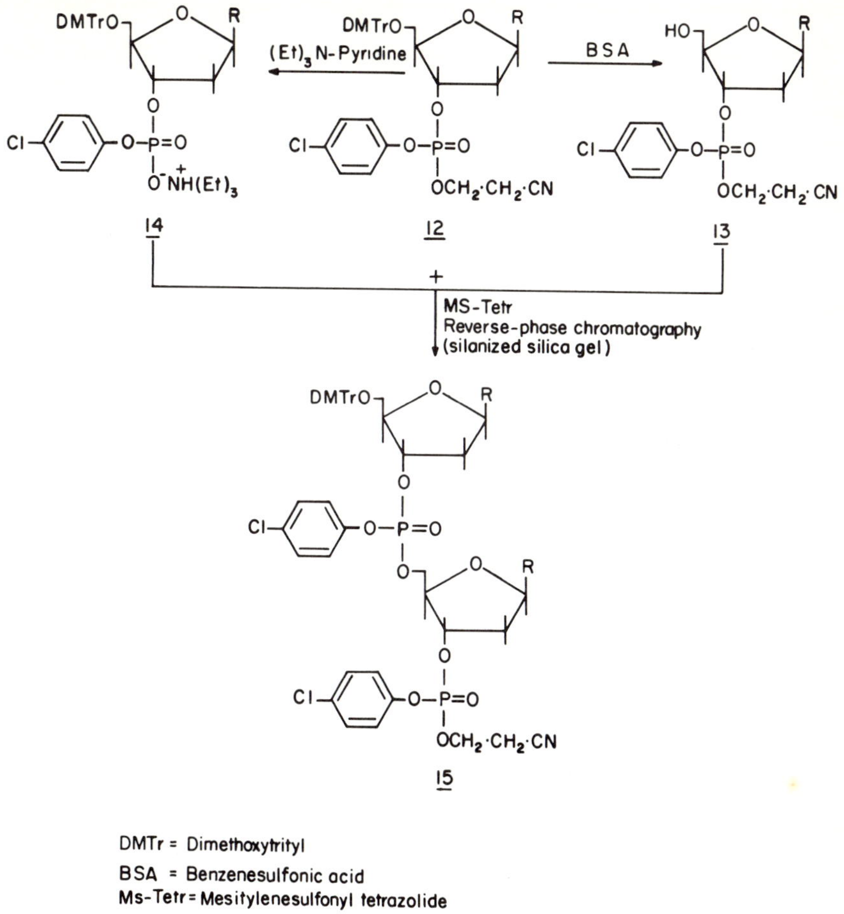
DMTrO
R
(Et)3 N-Pyridine
BSA
HO
Cl
O-P=O
ONH(Et)3
OCH2·CH2·CN
14
12
13
+
MS-Tetr
Reverse-phase chromatography
(silanized silica gel)
15
DMTr = Dimethoxytrityl
BSA = Benzenesulfonic acid
Ms-Tetr = Mesitylenesulfonyl tetrazolide

Fig. 2. Phophotriester approach to oligonucleotide synthesis.

Phosphite triester method. This method of oligonucleotide synthesis, outlined in Fig. 4, was first described by Letsinger and Lunsford.[20] In this method the nucleotides are coupled together by reaction with phosphodichloridite. The resulting phosphite esters are then oxidized to phosphotriester by treatment with aqueous iodine. The reactivity of the phosphite bond formation has been exploited to adapt this method for automated DNA synthesizers (see solid phase methods).

186

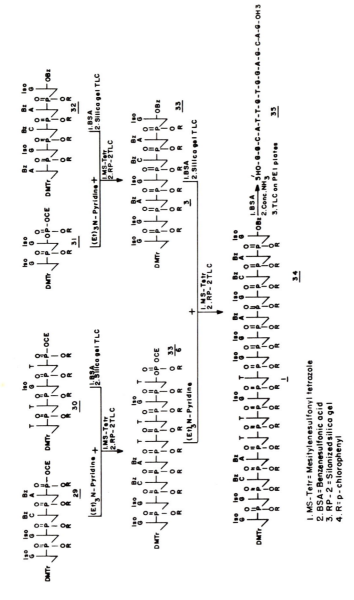

Fig. 3. Synthesis of a pentadecamer by the modified phosphotriester method.

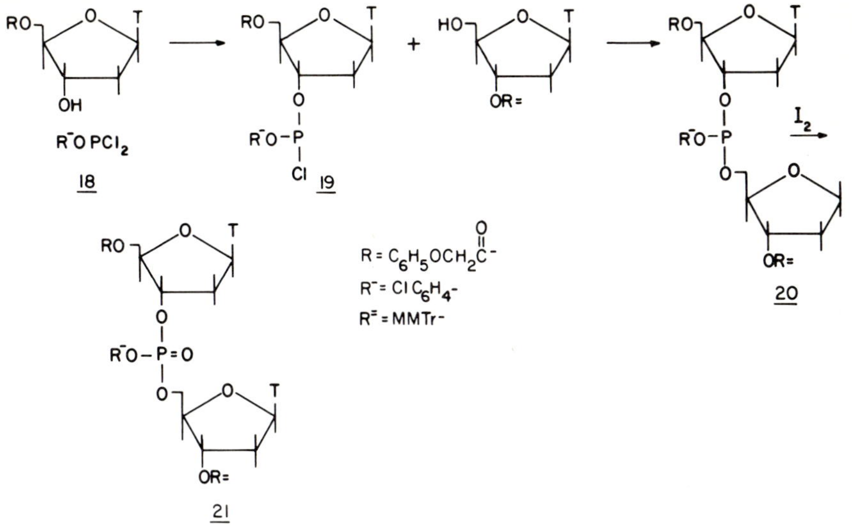

Fig. 4. Phosphite triester approach to oligonucleotide synthesis.

Solid phase methods. Solid phase synthetic methods have been used extensively in polypeptide synthesis. Gait and Sheppard reported the synthesis of an oligonucleotide in 1976 using the phosphodiester chemistry, but the yields were not very good.[21] Recent improvements in the solution methodology of oligonucleotide synthesis and an enormous need for these oligonucleotides in biological work has aroused a great interest in developing the solid phase methodology for the rapid synthesis of oligonucleotides with an ultimate aim of automation. All the solid phase methods being developed recently use either the phosphotriester method or the phosphite triester method for forming the internucleotide bond. The oligonucleotide chain is synthesized from 3' → 5' direction. The 3'-end nucleoside, suitably protected, is linked to a solid support-like silica gel, polystyrene or polydimethylacrylamide via a succinyl ester spacer. The support containing the 3'-ester of nucleoside is treated with an appropriate deblocking group to selectively deblock the 5'-OH group; the solid support is now ready for synthesis by either the phosphotriester method or the phosphite triester method.

In the phosphotriester solid support synthesis, the detritylated support in a pyridine solution is incubated with suitably protected nucleoside 3'-phosphate diester and a condensing reagent like the isopropyl benzene sulfonyl tetrazole. After 2-3 hours of reaction time, the reagents are removed from

the resin and the resin rinsed with pyridine followed by washing with chloroform methanol. The resin is then subjected to detritylation reaction and the
whole cycle is repeated until the desired chain length is achieved. In this
method one can also use nucleotide blocks synthesized by solution phosphotriester method.[22-24]

In the phosphite-based solid phase synthesis method, the resin containing
the mononucleoside ester is treated with suitably protected nucleoside, 3'-
phosphochlorindites[25] or nucleoside 3'-phosphoamidites.[26] The resulting phosphite bond is oxidized by aqueous iodine. The support is then subjected to
detritylation step followed by a drying step; the whole cycle is repeated until
the desired sequence is synthesized.

Once a desired chain of nucleotides has been synthesized it can be removed
from the support by hydrolysis with ammonium hydroxide. Of the two methods
being used, the phosphite method is faster than the triester method; each cycle
of nucleoside addition takes only 30 minutes in the phosphite method, as compared to 2-3 hours in the triester method. But in using the triester method,
one can add either mononucleotides or block of oligonucleotides, thereby one
can make longer DNA fragments with fewer cycles. By using block condensation
on solid support, overall synthesis on the solid support can be faster and the
products easier to isolate in a pure form.

APPLICATIONS OF SYNTHETIC OLIGONUCLEOTIDES

As mentioned in the introduction to this article, synthetic oligonucleotides have become very powerful tools in recombinant DNA research. Oligonucleotide synthesis has become an integral part of major molecular biology and
molecular genetics groups. In the following section we will discuss some
examples of the application of these compounds in recombinant DNA research.

SYNTHESIS OF REGULATORY SIGNALS

Regulation of gene expression is controlled by information residing in
segments of DNA. The first synthetic regulatory element was a 21 nucleotide-
long duplex DNA corresponding to the sequence of lactose operator of E. coli.[3]
This duplex DNA as well as a synthetic 17 nucleotide-long sequence were shown to
have properties similar to the natural lactose operator in vitro and in vivo.[27]
Synthesis of lactose operator sequences with altered bases at various locations
has been used for identifying the various contact points between the lac
operator and the lac repressor.[28] Similar studies are being carried out in
various laboratories to synthesize promoter and ribosome binding site sequences.

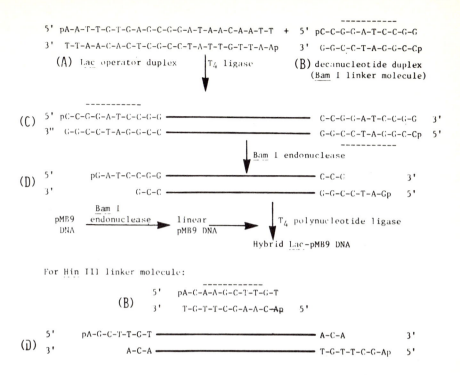

5' pA-A-T-T-G-T-G-A-G-C-G-G-A-T-A-A-C-A-A-T-T   +   5' pC-C-G-G-A-T-C-C-G-G

3' T-T-A-A-C-A-C-T-C-G-C-C-T-A-T-T-G-T-T-A-Ap   3' G-G-C-C-T-A-G-G-C-Cp

    (A) Lac operator duplex | T₄ ligase   (B) decanucleotide duplex
    (Bam I linker molecule)

(C)
5' pC-C-G-G-A-T-C-C-G-G ————————————— C-C-G-G-A-T-C-C-G-G   3'
3" G-G-C-C-T-A-G-G-C-C ————————————— G-G-C-C-T-A-G-G-C-Cp   5'

Bam I endonuclease

(D)
5'   pG-A-T-C-C-G-G ————————————— C-C-G   3'
3'   G-G-C ————————————— G-G-C-C-T-A-Gp   5'

pMB9 DNA — Bam I endonuclease → linear pMB9 DNA → T₄ polynucleotide ligase

Hybrid Lac-pMB9 DNA

For Hin III linker molecule:

(B)
5' pA-C-A-A-G-C-T-T-G-T
3' T-G-T-T-C-G-A-A-C-Ap   5'

(D)
5'   pA-G-C-T-T-G-T ————————————— A-C-A   3'
3'   A-C-A ————————————— T-G-T-T-C-G-Ap   5'

Fig. 5. A method for introducing restriction endonuclease site at the termini of a double-stranded DNA.

OLIGONUCLEOTIDES AS LINKERS AND ADAPTORS

The most common way of manipulating DNA in recombinant DNA research is through the restriction enzyme recognition sequences. Synthetic linkers and adaptors have made it possible to introduce a variety of restriction sites at various locations in the cloning vehicles or at the termini of the passenger DNA.[8,9] An example of introducing restriction sites at the termini of a passenger DNA for cloning is shown in Fig. 5. In addition to introducing restriction sites, one can also convert one site into another with the help of synthetic oligonucleotides.[10]

SYNTHESIS OF STRUCTURAL GENES

Khorana et al. first used chemical synthesis of oligonucleotides followed by enzymatic ligation to form two transfer RNA genes, one for yeast alanine tRNA and the other for *E. coli* tyrosine suppressor tRNA.[17] Recent advances in

190

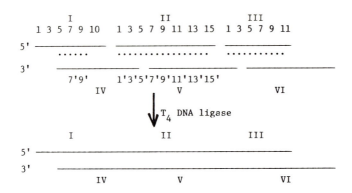

Fig. 6. Strategy for chemical synthesis of duplex DNA. Each dot represents a hydrogen bond between complementary nucleotides (such as 7 and 7').

recombinant DNA and the desire to manufacture useful mammalian proteins in large amount has put a great demand on the chemistry of oligonucleotides. In principle, for any protein of known amino acid sequence, one can design a double-stranded DNA which has a potential for coding that particular protein based on the genetic code. Oligonucleotides of chain length 10-15 are synthesized to cover the entire length of the double-stranded DNA. The oligo-nucleotides are designed in such a way that the break points in the two comple-mentary strands of DNA are staggered (Fig. 6). DNA ligase is used to seal these nicks in DNA, and then the DNA can be ligated to a suitable vector. After transformation of competent cells, the protein coded by this cloned DNA can be made (expressed) in the cells. Using this strategy, a number of "genes" for various mammalian peptides have been chemically synthesized, cloned and expressed in E. coli cells, e.g., somatostatin,[29] human insulin,[30] human proinsulin,[31] leucocyte interferon,[32] and secretin.[33]

ADAPTING cDNA SEQUENCES FOR EXPRESSION

The most common way of obtaining coding sequences for various mammalian structural genes is by cloning the double-stranded cDNA synthesized from enriched mRNA. Synthetic oligonucleotides have been used extensively for adapt-ing these cDNA for expression in bacterial hosts. Mammalian sequences obtained by cDNA synthesis contain the entire primary coding sequence including the coding sequences for the leader peptide. Unlike the mammalian systems which process away the leader peptide, the bacteria may not be able to process the

(A)

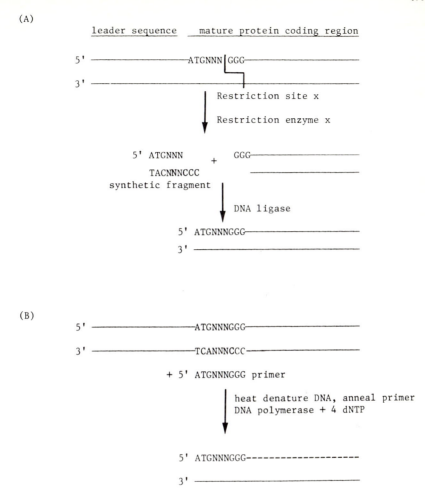

Fig. 7. Scheme for removing leader peptide. (A) A gene that includes the coding sequence for both the leader peptide and the mature protein coding region is digested by a restriction enzyme x which removes the leader sequence plus a short segment of the mature protein coding sequence. A synthetic fragment, which includes the ATG initiation codon and the deleted portion of the mature protein coding sequence, is ligated to the major portion of the mature protein coding sequence to complete the gene without extraneous sequence at the left-hand end. (B) A synthetic primer starting with a sequence 5' ATGNNNGGG is annealed to the denatured gene. *E. coli* DNA polymerase is used to extend the primer to copy the gene. At the same time, the 3'-5' exonuclease activity of the DNA polymerase degrades the 3' single-stranded region of the lower strand to give rise to a double-stranded DNA starting with ATG.

mammalian leader sequences. Thus, the protein synthesized by the bacteria from a cDNA will be different from the mature proteins synthesized by the mammalian cell. Synthetic oligonucleotides have been used in one of two possible ways to remove the sequences for these leader peptides. In one scheme, the cDNA is treated with a restriction enzyme so as to remove all of the coding sequence for the leader peptide but preserve as much of the mature protein coding sequence as possible. The portion of the mature protein coding sequence lost in this manipulation is replaced by synthetic oligonucleotide fragments. This scheme (Fig. 7A) was first used for the expression of human growth hormone[34] and recen recently for γ interferon expression.[35] In the other scheme Fig. 7B), an oligonucleotide having an ATG at the 5'-end followed by the sequence coding for the first few amino acids of the N-terminal of the mature protein, is used as a primer in the presence of DNA polymerase I. DNA polymerase catalyzed reaction in the presence of deoxynucleoside triphosphates extends the primer and simultaneously the 3' → 5' exonuclease activity of the polymerase degrades the 3'-end of the other strand giving rise to a double-stranded DNA starting with ATG followed by the coding sequence of the mature protein. This scheme has been successfully utilized for expression of a number of genes including β interferon.[36]

SYNTHETIC OLIGONUCLEOTIDES AS HYBRIDIZATION PROBES

Synthetic oligonucleotides have been used as powerful tools for screening the desired sequences in a cloning experiment. By using radioisotopically labeled probes, one can screen bacterial colonies or phage plaques for the presence of DNA complementary to the probe. The probes can also be used to detect the complementary RNA or DNA fractionated on gels.[11] The oligonucleotide can be synthesized based on the amino acid sequence of the protein being studied. Because of the degeneracy of the genetic code, one often selects a stretch of 4-5 amino acids in the protein such that the corresponding oligonucleotide sequence is the least ambiguous.[37] One may have to synthesize several oligonucleotide probes to cover the different possible sequences. A probe of about 12-16 in length can give specificity desired in the screening experiments. A number of genes have been identified using such probes, e.g., yeast iso-1 cytochrome c,[38,11] gastrin,[39] insulin,[40] interferons,[41,42] HLA,[43,44] and β-2 microglobulin.[45] Recently Wallace et al. have established hybridization conditions to use synthetic oligonucleotides to screen for genes with perfect match or one-nucleotide mismatches.[46]

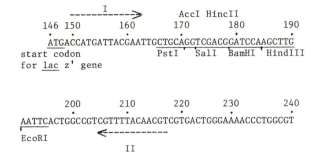

Fig. 8. Use of two synthetic primers for sequencing DNA cloned into pWR2 plasmid. A portion of the sequence near the polylinker region[53] of pWR2 plasmid is shown. Nucleotide 146 is the start codon for lac z' gene. The polylinker region (nucleotides 166-195) interrupted codon number 4 (thr) and number 6 (ser) of the lac z' gene[53]. DNA cloned into restriction site within this region usually give rise to white colonies whereas those containing only pWR2 DNA form blue colonies on indicator plates. The DNA cloned into any of these sites can be easily sequenced by using the synthetic primer I, to determine the sequence of one DNA strand, and the synthetic primer II to determine the sequence of the complementary strand.[52] Primer I has the sequence of 5' d(A-C-C-A-T-G-A-T-T-A-C-G-A-A-T-T), which hybridizes to the sequence complementary to that shown in this figure. Primer II has a sequence of 5' d(C-A-C-G-A-C-G-T-T-G-T-A-A-A-A-C), which is complementary to nucleotides 221-206 and can bind to pWR2 plasmid and be extended leftward.

SYNTHETIC OLIGONUCLEOTIDES FOR DNA SEQUENCING

Synthetic oligonucleotides are being used extensively in the sequencing of DNA by Sanger's dideoxynucleotide sequencing method.[47] The availability of single stranded M13 cloning vectors and the methodology specifically designed for sequencing[48,49] has made the primer-extension sequencing method attractive. Several synthetic primers suitable for the M13 dideoxynucleotide system have been reported.[49,50] Other primers for sequencing double-stranded DNA have been synthesized.[51,52] Two universal primers (16-mers) have been designed for sequencing any DNA fragments cloned into the polylinker region of pWR2 (Fig. 8), a double-stranded plasmid.[52]

It is essential that the sequence of the synthetic oligonucleotide be verified by sequence analysis. Either the mobility-shift method[54] or the chemical method can be used.[55]

194

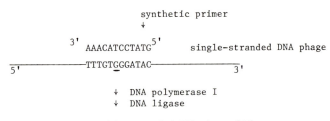

synthetic primer

↓

3' AAACATCCTATG 5'   single-stranded DNA phage

5' ————————————TTTGTGGGATAC———————————— 3'

↓ DNA polymerase I
↓ DNA ligase

double-stranded DNA phage (RF)

Fig. 9. A scheme for site specific mutagenesis. A synthetic oligodeoxynucleo-
tide (12-mer) with a single mismatched nucleotide (underlined) is annealed to
a circular single-stranded DNA phage. The primer is extended by repair
synthesis, catalyzed by *E. coli* DNA polymerase (large fragment), to complete
the second strand of DNA. The nick is joined by DNA ligase to give a closed
circular duplex DNA. Transfection of *E. coli* spheroplasts yields a phage
population with approximately 50% being the mutant phage, which contains a
single mismatched nucleotide at the location of the synthetic primer.

SYNTHETIC NUCLEOTIDES FOR SITE-SPECIFIC MUTAGENESIS

Synthetic oligonucleotides provide a very direct method for site-
specific mutagenesis. The mutation to be introduced can be base substitution,
insertion or deletions, and can be introduced at any site. The method involves
using a synthetic oligonucleotide containing the desired change of nucleotides
as primer and the DNA to be mutated as template (Fig. 9). The extended primer
strand then corresponds to the desired mutant sequence. The double-stranded
DNA contains a region of mismatched nucleotide and is used for transformation
or transfection.[16] The desired mutant can be screened for by phenotypic
selection or applying the same oligonucleotide designed for mutagenesis as a
hybridization probe. The site specific mutagenesis method is being used to
answer important biological questions. For example, Inouye et al. studied
the role of positive charge on the amino terminal of the signal peptide of *E.
coli* lipoprotein for secretion across the membrane.[56] Chang and Chang are
exploring the mechanism of anchoring of β-lactamase in *Bacillus subtilis*
membrane.[57] Kan et al. have constructed a functional human suppressor tRNA
gene for possible gene therapy for β-thalassaemia.[58]

In addition to the examples cited above, synthetic oligonucleotides have
been used for a number of other studies such as the inhibition of RSV replication,[59] the uptake of DNA by competent *Haemophilus* cells,[60] and the conformational analysis of DNA.[61,62]

ACKNOWLEDGEMENTS

This work was supported by research grant GM 27365 from the National Institutes of Health, USPHS. This paper is NRCC No. 20522.

REFERENCES

1. Watson, J. D. and Crick, F. H. C. (1953) Nature, 171, 737.
2. Khorana, H. G., et al. (1966) Cold Spring Harbor Symp. Quant. Biol., 31, 39.
3. Bahl, C. P., et al. (1976) Proc. Natl. Acad. Sci. USA, 73, 91.
4. Bahl, C. P., et al. (1977) Proc. Natl. Acad. Sci. USA, 74, 966.
5. Goeddel, D. V., Yansura, D. G., and Caruthers, M. H. (1977) Nuc. Acids. Res., 4, 3039.
6. Wu, R., Bahl, C. P., and Narang, S. A. (1978) Proc. Nucl. Acid Res. and Mol. Biol., 21, 101.
7. Itakura, K. and Riggs, A. D. (1980) Science, 209, 1401.
8. Bahl, C. P., et al. (1976) Gene, 1, 81.
9. Scheller, R. H., et al. (1977) Science, 196, 177.
10. Bahl, C. P., et al. (1978) Biochem. Biophys. Res. Comm., 81, 695.
11. Szostak, J. W., et al. (1979) Methods in Enzymology, 68, 419.
12. Wu, R., et al. (1972) Bull. Inst. Pasteur, 70, 203.
13. Padmanabhan, R., Padmanabhan, R., and Wu, R. (1972) Biochem. Biophys. Res. Commun., 48, 1295.
14. Sanger, F., et al. (1973) Proc. Natl. Acad. Sci. USA, 70, 1209
15. Sanger, F. and Coulson, A. R. (1975) J. Mol. Biol., 94, 441.
16. Smith, M. and Gillam, S. (1981) Genetic Engineering, 3, 1.
17. Khorana, H. G. (1979) Science, 203, 614.
18. Narang, S. A. (1983) Tetrahedron, 39, 3.
19. Narang, S. A., Hsiung, H. M., and Brousseau, R. (1979) Methods in Enzymology, 68, 90.
20. Letsinger, R. L. and Lunsford, W. B. (1976) J. Am. Chem. Soc., 98, 3655.
21. Gait, M. J. and Sheppard, R. C. (1976) J. Am. Chem. Soc., 98, 8514.
22. Miyoshi, K. and Itakura, K. (1980) Tetrahedron Lett., 29, 4159.
23. Gait, M. J., et al. (1980) Nucl. Acids Res., 8, 1081.
24. Dembek, P., Miyoshi, K., and Itakura, K. (1981) J. Am. Chem. Soc., 103, 706.
25. Alvardo-Urbina, G., et al. (1981) Science, 214, 270.
26. Mateucci, M. D. and Caruthers, H. M. (1981) J. Am. Chem. Soc., 103, 3186.
27. Wu, R., Bahl, C. P., and Narang, S. A. (1978) in Current Topics in Cellular Regulation, 13, 137.
28. Fisher, E. F. and Caruthers, M. H. (1979) Nucl. Acids Res., 7, 401.
29. Itakura, K., et al. (1977) Science, 198, 1056.
30. Crea, R., et al. (1978) Proc. Natl. Acad. Sci. USA, 75, 5765.
31. Brousseau, R., et al. (1982) Gene, 17, 279.
32. Edge, M. D., et al. (1981) Nature, 292, 756.
33. Suzuki, et al. (1982) Proc. Natl. Acad. Sci. USA, 79, 2475.
34. Goeddel, D. V., et al. (1979) Nature, 281, 544.
35. Gray, P. W., et al. (1982) Nature, 295, 503.
36. Goeddel, D. V., et al. (1980) Nucl. Acids Res., 8, 4057.
37. Wu, R. (1972) Nature, 236, 198.
38. Montgomery, D. L., et al. (1978) Cell, 14, 673.
39. Noyes, B. A., et al. (1979) Proc. Natl. Acad. Sci. USA, 76, 1770.
40. Chan, S. J., et al. (1979) Proc. Natl. Acad. Sci. USA, 76, 5036.
41. Houghton, M., et al. (1980) Nucl. Acids Res., 8, 1813.
42. Goeddel, D., et al. (1980) Nature, 287, 411.
43. Sood, A. K., Perira, D., and Weissman, S. M. (1980) Proc. Natl. Acad. Sci. USA, 78, 616.

44. Stetler, D., et al. (1983) Proc. Natl. Acad. Sci. USA, in press.
45. Suggs, S., et al. (1981) Proc. Natl. Acad. Sci. USA, 78, 6613.
46. Wallace, R. B., et al. (1981) Nucleic Acids Res., 9, 879.
47. Sanger, F., Nicklen, S., and Coulson, A. R. (1977) Proc. Natl. Acad. Sci. USA, 74, 5463.
48. Heidecker, G., Messing, J., and Gronenborn, B. (1980) Gene, 10, 69.
49. Messing, J., Crea, R., and Seeburg, P. H. (1981) Nucleic Acids, 9, 309.
50. Wu, R., et al. (1980) Miami Winter Symposia, 17, 419.
51. Wallace, R. B., Johnson, M. J., and Itakura, K. (1981) Gene, 16, 21.
52. Guo, L. H. and Wu, R. (1983) Methods in Enzymology, 100, 60.
53. Rüther, U., et al. (1981) Nucleic Acids Res., 9, 4089.
54. Tu, C. D., et al. (1976) Anal. Biochem., 74, 73.
55. Maxam, A. M. and Gilbert, W. (1977) Proc. Natl. Acad. Sci. USA, 74, 560.
56. Inouye, S., et al. (1982) Proc. Natl. Acad. Sci. USA, 79, 3438.
57. Chang, S.-Y. and Chang, S. (1983) in preparation.
58. Temple, G. F., et al. (1982) Nature, 296, 537.
59. Zemenick, P. C. and Stephenson, M. L. (1978) Proc. Natl. Acad. Sci. USA, 75, 280.
60. Danner, D. B., Smith, H. O., and Narang, S. A. (1982) Proc. Natl. Acad. Sci. USA, 79, 2393.
61. Wang, A. H. J., et al. (1982) Proc. Natl. Acad. Sci. USA, 79, 3968.
62. Drew, H. R., Samson, S., and Dickerson, R. E. (1982) Proc. Natl. Acad. Sci. USA, 79, 4040.

Biochemical and Biophysical Studies of Proteins and Nucleic Acids,
Lo, Liu, and Li, eds.

CURRENT PROGRESS IN NMR STUDIES OF NUCLEIC ACIDS:

CONFORMATIONS OF OLIGONUCLEOTIDES

LOU-SING KAN, DORIS M. CHENG, AND PAUL O. P. TS'O
Division of Biophysics, The School of Hygiene and Public Health,
The Johns Hopkins University, Baltimore, Maryland 21205.

ABSTRACT

The proton nuclear magnetic resonance (NMR) spectroscopy has been used to
study the conformations of nucleic acids. The target nucleic acids are synthetic oligonucleotides, one RNA- and three DNA-types. The NMR data (chemical
shifts and coupling constants) of these oligomers have been extensively
analyzed based on the chemical shift theory (ring-current anisotropic effect,
atomic anisotropic effect and polarized effect) and the Karplus theory, respectively. The results indicate that the r-A-A-G-C-U-U helix resembles the A'-
and d-C-C-A-A-G-C-T-T-G-G helix and d-C-G-C-G and d-C-G-C-G-C-G helices resemble
B-type DNA.

INTRODUCTION

Both ribonucleic acids (RNA) and deoxyribonucleic acids (DNA) play important
roles in genetic functions. Thus, the aspects such as structure, conformation,
and dynamics which lead us to understand how RNA and DNA function normally
and/or abnormally, are of prime importance as well as interesting to chemists,
biochemists, molecular biologists, and others. Since Watson and Crick made
the first proposal of a double helical structure of DNA in 1953,[1] a great
advance has been made in understanding of nucleic acids by fiber methods.[2]
However, it is impossible to solve the three-dimensional structure of a macro-
molecule from a fiber diffraction pattern without making a large number of
simplifying assumptions. For this reason, short oligoribo- and oligodeoxyribo-
nucleotide helices with defined sequences are excellent models for studies of
RNA and DNA conformations. Three stable helices have been crystallized and
studied by X-ray diffraction techniques.[3-6] These studies provide information
on conformation of these helices at the atomic level of resolution. In order
to understand the properties of these short helices in aqueous solution, the
high resolution nuclear magnetic resonance (NMR) spectroscopy is utilized.
Correlation of nucleic acid conformation in the crystalline and solution states
can be made by comparing the X-ray and NMR results. While these helices are
sufficiently short to yield X-ray diffraction data and NMR signals for high

SYNTHESIS OF ApApGpCpU*pU*

$$ppA + ppG \xrightarrow{\text{PNPase}} poly(A,G) \quad (\text{input ratio 3A:1G})$$

$$poly(A,G) \xrightarrow{\text{RNase } T_1} A_nG_p \xrightarrow{\text{DEAE-seph}} A_2Gp \xrightarrow{\text{BAPase}} A_2G$$

$$A_2G + ppC \xrightarrow[\text{RNase A}]{\text{PD PNPase}} A_2GCp \xrightarrow{\text{BAPase}} A_2GC$$

$$A_2GC + ppU* \xrightarrow{\text{PD PNPase}} A_2GCU*_n \xrightarrow{\text{DEAE-seph}} A_2GCU*_2$$

(U* = ^{3}H-labelled uridine at 25 µc/mmole)

Chart 1. Where (PD) PNPase: (primer dependent) polynucleotide phosphorylase; RNase T_1 (and A): ribonuclease T_1 (and A); BAPase: bacteria alkaline phosphatase; seph: sephadex column.

resolution investigation, they are long enough for studies on drug-nucleic acid helix interaction (for RNA helices: 7-14; for DNA: 15-21; for DNA-RNA hybrid: 22; enzyme-nucleic acid interactions: 23).

In this communication, the NMR results of four oligonucleotides which all have self-complementary sequences: r-A-A-G-C-U-U (a), d-C-C-A-A-G-C-T-T-G-G (b), d-C-G-C-G (c), and d-C-G-C-G-C-G (d) have been reported, where r represents RNA and d is for DNA. This is why uridine (U) appears in compound a and thymidine (T) appears in others.

METHODS

Syntheses

The detailed synthetic procedures of these oligomers have been published elsewhere. The following are brief outlines.

i. r-A-A-G-C-U-U (a)

The synthesis of a was accomplished by enzymatic method,[9] and is briefly outlined in Chart 1.

ppA and ppG were copolymerized with PNPase to produce poly (A,G) (step 1). Then the polymer was hydrolyzed with RNase T_1. The resulting A_nGp oligomer was then separated on a DEAE-Sephadex-Cl⁻ anion exchange column with a linear NaCl gradient in 7 M urea. The A_2Gp peak was selected and purified. The terminal phosphate of A_2Gp was removed with BAPase and A_2G was separated on a Sephadex G-15 column (step 2). A single cytidylate was esterified to the 3'-OH of A_2G in the presence of primer-dependent PNPase and RNase A. The

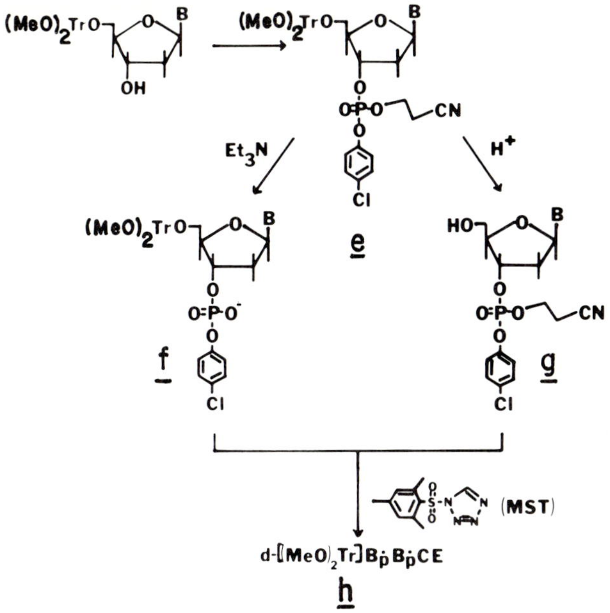

Chart 2. Where (MeO)$_2$Tr: dimethoxytrityl; CN: -OCH$_2$CH$_2$CN; B: protected base; p̣: p-chlorophenylphosphate; and MST: mesitylenesulfonyl tetrazole.

produced terminal phosphate of A$_2$GCp was again removed by BAPase (step 3). In the last step, A$_2$GC and ppU were copolymerized in the presence of PD PNPase to form a series of A$_2$GCU$_n$ oligomers. The mixture was separated on DEAE-Sephadex-Cl$^-$ column again with linear NaCl gradient in 7 M urea. Desired oligomer a was purified from urea and then desalted on Bio-Gel P-2 column until salt free.

The sequence integrity of a was carefully checked by enzymatic digestion as well as acidic hydrolysis. These results completely and redundantly support the integrity of the sequence r-A-A-G-C-U-U.

ii. DNA oligomers (b, c, and d)

On the other hand, all three DNA oligomers were synthesized chemically. The strategy of the synthesis is illustrated in Chart 2.

5'-OH protected nucleoside was first phosphorylated with phosphodiesters to become product e. All OH groups in phosphate have been esterified. Thus, this procedure is called the phosphotriester method. The protected groups of (MeO)$_2$Tr and CE can be selectively removed by treatment with acid and base, respectively. Therefore, compounds f and g are ready to be joined through 3'-5' linkage by a condensation reagent, MST. The product h is a fully protected dimer. So the CE or $(MeO)_2Tr$ of h can again be removed by base or acid. These dimers are ready to be joined to a fully protected tetramer.

The base components of f and g are not necessarily identical. Thus, hetero-base dimers can be synthesized in the same manner.

In summary, the above-mentioned technique was applied to make the DNA short helix. Therefore, the d-C-G-C-G was synthesized by linking two dimers, d-C-G-C-G-C-G was synthesized by dimer plus tetramer, and d-C-C-A-A-G-C-T-T-G-G was condensated by d-C-C-A-A-G-C and d-T-T-G-G. The strategy is to avoid . difficulties of separation between reactants and product. The synthetic procedures of these three oligomers were published elsewhere.[24,25]

NMR spectroscopy

Several NMR spectrometers with magnetic fields ranging from 23 to 140 kilo-gauss (namely, from 100 MHz to 600 MHz in the frequency of proton) were used. All these spectrometers are equipped with pulse units and minicomputers to perform the Fourier Transform. The digital resolution of each spectrum is 0.4 Hz/pt or better. All reported chemical shifts are in respect to the methyl resonances of sodium 2,2-dimethyl-2-silapentane-5-sulphonate.

GENERAL FEATURES OF THE ^{1}H NMR SPECTROSCOPY ON OLIGONUCLEOTIDE

All these short oligonucleotides are capable of forming double strands in aqueous solution at low salt condition.[24-26] Thus, they can be examined in both single-stranded and helical states by NMR method. The resonance lines are narrow at high temperature and broad at low temperature. There are dramatic, predominantly upfield changes in the chemical shifts of these resonances upon lowering the temperature. This may reflect a large increase in the shielding effects due to the stacking of the bases. Thus, these phe-nomena can be utilized for assignment, especially the sequential incremental assignment techniques.[9]

The unambiguous assignment of the resonance lines is the most important criteria of the conformational analyses of these or any oligomers。 In general, the assignment can be achieved by the following NMR features or techniques:

1. spectral pattern
2. chemical shift values
3. deuterium exchange
4. paramagnetic ion doping

5. relaxation times

6. thermal perturbation

7. double resonances--tickling, decoupling and
   nuclear Overhauser effect (NOE)

8. sequence comparison

The first seven techniques are directly related to NMR itself and have been used (except NOE) for a long time by many observers.[28,29] The NOE is an effective assignment tool which has been widely used recently for oligonucleotides (unpublished data).[29] The eighth method compares the NMR spectra between two oligonucleotides with very similar sequences. This method was the first used successfully to assign the nonexchangeable base proton resonances of a, and recently, all the nonexchangeable proton resonances (base and sugar) of two deoxypentanucleotides, d-C-C-A-A-G and d-C-T-T-G-G[30] and c, and d. Thus, all the nonexchangeable base proton resonances of these four oligonucleotides can be unambiguously assigned.

One of the advantages of the NMR method is that multiple parameters can be generated by NMR spectra. The chemical shift changes reflect the relative positions between two overlapped based in the oligonucleotide. The magnitude of upfield or downfield shifts of resonances can be calculated by ring-current anisotropic effect.[31] Thus, the mode of base-base stacking can be illustrated by computer graphics.[32,33]

The coupling constants, both homonuclear and heteronuclear, are especially useful for revealing the structure of sugar-phosphate backbone of nucleic acids.[28,30] However, the spectral pattern becomes much more complicated as the chain length of the oligonucleotides increases. The coupling constants may be extracted from these complex spectra by use of the two-dimensional NMR (2D-J) techniques[34] (unpublished data).[35] At present, the conformation of sugar-phosphate backbone can be solved up to six nucleotidyl in length.[30,36]

The NOE is found not only useful for assignment but also a potential method to calculate the distances between protons. However, the experiment has to be carried out under extreme precautions to avoid any spin-diffusion phenomena.[21] Both theoretical treatment and experiment of NOE in oligonucleotides are under development in our laboratory.

The NMR spectrum can also generate two more parameters, namely, spin-lattice and spin-spin relaxation times (T_1 and T_2). These two parameters are closely related to the dynamics of molecules, which is not in the scope of the manuscript and will not be discussed here. However, it should be noted that mechanisms of NOE are similar to those of relaxation times. For this reason, relaxation times are also under investigation in our laboratory.

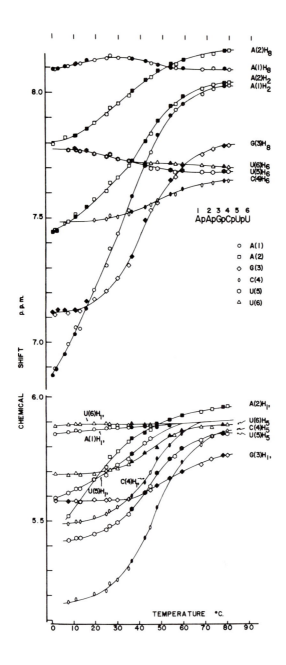

Fig. 1. The plot of chemical shifts of base and H_1, protons of A_2GCU_2 in D_2O (10 mM in strand concentration, 0.01 M sodium phosphate buffer, pD = 7.0, 0.07 M Na^+) vs. temperature. All chemical shifts are expressed in reference to DSS. The solid symbols represent the data from the 100 MHz spectrometer and the open symbols represent the data from the 220 MHz spectrometer.

RESULTS AND DISCUSSIONS

r-A-A-G-C-U-U (*a*)

The profiles of chemical shifts of all base and C_1, proton resonances of *a* versus temperature are illustrated in Fig. 1. All seventeen resonances have been assigned as described in the previous section. Thus, the differences of chemical shift values from mononucleotide to helical states can be calculated and tabulated in Table 1. These values can be compared to the calculated values based on the model constructed by X-ray diffraction data of fibers. The

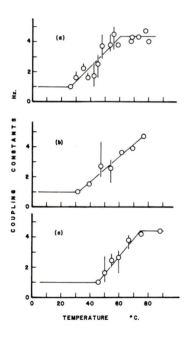

Fig. 2. The H_1,-H_2,, coupling constants, $J_{1'-2''}$ for the six ribose H_1, resonances of A_2GCU_2 as a function of temperature and counterion concentration: (a) A^1U^6; (b) A^2U^5; and (c) G^3C^4 base pairs.

helical models of a were constructed from the coordinates of A-RNA (11 bases
per helical turn), A'-RNA (12 bases per helical turn), and B-DNA (10 bases per
helical turn) given by Arnott and co-workers.[2,37,38] Thus, the theoretical
values of the chemical shift changes can be calculated based on these models
(Table 1) by comparing the predicted and observed values in Table 1. The difference of chemical shift values of base protons in the duplex calculated,
assuming the A'-RNA geometry agrees (within $\pm$ 0.1 ppm) with the observed values,
is much more accurate than those calculated based on the B-DNA geometry.

   The conformation of sugar puckering can be revealed by the $H_{1'}-H_{2'}$ coupling
constant value $(J_{1'-2'})$ as described in the previous section. In general, the
$J_{1'-2'}$ data on the nucleosides and nucleotides indicate their furanose conformation in aqueous solution is not frozen in a given conformation but exists in
a dynamic equilibrium among several forms. The results from the $J_{1'-2'}$ of
each residue of a in duplex is less than 1.5 Hz (Fig. 2). Therefore, the
equilibrium distribution of furanose conformations in the short ribosyl helix
must have very small population of $C_{2'}$-endo form. The $J_{1'-2'}$ values indicate
the $C_{3'}$-endo is the predominant form in a duplex.[39] Thus, the $J_{1'-2'}$ in a
helix provides direct verification that the RNA double-stranded helix in aqueous
solution has a conformation similar to that of A'-RNA form which is determined
by X-ray diffraction study of fibers.

d-C-C-A-A-G-C-T-T-G-G (b)

   The profiles of all nonexchangeable base proton resonances versus temperature from 0°C to 90°C are shown in Fig. 3. Since the change in chemical shift
faithfully reflects the change of the magnetic environment of each proton, this
change can serve as a sensitive monitor of conformational change during helix
melting. For example, the sequence of b contains the sequence of a. The
proton chemical shift profiles of -A-A-G-C-T-T- are different than those in
Fig. 1. This result clearly implies the different structure of DNA to RNA.

   The chemical shift changes from mononucleotides to b helical states can be
calculated as described in the previous section. The predicated values based
on X-ray diffraction data from fibers on the same sequence can also be
calculated.

   However, three categories of through-space magnetic effects have been
identified for the theoretical and experimental investigations: (i) the ring-
current anisotropic effect (RC)--this is the only effect considered on helix a;
(ii) atomic anisotropic effect (both diamagnetic (AD) and paramagnetic (AP);
and (iii) the polarization effect of the negatively charged phosphates (POL).

TABLE 1

THE CHEMICAL SHIFTS OF FORMATION OF 17 NONEXCHANGEABLE PROTONS IN
(1) THE A_2GCU_2 DUPLEX AS COMPARED WITH CALCULATED RING CURRENT SHIFTS
FROM TWO X-RAY CRYSTALLOGRAPHIC MODELS AND (2) THE A_2GCU_2 COILS
AT 90°C (δ IN PPM FROM DSS).

| | | | Duplex Form | | | | | |
| | | | $\Delta\delta_{calcd}$[c] | | $\Delta\delta_{obsd} - \Delta\delta_{calcd}$ | | Coil Form | |
	δ_{obsd}[a]	$\Delta\delta_{obsd}$[b]	A'	B	A'	B	δ[d]	$\Delta\delta$[e]
$G(3)H_8$	7.20	0.93	0.85	0.33	0.08	0.60	7.91	0.13
$C(4)H_6$	7.56	0.48	0.42	0.21	0.06	0.27	7.77	0.15
$C(4)H_5$	5.14	0.97	1.07	0.66	-0.10	0.31	5.89	0.19
$A(2)H_2$	7.50	0.76	0.64	0.44	0.12	0.32	8.16	0.13
$A(2)H_8$	7.84	0.67	0.79	0.25	-0.12	0.45	8.28	0.27
$U(5)H_6$	7.95	0.07	0.15	0.18	-0.08	-0.11	7.79	0.11
$U(5)H_5$	5.43	0.51	0.46	0.31	0.05	0.20	5.87	0.06
$A(1)H_2$	6.89	1.36	1.05[f]	1.10[f]	0.31[f]	0.26[f]	8.16	0.12
$A(1)H_8$	8.16	0.22	0.00[f]	0.02[f]	0.22[f]	0.20[f]	8.20	0.16
$U(6)H_6$	7.95	0.07	0.03	0.03	0.04	0.04	7.81	0.09
$U(6)H_5$	5.69	0.25	0.20	0.09	0.05	0.16	5.90	0.03
$G(3)H_1'$	5.59	0.35	0.08	0.17	0.27	0.18	5.77	0.13
$C(4)H_1'$	5.48	0.50	0.04	0.07	0.46	0.43	5.92	0.04
$A(2)H_1'$	5.45	0.69	0.10	0.27	0.59	0.42	5.98	0.15
$U(5)H_1'$	5.60	0.38	0.02	0.05	0.36	0.33	5.92	0.08
$A(1)H_1'$	5.86	0.27	0.00	0.22	0.27	0.05	5.92	0.19
$U(6)H_1'$	5.86	0.12	0.00	0.00	0.12	0.12	5.92	0.08

[a]These are the low temperature plateau values of δ at 10 mM strand concentration, pD 7.0, 1.07 M Na^+. If the 0.17 or 0.07 M Na^+ δ vs. T (°C) profiles level off at low temperature, their plateau values were averaged with the 1.07 M Na^+ number to generate the δ_{duplex} value.

[b]$\Delta\delta_{duplex} = \delta_{duplex} - \delta^{0=6°}_{mononucleotide}$ is the chemical shift of formation for a proton in the duplex. Appropriate mononucleotide values were selected from the following list (δ in ppm from DSS): pG-H_8, H_1', 8.126, 5.936, respectively; pC-H_6, H_5, H_1', 8.038, 6.110, 5.979; pA-H_2, H_8, H_1', 8.257, 8.512, 6.141; pU-H_6, H_5, H_1', 8.015, 5.935, 5.982; and Ap-H_2, H_8, H_1', 8.252, 8.381, 6.132, measured in 0.01 M sodium cacodylate buffer, pD 5.9 $\pm$ 0.1 in D_2O, 1 mM in the appropriate 5'-mononucleotide (a 2.0 mM sample of 3'-AMP was also measured). Little or no association of the mononucleotides is expected at these low concentrations. Temperatures were 0-5°C.

[c]Based on the A'-RNA and B-DNA geometries (Arnott et al., 1972; Arnott and Hukins, 1972) and ring current isoshielding contours provided by B. Pullman (private communication, see Discussion).

[d]These averages of the 90°C chemical shifts of protons in 0.07, 0.17, and 1.07 M Na^+ at 10 mM strand concentration pD 7.0.

[e]$\Delta\delta_{coil} = \delta_{coil} - \delta^{90°}_{mononucleotide}$ is the chemical shift of formation for a proton in the high temperature coil form. These mononucleotide values were used (see note b for buffer and nucleotide concentrations): pG-H_8, H_1', 8.041, 5.890; pC-H_6, H_5, H_1', 7.921, 6.084, 5.960; pA-H_2, H_8, H_1', 8.287, 8.547, 6.130; pU-H_6, H_5, H_1', 7.905, 5.930, 6.001; and Ap-H_2, H_8, H_1', 8.282, 8.359, 6.109.

[f]These calculated values are subject to a correction of $\sim$0.15 ppm due to shielding in end-to-end aggregates.

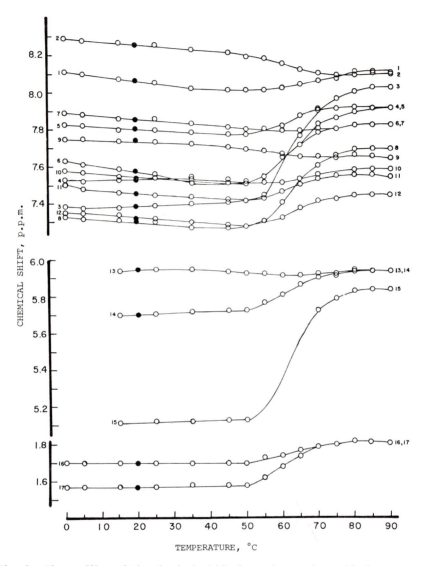

Fig. 3. The profiles of the chemical shift data of nonexchangeable base proton resonances of d-C-C-A-A-G-C-T-T-G-G versus temperature. The numbering system of peaks is 1 = A^4-H_8; 2 = A^3-H_8; 3 = A^4-H_2; 4 = A^3-H_2; 5 = G^{10}-H_8; 6 = G^5-H_8; 7 = G^9-H_8; 8 = C^6-H_6; 9 = C^1-H_6; 10 = C^2-H_6; 11 = T^7-H_6; 12 = T^8-H_6; 13 = C^1-H_5; 14 = C^2-H_5; 15 = C^6-H_5; 16 = T^7-CH_3; 17 = T^8-CH_3. The open circles represent data which were obtained at 360 MHz and filled circles are data obtained at 600 MHz.

All these calculated values are listed in Table 2 for helix b. The first two effects have been investigated and calibrated experimentally with a purine stacking in solution as model system.[40] For the base proton resonances of b, Table 2 shows the observed decameric shifts ($\Delta\delta_{10mer}$), the computed $\Delta\delta_{10mer}$ derived from four individual contributions (RC, AD, AP, and POL), the mean derivation between the observed and the computed $\Delta\delta_{10mer}$, and finally the standard derivation of the mean derivation. For the calculation based on B form, the mean derivation of difference of the experimental and theoretical $\Delta\delta$ values are the smallest (0.25 ppm for RC+AD+AP+POL). The mean derivations based on the A and A' conformation were 0.26 ppm for RC only; 0.30-0.41 ppm for RC+AD+AP; and 0.36-0.59 ppm for RC+AD+AP+POL. Thus, the mean derivations from the calculations based on B conformation are within the expected error range of 0.2 ppm and can be considered as a support to conformational assignment of this decamer.

The $C_{1'}$-H proton resonances of b helix at 600 MHz are well separated.[29] The identities of these $C_{1'}$-H resonances can be assigned by NOE technique with the aid of the known base proton resonances.[29] The $J_{1'-2'}$ and $J_{1'-2''}$ can be evaluated by computer simulation of assigned and well separated $C_{1'}$-H signals. They are 7.8, 7.9, 8.5, 7.3, 9.0, 9.2, and 9.6 Hz for G^{10}, C^2, T^7, C^1, A^3, A^4, and C^6, respectively. As mentioned in the previous section, the sugar conformation of an oligonucleotide can be evaluated based only on $J_{1'-2'}$ and $J_{1'-2''}$.[28,30] Thus, the percent of $C_{2'}$-endo form of the sugar puckering is calculated as 66, 72, 82, 84, 87, 77, and 71% for C^1, C^2, A^3, A^4, C^6, T^7, and G^{10}, respectively (whereas the $C_{1'}$-H signals of G^5, T^8, and G^9 are merged with other resonances and cannot be evaluated accurately). The result clearly shows that the percentage of $C_{2'}$-endo population increases from the terminal residue toward the inner base pair. This result also indicates that the sugar conformation of this short helix tends to resemble more of a "standard" B-DNA conformation.

d-C-G-C-G (c) and d-C-G-C-G-C-G (d)

The technique for the assignment of base proton resonances of c and d is similar to that of b as described previously. The features of chemical shift versus temperature profiles of c and d in low salt can be rational to B-DNA form.[36]

The following section will concentrate on the studies of chemical shift and coupling constants of deoxyribofuranose proton resonances of c and d helices.

TABLE 2

THE OBSERVED AND CALCULATED CHEMICAL SHIFT CHANGES OF ALL NONEXCHANGEABLE BASE PROTON RESONANCES OF d-CCAAGCTTGG HELIX. THE
CALCULATED VALUES ARE IN TERMS OF RING-CURRENT ANISOTROPIC EFFECT (RC), ATOMIC ANISOTROPIC EFFECTS (DIAMAGNETIC, AD AND PARA-
MAGNETIC, AP), AND POLARIZATION EFFECT (POL) BASED ON A, A', AND B-DNA STRUCTURES (SEE TEXT)

Proton #[a]	obs.	RC A	RC A'	RC B	AD A	AD A'	AD B	AP A	AP A'	AP B	RC+AD+AP A	RC+AD+AP A'	RC+AD+AP B	POL A	POL A'	POL B	SUM[b] A	SUM[b] A'	SUM[b] B
C¹-H₅ 13	0.09	-0.05	-0.04	0.04	0.00	0.00	-0.01	-0.02	-0.03	0.12	-0.07	-0.07	0.15	-0.42	-0.17	0.04	-0.49	-0.23	0.19
C¹-H₆ 9	0.12	-0.04	-0.04	0.09	0.00	0.00	-0.01	-0.03	-0.04	0.13	-0.07	-0.08	0.21	-0.08	0.09	0.32	-0.14	0.01	0.53
G¹⁰-H₈ 5	0.33	0.61	0.62	0.15	-0.01	-0.01	-0.01	0.27	0.29	0.08	0.87	0.89	0.22	0.42	0.41	0.39	1.29	1.30	0.62
C²-H₅ 14	0.39	0.24	0.32	0.19	-0.01	-0.01	-0.01	0.30	0.34	0.26	0.53	0.66	0.44	-0.07	0.27	0.35	0.46	0.92	0.79
C²-H₆ 10	0.52	0.06	0.02	0.15	-0.00	0.00	-0.00	0.14	0.16	0.07	0.20	0.18	0.22	-0.01	0.36	0.24	0.19	0.54	0.45
G⁹-H₈ 7	0.27	0.17	0.17	0.08	-0.01	-0.01	-0.01	0.28	0.29	0.12	0.44	0.46	0.20	-0.45	0.55	0.45	0.89	1.00	0.64
A³-H₂ 4	0.67	1.36	1.55	1.08	-0.01	-0.02	0.00	0.25	0.28	0.35	1.59	1.81	1.44	-0.01	0.06	-0.67	1.59	1.87	0.77
A³-H₈ 2	0.26	0.18	0.15	0.19	-0.01	-0.00	-0.01	0.24	0.27	0.12	0.42	0.42	0.29	0.29	0.64	0.42	0.70	1.05	0.71
T⁸-H₆ 12	0.51	0.06	0.05	0.06	-0.01	-0.01	-0.01	0.25	0.25	0.12	0.30	0.29	0.18	0.54	0.67	0.41	0.84	0.96	0.58
T⁸-CH₃ 17	0.35	0.13	0.20	0.01	-0.01	-0.02	-0.01	0.39	0.50	0.17	0.51	0.68	0.17	-0.12	-0.25	0.34	0.39	0.43	0.51
A⁴-H₂ 3	0.82	0.64	0.81	0.74	-0.01	-0.02	-0.00	0.26	0.30	0.31	0.89	1.10	1.04	0.02	0.08	-0.72	0.91	1.18	0.32
A⁴-H₈ 1	0.45	0.78	0.81	0.29	-0.01	-0.01	-0.01	0.23	0.23	0.11	0.99	1.03	0.39	0.38	0.63	0.42	1.37	1.66	0.81
T⁷-H₆ 11	0.31	0.17	0.13	0.06	-0.01	-0.00	-0.01	0.23	0.23	0.17	0.40	0.36	0.22	0.53	0.68	0.29	0.92	1.04	0.50
T⁷-CH₃ 16	0.22	0.38	0.59	0.09	-0.02	-0.02	-0.01	0.39	0.53	0.20	0.76	1.10	0.27	-0.14	-0.19	0.14	0.62	0.91	0.42
G⁵-H₈ 6	0.54	0.87	0.91	0.30	-0.02	-0.01	-0.02	0.24	0.24	0.17	1.10	1.14	0.45	0.40	0.57	0.44	1.50	1.71	0.89
C⁶-H₅ 15	0.98	0.88	1.23	0.37	-0.00	-0.00	-0.00	0.31	0.41	0.23	1.19	1.63	0.60	-0.44	-0.47	0.00	0.75	1.16	0.60
C⁶-H₆ 8	0.77	0.48	0.42	0.16	-0.01	-0.01	-0.01	0.20	0.18	0.16	0.67	0.59	0.32	0.45	0.60	0.24	1.12	1.19	0.56
x̄ₑ[c]		0.26	0.26	0.25							0.30	0.41	0.20				0.36	0.59	0.27
s[d]		0.16	0.32	0.14							0.23	0.29	0.20				0.34	0.36	0.14

a designated signal numbers, the same as in Fig. 1.
b SUM = RC + AD + AP + POL
c average error of all 17 signals
d standard deviations of averaged errors

209

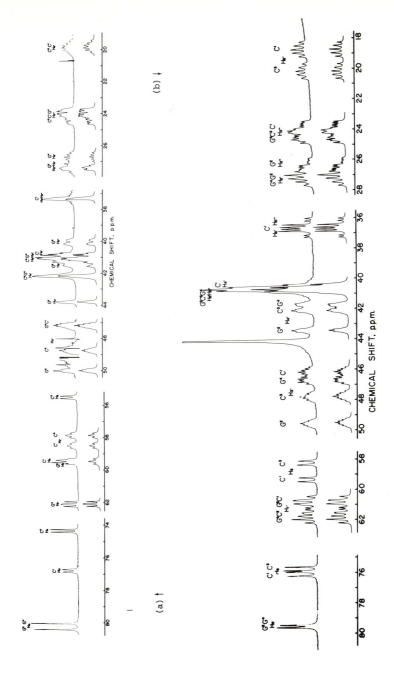

Fig. 4. (a) 300 MHz ^{1}H NMR spectrum of d-CpGpCpG at 70°C (200 scans); (b) 500 MHz ^{1}H NMR spectrum of d-CpGpCpG at 9°C (400 scans); 250 O.D./mL, 0.02 M NaHPO$_4$, pD 7.4 in D$_2$O. The lineshape simulation based on values listed in Tables 3 and 4 is shown at the bottom of the observed spectra.

By sequential homodecoupling technique, the complete set of $H_{1'}$, $H_{2'}$, $H_{2''}$, $H_{3'}$, $H_{4'}$, $H_{5'}$, and $H_{5''}$ resonances of individual sugar in a nucleotidyl unit can be interconnected and identified.[27,30] Thus, the identity of any one resonance from each set is known, which would then permit the assignment of the complete set to a specific residue.

Figure 4a and 4b show the complete proton NMR spectrum of c at single-stranded form (70°C) and at helical form (9°C) respectively. For c at 70°C (Fig. 4a), $H_{5'}$ and $H_{5''}$ of C^1 residue can be easily distinguished from $H_{5'}$ and $H_{5''}$ resonances of remaining residues by their unique coupling pattern, the absence of ^{31}P coupling. Similarly, $H_{3'}$ of G^4 residue can be identified by its unique coupling pattern (the absence of ^{31}P coupling). Therefore, these two sets of sugar proton resonances are unambiguously assigned. The remaining sets, C^3 and G^2 residues, can be assigned by comparing the $H_{2'}$ resonances with corresponding residue in the d-CpG and d-GpC dimers as shown in Fig. 4a.[32]

The chemical shift and coupling constant values of all sugar proton resonances of c at 9°C (i.e., ~100% in helical form); 30°C (~75% helix); 45°C (~50% helix) and 70°C (~100% in coil form) were analyzed by computer simulation and are summarized in Tables 3 and 4.

The sugar proton resonances of d-CpGpCpGpCpG were measured at 25°C by 500 MHz NMR spectrometer (Fig. 5). The assignments of all sugar proton resonances of the d helix were accomplished by comparison with the c helix as shown in Fig. 5.[32]

After the $H_{1'}$ resonances in d are identified, then the other sugar proton resonances can be assigned through the sequential decoupling technique.[30]

The chemical shifts and coupling constants of all sugar proton resonances of d-(CpGpCpGpCpG)$_2$ are summarized in Tables 3 and 4. Thus, the conformation of the deoxyribofuranose ring can be computed.

For the population of ^2E conformers, a representative sum of $J_{1',2'}$ plus $J_{3',4'}$ was obtained by averaging the summed J values from the tetramer or the hexamer at all temperatures. These representative sums for the c and d are 10.6 Hz and 11.1 Hz, respectively. The percentage populations of ^2E-conformers in c and d were computed from the above summed J values and the individual $J_{3',4'}$ values and are listed in Table 5. At 9°C in ~100% helical form, c helix shows a clear preference to populate in ^2E-conformation especially the internal G^2 and C^3 residues. When the temperature was elevated to 70°, i.e., c is now in single-stranded state, ^2E population is reduced and approaches to its monomeric value, such as 66% ^2E for d-Cp and 70% ^2E for pdG.[27]

TABLE 3

CHEMICAL SHIFTS OF THE SUGAR PROTONS OF d-CpGpCpG and d-CpGpCpGpCpG

IN D_2O, pD 7.4, 0.02 M $NaHPO_4$ AT DIFFERENT TEMPERATURE (PPM FROM DSS)

Compounds	Temperature °C	$H_{1'}$	$H_{2'}$	$H_{2''}$	$H_{3'}$	$H_{4'}$	$H_{5'}$	$H_{5''}$
CGCG C^1	9	5.85	1.98	2.40	4.71	4.08	3.74	3.73
C^2		5.94	2.73	2.75	5.00	4.38	4.14	4.00
C^3		5.78	2.00	2.45	4.87	4.22	4.12	4.10
G4		6.21	2.67	2.40	4.73	4.21	4.10	4.09
CGCG C^1	30	5.94	1.95	2.40	†	4.09	3.74	3.72
G^2		5.98	2.73	2.74	5.00	4.38	~4.13	~4.02
C^3		5.87	1.97	2.43	4.85	4.21	~4.13	~4.11
G^4		6.21	2.69	2.43	†	4.21	~4.12	~4.11
CGCG C^1	45	5.93	1.91	2.40	4.68	4.10	3.73	3.69
G^2		6.00	2.71	2.73	4.99	4.36	4.19	4.03
C^3		6.00	1.96	2.41	4.82	4.21	~4.13	~4.11
G^4		6.21	2.69	2.44	4.70	4.19	~4.10	~4.08
CGCG C^1	70	6.07	1.88	2.39	4.61	4.05	3.69	3.63
G^2		6.09	2.72	2.64	4.95	4.34	~4.10	~4.01
C^3		6.15	2.02	2.43	4.77	4.20	~4.08	~4.05
G^4		6.20	2.73	2.45	4.67	4.17	~4.07	~4.05
CGCGCG C^1	25	5.78	1.99	2.43	4.72	4.07	3.74	3.71
G^2		5.90	2.65	2.73	4.99	4.36	4.21	4.00
C^3		5.76	2.04	2.43	4.87	4.21	4.16	4.16
G^4		5.92	2.69	2.76	4.99	4.37	4.11	4.04
C^5		5.78	1.90	2.34	4.81	4.16	4.20	4.19
G^6		6.17	2.61	2.37	4.68	4.18	4.12	4.11

†Under HDO

TABLE 4

COUPLING CONSTANTS OF THE SUGAR PROTONS OF d-CpGpGpG and d-CpGpCpGpCpG IN D$_{2O}$, pD 7.4, 0.02 M NaHPO$_4$ AT DIFFERENT TEMPERATURE (J$_{H-H}$ or J$_{H-P}$ IN Hz)

Compounds	Temperature °C	1'2'	1'2"	2'2"	2'3'	2"3'	3'4'	4'5'	4'5"	5'5"	3'P	4'P	5'P	5"P
CGCG C^1	9	8.0	6.1	-14.0	6.7	2.6	3.0	3.1	4.0	-12.1	5.7	-	-	-
G^2		8.2	6.6	-14.0	6.4	2.7	2.0	2.4	2.0	-11.8	5.5	1.0	3.0	3.8
C^3		7.9	6.1	-14.0	6.8	2.5	2.7	3.0	3.0	-11.3	5.5	2.1	3.2	3.2
G^4		8.0	6.3	-14.0	6.0	2.6	2.5	3.1	3.1	-11.3	-	1.4	3.5	3.5
CGCG C^1	30	7.9	6.0	-14.0	6.8	2.5	3.1	3.3	4.4	-12.2	†	-	-	-
G^2		8.1	6.5	-14.0	6.5	2.7	2.3	~2.4	~1.9	-11.5	5.7	0.8	~3.1	3.7
C^3		8.0	6.2	-14.0	6.8	3.0	2.8	~3.1	~3.1	-11.4	5.6	2.1	~3.2	3.2
G^4		8.0	6.0	-14.0	6.5	2.4	2.7	~3.2	~3.2	-11.4	-	1.4	~3.5	3.5
CGCG C^1	45	7.8	6.0	-14.0	7.0	2.4	3.2	3.2	~4.4	-12.1	5.9	-	-	-
G^2		7.7	6.5	-14.0	6.3	2.7	2.3	2.4	1.8	-11.8	5.7	1.8	3.0	4.3
C^3		8.1	6.4	-14.0	7.1	3.4	2.8	~3.2	~3.2	-11.5	5.8	2.5	~3.3	~3.3
G^4		7.2	6.9	-14.0	6.6	3.5	3.0	~3.3	~3.3	-11.5	-	1.8	~3.5	~3.5
CGCG C^1	70	7.3	6.3	-14.0	6.5	2.9	3.4	4.0	4.8	-12.1	6.0	-	-	-
G^2		7.0	6.4	-14.0	6.3	2.5	2.8	~2.4	~2.3	-11.6	5.7	2.0	3.0	~4.3
C^3		8.1	6.3	-14.0	6.8	3.5	3.4	~3.3	~3.3	-11.4	6.0	1.9	3.4	~3.4
G^4		7.0	6.6	-14.0	6.4	3.9	3.3	~3.4	~3.4	-11.4	-	1.4	3.6	3.6
CGCGCG C^1	25	8.1	6.2	-14.0	6.2	2.9	3.2	3.6	5.0	-11.8	5.9	-	-	-
G^2		8.6	5.5	-14.3	6.3	1.7	2.8	~3.5	2.0	-11.6	5.8	2.0	3.8	4.0
C^3		8.5	5.6	-14.0	6.7	3.1	2.7	~2.8	~2.8	-11.3	5.8	2.0	3.0	3.0
G^4		8.5	5.5	-14.3	6.3	1.7	2.8	~3.5	2.0	-11.5	5.8	1.9	3.8	4.2
C^5		8.3	6.4	-14.0	6.9	2.5	2.6	~2.8	~2.8	-11.3	5.8	2.0	3.2	3.2
G^6		8.0	6.0	-14.0	6.3	3.1	2.7	~3.0	~3.0	-11.3	-	2.0	3.3	3.3

†Under HDO.

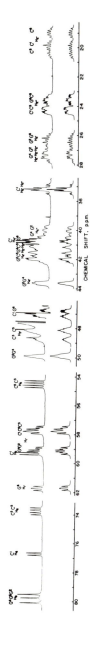

Fig. 5. 500 MHz ^{1}H NMR spectrum of d-CpGpCpGpGpCpG at 25°C (600 scans) 350 O.D./ml, 0.02 M NaHPO$_4$, pD 7.4 in D$_2$O. The lineshape simulation based on values listed in Tables 3 and 4 is shown at the bottom of the observed spectrum.

Similarly, there is a clear preference for d helix to populate in ^{2}E-type conformer (Table 5). Internal C^3 and C^5 residues have a slightly higher %^{2}E than the terminal C^1 residue ($\sim$ 5%-7%).

It should be noted that even at the 100% helical form, and even for the internal base pairs, the percent of ^{2}E conformation does not exceed 81%, a value slightly less than 87% estimated for the center GC base pair of the b helix at 20°C, which is 40°C below the Tm as described in the previous section. It is not certain how large this percentage can be enhanced by increasing the chain-length or by lowering the temperature. It is, however, very unlikely that the furanose conformation of these helices will ever assume a 100% ^{2}E form. Therefore, while this data totally reject the A conformation with a ^{3}E conformation for these helices, the observation does not support either a "typical" B conformation previously derived from fiber diffraction pattern.[37]

Similarly, the backbone conformations of c and d helix can also be computed by the coupling constant data in Table 4. The population distribution of conformers about the $C_4{}'$-$C_5{}'$ bond (%gg) and $C_5{}'$-$O_5{}'$ bond (%g'g') for c and d are compiled in Table 5. As shown in Figs. 4 and 5, $H_4{}'$, $H_5{}'$, and $H_5{}''$ resonances of each residue are extensively overlapped with each other, therefore a precise determination of coupling constant values between $H_4{}'$, $H_5{}'$, and $H_5{}''$ and ^{31}P became difficult. Hence, the computed populations of gg and g'g' conformers for both oligomers are less accurate as compared to the calculated ^{2}E population. Generally, the data show a clear preference for gg and g'g' conformers for both c and d helices. Elevation of temperature causes a reduction in gg populations. This is clearly indicated in C^1 residue of c whose $J_4{}'_5{}'$ and $J_4{}'_5{}''$ coupling constants can be accurately determined since $H_5{}'$ and $H_5{}''$ resonances of C^1 residue are clearly resolved at the most upfield region (Fig. 4).

As shown in Table 5, the values of percent gg of free 3' ends are much higher than those at free 5' ends. This observation indicates that the rotation of the $C_4{}'$-$C_5{}'$ bond at the 5' ends is less restricted than that at the 3' ends. The same result has been found previously in our studies of the oligomers.[30] The rotamer distribution about the $C_3{}'$-$O_3{}'$ bond (ϕ') for c and d is listed in Table 5. This computation was made on the basis of a two-rotamer model.[27,41] Data shows that both deoxy-oligomers have similar ϕ' values, i.e., $\sim$195°/285°. Since no ^{J}H$_2{}'$-P four-bond couplings were observed in both short oligomers, it appears that the torsion about the $C_3{}'$-$O_3{}'$ bond is restricted to a domain around $\phi' \simeq 195°$. Temperature or helix-coil transition has little effect on the average torsion angle of this $C_3{}'$-$O_3{}'$ bond.

TABLE 5

CONFORMATION AND POPULATION DISTRIBUTION OF CONFORMERS OF THE SUGAR
BACKBONE OF d-CpGpCpG AND d-CpGpCpGpCpG AT DIFFERENT TEMPERATURES[a]

Compounds	Temperature °C	% Helix	% ^{2}E	% gg	% g'g'	ϕ'
CGCG C^1	9	100	72	68	–	195°/285°
G^2			81	96	88	194°/286°
C^3			75	79	89	194°/286°
G^4			76	77	87	–
C^1	30	75	71	62	–	Under HDO
G^2			78	97	88	195°/285°
C^3			74	77	89	195°/285°
G^4			75	75	87	–
C^1	45	50	70	63	–	196°/284°
G^2			78	98	85	194°/286°
C^3			74	75	88	194°/286°
G^4			72	73	87	–
C^1	70	0	68	51	–	196°/284°
G^2			74	93	85	195°/285°
C^3			68	73	87	196°/284°
G^4			69	71	86	–
CGCGCG C^1	25	100	71	53	–	196°/284°
G^2			75	85	83	195°/285°
C^3			76	83	91	195°/285°
G^4			75	85	82	195°/285°
C^5			78	83	89	195°/285°
G^6			76	79	88	–

[a] $\%^2E = 100 - [J_{3'4'}/(J_{1'2'} + J_{3'4'})] \times 100$; accurate to $\pm$ 1%-2%.

$\%gg = [(13.7 - \Sigma)/9.7] \times 100$; $\Sigma = J_{4'5'} + J_{4'5''}$; accurate to $\pm$ 6%-7%.

$\%g'g' = [(25 - \Sigma')/20.8] \times 100$; $\Sigma' = J_{5'p} + J_{5''p}$; accurate to $\pm$ 6%-7%.

$^3J_{HP} = 18.1 \cos^2 \theta_{HP} - 4.8 \cos \theta_{HP}$; $\phi' = 240° \pm \theta$; accurate to $\pm$ 3°.

CONCLUSION

In this communication, four oligonucleotides have been studied thoroughly. The identities of base, C_1, and NH-N proton resonances of all four compounds were assigned unambiguously. Thus, the mode of base-base stacking, and the conformation of the furanose ring of these four oligonucleotides can be determined. Furthermore, all sugar proton resonances of c and d were also assigned and their coupling patterns have been analyzed. Thus, phosphate-sugar backbone conformation of DNA-type oligonucleotides were also explored. We found the RNA-type oligonucleotide (a) more resembles A (A')-type DNA; and DNA-type oligomers in low salt condition more resemble the B-DNA in aqueous solution.

The NMR technique has been extensively demonstrated, through the studies of ^{1}H NMR of four oligonucleotides, as a very powerful method to reveal the conformations of nucleic acids in aqueous solution. Based on these observations, the fundamental work of this field has been laid down. Thus, more advanced aspects which are more relevant to biomedicine such as nucleic acids-protein interaction and nucleic acids with lesions can be approached. These studies are in progress in our laboratory.

ACKNOWLEDGEMENTS

The authors are indebted to Drs. Paul S. Miller, Krishna Jayaraman, Eldon E. Leutzinger, and Philip N. Borer for their contributions to this work, and the National Science Foundation and the National Institutes of Health for their financial support.

REFERENCES

1. Watson, J. D. and Crick, F. H. C. (1953) Nature, 171, 737, 964.
2. Arnott, S., et al. (1973) J. Mol. Biol., 81, 107.
3. Wang, A. H.-J., et al. (1979) Nature (London), 282, 680.
4. Crawford, J. L., et al. (1980) Proc. Natl. Acad. Sci. USA, 77, 4016.
5. Drew, H., et al. (1980) Nature (London), 286, 567.
6. Dickerson, R. E. and Drew, H. R. (1981), J. Mol. Biol., 149, 761.
7. Bubienko, E., Uniak, M. A., and Borer, P. N. (1981), Biochemistry, 20, 6987.
8. Arter, D. B., et al. (1974) Biochem. Biophys. Res. Commun., 61, 1089.
9. Borer, P. N., Kan, L.-S., and Ts'o, P.O.P. (1975) Biochemistry, 14, 4847.
10. Hughes, D. W., et al. (1978) Can. J. Chem., 56, 2243.
11. Romaniuk, P. J., et al. (1978) Can. J. Chem., 56, 2249.
12. Romaniuk, P. J., et al. (1978) J. Am. Chem. Soc., 100, 3971.
13. Romaniuk, P. J., et al. (1979) Biochemistry, 18, 5109.
14. Alkema, D., et al. (1981) J. Am. Chem. Soc., 103, 2866.
15. Cross, A. D. and Crothers, D. M. (1971) Biochemistry, 10, 4015.
16. Patel, D. W. (1979) Stereodynamics of Molecular Systems, Sarma, R. H., ed., Progress Press, p. 397.
17. Patel, D. W., et al. (1982) Biochemistry, 21, 428.

18. Patel, D. W., et al. (1982) Biochemistry, 21, 437.
19. Patel, D. W., et al. (1982) Biochemistry, 21, 445.
20. Patel, D. W., et al. (1982) Biochemistry, 21, 451.
21. Early, T. A. and Kearns, D. R. (1980) Nucleic Acids Res., 8, 5795.
22. Pardi, A., Martin, F. H., and Tinoco, I., Jr. (1981) Biochemistry, 20, 3986.
23. Miller, P. S., et al. (1982) Biochemistry, 21, 5468.
24. Uesugi, S., Shida, T., and Ikehara, M. (1981) Chem. Pharm. Bull., 29, 3573.
25. Miller, P. S., et al. (1981) Biochemistry, 19, 4688.
26. Borer, P. N., et al. (1973) J. Mol. Biol., 80, 759.
27. Ts'o, P. O. P. (1974) Basic Principles in Nucleic Acid Chemistry,
    New York: Academic Press, Vol. I, p. 453; Vol. II, p. 305.
28. Cheng, D. M. and Sarma, R. H. (1977) J. Am. Chem. Soc., 99, 7333.
29. Kan, L.-S., et al. (1982) Biochemistry, 21, 6723.
30. Cheng, D. M., et al. (1982) Biochemistry, 21, 621.
31. Giessner-Prettre, C. and Pullman, B. (1970) J. Theor. Biol., 27, 87.
32. Kan, L-S., et al. (1979) Comp. Prog. Biomed., 10, 16.
33. Giessner-Prettre, C., et al. (1981) Comp. Prog. Biomed., 13, 167.
34. Nagayama, K., et al. (1977) Biochem. Biophys. Res. Commun., 78, 99.
35. Kan, L.-S., Cheng, D. M., and Cadet, J. (1982) J. Mag. Reson., 48, 86.
36. Cheng, D. M., et al. (1983) Biopolymers, in press.
37. Arnott, S. and Hukins, D. W. L. (1972) Biochem. Biophys. Res. Commun., 47,
    1504.
38. Arnott, S. and Hukins, D. W. L. (1973) J. Mol. Biol., 81, 93.
39. Hruska, F. E. and Danyluk, S. S. (1968) J. Am. Chem. Soc., 90, 3266.
40. Cheng, D. M., et al. (1980) J. Am. Chem. Soc., 102, 525.
41. Alderfer, J. L. and Ts'o, P. O. P. (1977) Biochemistry, 16, 2410.

THE METAL CENTERS OF CYTOCHROME C OXIDASE: STRUCTURE AND FUNCTION

SUNNEY I. CHAN, CRAIG T. MARTIN, HSIN WANG, DAVID F. BLAIR,
JEFF GELLES, GARY W. BRUDVIG, AND TOM H. STEVENS
A. A. Noyes Laboratory of Chemical Physics, 127-72
California Institute of Technology, Pasadena, California 91125

ABSTRACT

Progress toward elucidation of the structures of the metal centers in cytochrome c oxidase will be reviewed. Our studies are based primarily on low-temperature electron paramagnetic resonance (EPR) spectroscopy. We have used nitric oxide (NO) extensively to probe the O_2 reduction site of the enzyme. In addition, we have isolated auxotrophs of $Saccharomyces$ $cerevisiae$ in order to incorporate isotopically substituted amino acids into the yeast protein. The latter approach, in conjunction with low-temperature EPR and electron nuclear double resonance (ENDOR) spectroscopy, has furnished unambiguous information on the structure of two of the four metal centers. The implications of these structural results with respect to the mechanism of dioxygen reduction by cytochrome c oxidase will be discussed.

INTRODUCTION

Cytochrome c oxidase mediates the transfer of electrons from reduced cytochrome c to molecular oxygen, reducing dioxygen to two molecules of water. The overall reaction is given by:

$$4 \text{ ferrocytochrome } c + O_2 + 4H^+ \longrightarrow 4 \text{ ferricytochrome } c + 2H_2O$$

The free energy for this reaction at physiological pH is -50 kcal per mole of O_2 reduced.[1] The reduction of O_2 is extremely efficient, with the turnover rate approaching 400 sec^{-1}.[2]

Cytochrome c oxidase is a transmembrane protein embedded in the inner membrane of the mitochondrion. All indications are that the oxidase receives its reducing equivalents from cytochrome c in the intermembrane space, i.e., from the cytosol side, and the four H^+ that are used up per turnover originate from the matrix. In this manner, four negative charges are transferred across the inner membrane and a transmembrane electrochemical potential gradient is established. Evidence is mounting that the enzyme is also a proton pump[3] with protons pumped across the inner membrane from the matrix to the cytosol side

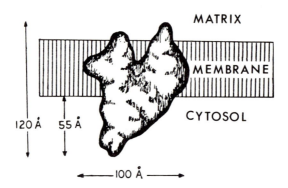

Fig. 1. Three-dimensional folding and assembly of oxidized cytochrome c oxidase as revealed by electron microscopy imaging.

during the electron transport and/or O_2 reduction process. The number of vectorial protons (four to eight) pumped per turnover, however, remains elusive.[3,4]

Because of the importance of cytochrome c oxidase in cellular respiration and energy conservation, the structure of the protein has naturally received considerable attention. Impressive progress has been made toward deciphering the primary sequence[5-7] and the subunit structure[8] of the enzyme. A glimpse of the three-dimensional folding and assembly of the protein (Fig. 1) has also emerged from electron microscopy imaging studies.[9] The structures of the four metal centers that are intimately associated with the electron transfer, O_2 reduction, and possibly proton pumping, are beginning to unfold.[10-13] Information on the location of the metal centers in the enzyme and their relative proximity to one another is expected to be forthcoming.

In this report, we will review our efforts toward elucidating the structure and function of the four metal centers in cytochrome c oxidase. Our studies are based largely on low-temperature electron paramagnetic resonance (EPR) spectroscopy. We have used nitric oxide (NO) extensively to probe the structure of the O_2 reduction site. In addition, we have manipulated the yeast system *Saccharomyces cerevisiae* in order to incorporate isotopically substituted derivatives by EPR and electron nuclear double resonance (ENDOR) spectroscopy. These approaches have provided unambiguous information on the ligands of two of the four metal centers. Although our work is still incomplete, the structural results obtained to date have already allowed us to gain some insight into the mechanism of dioxygen reduction by this enzyme. We feel that an important beginning has been made and are optimistic that a molecular picture of the structure and function of cytochrome c oxidase is on the horizon.

THE METAL CENTERS

Cytochrome c oxidase contains two heme a's and two copper ions, all of which are inequivalent. Cytochrome a_3 and Cu_B form the oxygen reduction site, while cytochrome a and Cu_A participate in electron transfer. We now describe in turn results relating to the structure of each of these metal centers.

Cytochrome a

There seems to be general agreement that cytochrome a accepts electrons from ferrocytochrome c and transfers them to the Cu_A center. The cytochrome c binding site(s) is known to be located on the cytosol side of the inner mitochondrial membrane.[14] The reduction potential of cytochrome a is similar to that of ferricytochrome c,[15] which is expected if cytochrome a is the point of entry of the electrons into the protein and degradation of the redox free energy is to be minimized in this electron transfer step.

All indications are that cytochrome a is six-coordinate and low spin, with nitrogens from two neutral imidazoles as axial ligands. Such a structure would be consistent with the proposed electron transfer function of cytochrome a. The magnetic circular dichroism (MCD) and EPR spectra of cytochrome a demonstrate clearly that cytochrome a is low spin.[16,17] There have been two lines of evidence supporting two histidine imidazoles as axial ligands. Babcock et al.[18] have compared the optical and resonance Raman spectra of cytochrome a with those of related low-spin heme a models. Similarly, Peisach[19] has compared the EPR g-values of cytochrome a with those of low-spin heme a models with known axial ligands. The Peisach compilation of g-values is reproduced in Table 1. The g-values (3.0, 2.21, 1.45) of cytochrome a (Fig. 2) are reproduced only by the bis neutral-imidazole complex of heme a.

Copper A

Copper A is the low potential copper in cytochrome c oxidase, and it has been suggested that the functional role of this metal center is to transfer electrons from cytochrome a to the dioxygen reduction site. Its reduction potential is similar to that of cytochrome a.[15] Since the first reduction potential drop occurs between Cu_A and the O_2 reduction site, Chan et al.[20] have proposed that this redox energy may be conserved and that Cu_A might serve as the proton pump of the enzyme.

TABLE 1

THE RELATION BETWEEN EPR PARAMETERS AND STRUCTURE
OF LOW SPIN FERRIC HEME COMPOUNDS[*]

Compound	Axial ligands		g values		
$\mathit{Glycera}$ Hb MeNH$_2$	imid	RNH$_2$	3.30	1.98	1.20
Leg Hb pyridine	imid	pyr	3.26	2.10	0.82
Cytochrome c	imid^O	met	3.07	2.26	1.25
Bis imid heme	imid^O	imid^O	3.02	2.24	1.51
Bis imid$^-$ heme	imid$^-$	imid$^-$	2.80	2.26	1.72
MbOH	imid$^-$	OH$^-$	2.55	2.17	1.85
Cyt. P-450$_{cam}$	imid^O	RS$^-$	2.45	2.26	1.92
Cyt. c oxidase	imid^O	imid^O	3.0	2.2	1.5

[*]
Taken from reference 19.

The structure of the Cu$_A$ center has been of considerable interest because
this copper is highly unusual.[21] The EPR signal of the Cu$_A$ center (Fig. 2)
is atypical of Cu(II) in that no copper hyperfine splittings are clearly
resolved and the g-values (2.18, 2.03, 1.99) are quite low; in fact, one g-
value is below that of the free electron. Two mechanisms have been proposed
to explain the unusual EPR properties. One possibility is that the unpaired
electron spin resides primarily on an associated ligand.[22] This mechanism
calls for extensive electron charge delocalization from the involved associated
ligand onto the copper ion. The g-values for the Cu$_A$ center EPR signal are,

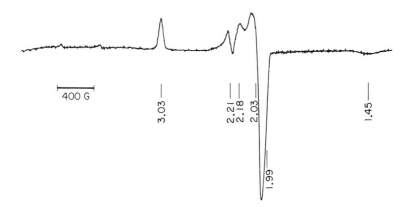

400 G

3.03   2.21 2.18 2.03   1.45

1.99

Fig. 2. X-band EPR spectrum of oxidized beef heart cytochrome c oxidase at 10 K.

in fact, typical of those of thiyl radicals,[23] and it was originally suggested that this EPR signal might be due to a disulfide interacting with a copper ion[24] or due to a sulfur radical.[25] A second possibility is that the orbital containing the unpaired electron is a copper hybrid 3d orbital with strong admixtures of 4s and 4p character.[26] In this case, the unpaired electron would reside primarily on the copper, and the unusual EPR properties would result from distorting the copper into a near tetrahedral geometry which allows mixing of copper 4s and 4p orbitals with the 3d ground state.

Recent ENDOR studies[27] have revealed a small copper hyperfine interaction associated with the Cu_A center, but this copper hyperfine interaction is nearly isotropic. This result indicates that both the isotropic hyperfine interaction and the anisotropic distributed dipole interaction between the unpaired electron and the copper nucleus are small. The implication is that *either* the unpaired electron spin density resides primarily on an associated ligand sufficiently removed from the copper ion, *or* the unpaired electron is localized on a copper ion with cubic (or higher symmetry) coordination. However, the Cu ENDOR results, i.e., the small anisotropy in the copper hyperfine interaction, place limits on the extent of any d-p mixing. It is possible to ascertain the relative amounts of 3d and 4p character in a hybrid orbital necessary to eliminate the anisotropic copper hyperfine interaction by calculating the distributed dipole interaction of an electron in a 3d or 4p orbital with the copper nucleus. The calculations show that it is necessary to include about three times as much $4p(z)$ character as $3d(x^2 - y^2)$ or $3d(xy)$ character to obtain an isotropic copper hyperfine interaction.[28] Such a large mixing is unreasonable in view of the fact that the $3d^8 4p$ configuration lies about 125,000 cm^{-1} above the $3d^9$ configuration for divalent copper.[29] Thus, although mixing of copper 3d and 4p orbitals could, in principle, account for the observed EPR properties of the Cu_A center, this possibility is unreasonable from the standpoint of energetics.

We subscribe to the view that the EPR properties of the Cu_A center are best accounted for by delocalization of the unpaired electron spin onto an associated ligand. In fact, X-ray absorption edge data indicate that Cu_A in the oxidized protein is in a highly covalent environment[30] and might even be reduced[31] to Cu(I). Extensive delocalization of spin onto an associated ligand in the oxidized enzyme would predict that Cu_A remains mainly in the Cu(I) state both in the oxidized and reduced forms of the Cu_A center, since upon reduction an associated ligand would be the actual electron acceptor rather than the Cu ion itself. This prediction seems to be born out by the observation that the

"1s-4s" transition in the X-ray edge spectrum of Cu_A does not change substantially when the enzyme is reduced.[31]

If our picture of the structure of the Cu_A center is correct, the question
remains as to which ligands among the various amino acid residues can render
to Cu_A the unusual EPR properties manifested by this metal center. Cysteines
come to mind; however, at least two cysteines must be involved since ligation
of one cysteine to a divalent copper does not lead to sufficient electron
transfer from the cysteine sulfur to account for the observed EPR properties
(cf. "blue coppers"[32]). However, it is well known that ligation of one
cysteine to a $3d^9$ copper renders the metal center more susceptible to reduction
by a second cysteine sulfur.[24]

These considerations led us to propose some years ago[22] that the Cu_A center
in cytochrome c oxidase consisted of a Cu(I) ion ligated by two cysteine
sulfurs, one a cysteinate and the other a sulfur radical. We have now undertaken experiments to test this model by incorporating cysteine substituted
with ^2H at the β-methylene carbon ((^2H)Cys) into yeast cytochrome c oxidase
using a cysteine auxotroph of the yeast system $Saccharomyces\ cerevisiae$. As
part of a larger overall effort to define the ligands of the various metal
centers, a histidine auxotroph was also isolated and used to incorporate
histidine substituted with ^{15}N at both imidazole ring positions ((^{15}N)His) into
the yeast enzyme. The strategy here is to exploit the differences in nuclear
spin or nuclear magnetic moment of isotopes to identify magnetic interactions
between the unpaired electron spin and the atom in question. Thus, perturbation of the EPR spectrum for the metal center by the isotopic substitution,
or in the case where the hyperfine interaction is too small to be discernible
in the EPR, modification of the ENDOR spectrum, may be used to obtain unambiguous information about the involvement of a particular amino acid at the site.

In the specific case of Cu_A, we have confirmed the involvement of at least
one histidine and one cysteine as ligands. We have compared the EPR spectra of
native, (^2H)Cys, and (^{15}N)His yeast cytochrome c oxidase. Although subtle
differences were noted between the EPR spectra of the native and the isotopically substituted yeast proteins, the recent ENDOR studies,[12] carried out in
collaboration with Dr. C. P. Scholes of SUNY at Albany, were unequivocal in
this conclusion. The Cu_A ENDOR spectra of the (^2H)Cys and (^{15}N)His yeast
oxidase are compared with that of the native yeast enzyme in Fig. 3. We note
that the spectrum observed for the native yeast oxidase is very similar to that
of the beef heart protein.[33] The two signals seen at 21.7 and 19.7 MHz are due
to strongly coupled protons and correspond to proton hyperfine couplings of
16.2 and 12.2 MHz. That these protons have origin in the β-CH_2 of cysteine may

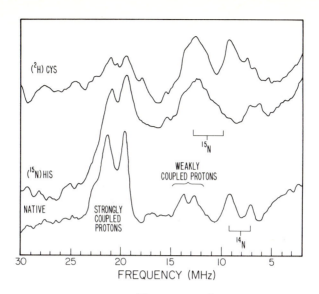

Fig. 3. ENDOR spectra of native, (^{15}N)His and (^2H)Cys yeast cytochrome c oxidase observed at $g = 2.04$, microwave frequency 9.12 GHz and temperature 2.1 K.

be ascertained by noting the intensity change in the ENDOR signal in this region upon the deuterium substitution. The two signals seen at 7.1 and 9.2 MHz in the native enzyme are split by twice the characteristic ^{14}N Zeeman energy $(2\nu(^{14}N) = 2.00$ MHz$)$ and can be assigned to a ^{14}N with hyperfine coupling of 16 MHz. This assignment was confirmed by the ENDOR spectrum of the (^{15}N)His yeast enzyme. Upon comparison of the ENDOR spectra for the (^{15}N)His and native yeast protein, we see that only the intensities of the ^{14}N ENDOR signals at 7 and 9 MHz are substantially reduced by the isotopic substitution of the imidazole ring nitrogens. On the basis of the difference in nuclear magnetic moments between the two nitrogen isotopes, substitution of the ^{14}N nucleus by ^{15}N would replace the ^{14}N ENDOR signals at 7 and 9 MHz by ^{15}N ENDOR signals at 10 and 13 MHz, respectively. The increased intensity observed in the 10-15 MHz region in the ENDOR spectra of the (^{15}N)His protein is consistent with the appearance of these ^{15}N ENDOR signals. From these observations, it is clear that there is at least one histidine and one cysteine ligand to Cu_A in cytochrome c oxidase.

The issue of whether there are one or two cysteines associated with Cu_A remains unsettled. In principle, the number of cysteines could be inferred from the number of proton ENDOR signals and/or the magnitudes of the proton hyperfine interactions affected by the deuterium substitution. In our experiments, only the two strongly coupled proton ENDOR signals clearly disappear upon deuterium substitution, although the signal-to-noise in the 10-15 MHz region of the ENDOR spectra does not allow us to rule out the existence of weakly coupled β-CH$_2$ protons should there exist a second cysteine ligand. In any case, from the well-known equations describing hyperconjugation in sulfur radicals,[34] and by comparison with model compounds,[23] we can estimate the unpaired spin density (ρ_S^{π}) in the sulfur 3p(z) orbital(s). If we assume that the strongly coupled proton hyperfine interactions arise from two protons on two *different* cysteine ligands to Cu$_A$, one cysteine sulfur would have ρ_S^{π} ranging from 0.14 to 0.37, while for the other cysteine sulfur ρ_S^{π} would range from 0.23 to 0.52, depending on the choice of dihedral angle between the strongly coupled β-CH$_2$ proton and the 3p(z) orbital of the sulfur in each case. We make the assumption here that the other proton on each methylene carbon exhibits a negligible (less than 3 MHz) hyperfine coupling. A simpler situation is that these hyperfine couplings arise from protons on the *same* cysteine ligand. In this case, only two distinct orientations of the β-CH$_2$ protons relative to the sulfur 3p(z) orbital are possible. In the one case, the resultant unpaired spin density on sulfur would be 0.23, and in the other, 0.83. The latter value is more in agreement with the X-ray absorption edge and EXAFS data on the Cu$_A$ site. It is important to note that the assignment of the strong proton hyperfine couplings to methylene protons on the same cysteine ligand does not preclude the existence of a second cysteine ligand to Cu$_A$. In fact, if ρ_S^{π} is as high as 0.83, a second cysteine ligand is necessary to facilitate this large delocalization of spin from the copper to the first cysteine (i.e., a delocalization of charge from cysteine to copper). If this were so, the magnitude of the hyperfine coupling to these more distant protons would be expected to be very small.

Experiments are currently in progress to incorporate cysteine substituted with ^{13}C at the methylene carbon ((^{13}C)Cys) into yeast cytochrome c oxidase using the cysteine auxotroph which we have isolated. In contrast to the hyperfine interaction of the β-methylene protons, the ^{13}C hyperfine interaction is relatively insensitive to the dihedral angle between the C-H bond and the sulfur 3p(z) orbital bearing the unpaired electron and hence should provide a more direct indication of ρ_S^{π}.

Cytochrome a_3-Cu$_B$

Cytochrome a_3-Cu$_B$ constitutes the oxygen reduction site of the enzyme. The structure of this site has remained elusive because antiferromagnetic coupling between the high-spin heme and the d^9 copper yields an S = 2 ground state and renders the site EPR silent in the oxidized enzyme.[35] Even in the reduced enzyme, where the copper is d^{10}, the high-spin ferrous heme has an S = 2 ground state. To obtain EPR signals from the cytochrome a_3-Cu$_B$ site, it is necessary to convert the site into a half-integral spin system which has a ground state Kramer's doublet.[36] Indeed, upon partial reduction of oxidized cytochrome c oxidase, one observes at neutral pH, a high-spin heme EPR signal at g = 6, and at higher pH's, a low-spin heme EPR signal at g = 2.6.[37] The high-spin heme species has been assigned to a cytochrome $a_3^{+3} \cdot$ H$_2$O species in which the antiferromagnetic coupling between cytochrome a_3 and Cu$_B$ has been eliminated by reduction of Cu$_B$. Increasing the pH merely leads to deprotonation of the bound water to form the low-spin cytochrome $a_3^{+3} \cdot {}^-$OH species. These results clearly demonstrate that the reduction potential of Cu$_B$ is higher than that of cytochrome a_3, at least under these low-temperature conditions. Thus, simple reductive titration of the enzyme allows only the detection of an EPR signal from cytochrome a_3. We recently showed, however, that the binding of NO to the site under various conditions allows EPR signals to be observed from both cytochrome a_3 and Cu$_B$.[10]

It is well known that cytochrome a_3 can bind a variety of ligands including NO, CO, CN$^-$, F$^-$, and HCOO$^-$;[38] we have shown that NO can bind to Cu$_B$ as well.[10,39] Depending on the conditions, three complexes of NO with cytochrome c oxidase can be prepared: 1) the reduced enzyme plus NO, which exhibits a well-characterized nitrosylferrocytochrome a_3 EPR signal near g = 2;[40,41] 2) the oxidized enzyme plus NO, which exhibits a high-spin ferricytochrome a_3 EPR signal at g ~ 6;[10] and 3) the one-quarter reduced NO-bound enzyme, which exhibits a triplet (S = 1) EPR signal due to the interaction of nitrosylferrocytochrome a_3 and Cu$_B^{+2}$.

The EPR signal observed when NO is added to reduced cytochrome c oxidase is typical of that observed for other nitrosylferroheme proteins such as cytochrome c, hemoglobin, or cytochrome c peroxidase.[42] In particular, nitrogen super-hyperfine splittings are observed for both the NO nitrogen and an endogenous nitrogen bound axially to cytochrome a_3 opposite the NO. The g-values for this complex as well as the magnitudes of the ^{14}N hyperfine splittings indicate that the endogenous nitrogen ligand of cytochrome a_3 is most likely from an imidazole nitrogen. Unambiguous identification of histidine as the endogenous

228

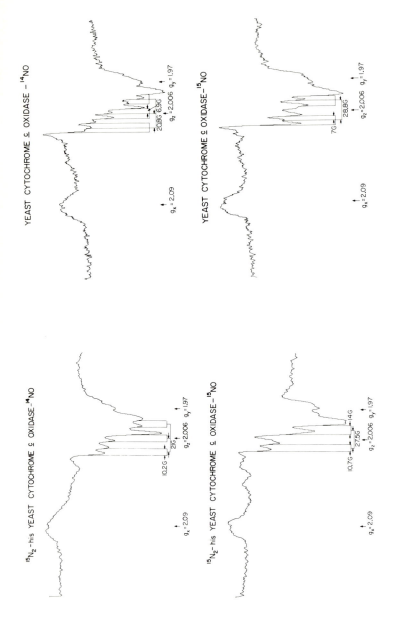

Fig. 4. X-band EPR spectra of NO-bound reduced native and (^{15}N)His yeast cytochrome c oxidase at 30 K.

axial ligand to cytochrome a_3[11] was obtained from the nitrosylferrocytochrome a_3 complex prepared from the (^{15}N)His yeast oxidase described earlier.

When ^{14}NO is bound to ferrocytochrome a_3 of the yeast enzyme, the g = 2.006 component of the EPR signal of the complex exhibits a nine-line hyperfine pattern which can be interpreted in terms of the superposition of three sets of three lines arising from two nonequivalent nitrogens (I = 1) interacting with the unpaired electron. The larger of the two hyperfine coupling constants is 20.8 G and the smaller 6.9 G. When ^{15}NO is used in this experiment, the ^{15}NO-bound protein exhibits an EPR spectrum with g-values identical to those of the ^{14}NO-bound species, but the g=2.006 component shows a hyperfine pattern consisting of two sets of three lines. This pattern is consistent with the presence of one ^{14}N and one ^{15}N nitrogen bound axially to cytochrome a_3 with a 28.2 G splitting for the ^{15}N and a 7.0 G splitting for the ^{14}N ligand. These spectra of the NO-bound native yeast protein are compared with those of the ^{14}NO- and ^{15}NO-bound (^{15}N)His yeast cytochrome c oxidase in Fig. 4. It is apparent from this comparison that the hyperfine patterns have been altered upon (^{15}N)His substitution. The ^{15}NO-bound (^{15}N)His protein hyperfine pattern consists of two sets of doublets, with a ^{15}NO nitrogen splitting of 27.5 G and a splitting of about 12 G for the ^{15}N nitrogen of the histidine. The ^{14}NO-bound (^{15}N)His protein hyperfine pattern consists of three sets of doublets, with splittings of 21 G and 10.2 G for the ^{14}NO and (^{15}N)His nitrogen, respectively. Thus, the substitution of (^{15}N)His for (^{14}N)His in cytochrome c oxidase results in the involvement of an (I = 1/2) ^{15}N nucleus rather than an (I = 1) ^{14}N nucleus in the nitrosylferrocytochrome a_3 EPR signal. These studies provide unequivocal identification of histidine as the endogenous fifth ligand to cytochrome a_3.

In the course of our work, we have shown that NO also binds to the *oxidized* enzyme.[10] Unlike the reduced enzyme, in which NO binds to ferrocytochrome a_3, the site of interaction with the oxidized enzyme appears to be Cu_B. The addition of NO to the oxidized enzyme results in the disruption of the antiferromagnetic coupling between cytochrome a_3 and Cu_B, as evidenced by the appearance of the high-spin heme EPR signal (Fig. 5). This EPR signal is similar to the g = 6 signal seen upon partial reduction of the oxidized enzyme.[37] However, in the present case, Cu_B does not appear to be reduced as NO can be reversibly photolyzed from the site at low temperatures to re-establish the antiferromagnetic couple between ferricytochrome a_3 and Cu_B^{+2}.[43] The action spectrum of this photodissociation does not correspond to the absorption spectrum of heme a, but rather it is indicative of a divalent copper. Further evidence comes from the addition of CN$^-$ to the oxidized enzyme plus NO species, which converts

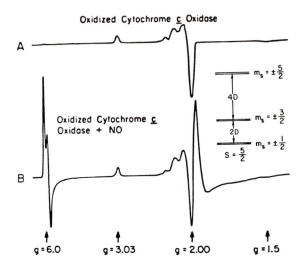

Fig. 5. X-band EPR spectra of native oxidized beef heart cytochrome c oxidase in the absence and presence of NO. (Inset): Energy level diagram of the S = 5/2 spin manifold.

cytochrome a_3 to a low-spin heme with EPR signals characteristic of the cyano-ferricytochrome a_3 species.[10] From these results, it is clear that NO binds to Cu_B to yield an EPR-silent Cu_B^{+2}-NO species.

The NO complex with the oxidized enzyme can be used to ascertain whether the histidine identified earlier as the axial ligand to cytochrome a_3 in the reduced protein remains coordinated to the heme in the oxidized enzyme. The high-spin heme EPR signal seen at g ∼ 6 upon binding of NO to Cu_B^{+2} (Fig. 5) arises from the $M_S = \pm 1/2$ component (Kramer's doublet) of the S = 5/2 spin manifold when the applied field is in the plane of the porphyrin ring. The intensity of these signals is temperature dependent and can be used to measure the zero field splitting D for the high spin heme, which could, when compared with those of known hemes, in turn be used to infer the nature of the fifth ligand. In the absence of exogenous ligands, we measured D to be 9 cm^{-1}, and in the presence of F$^-$, D ≈ 6 cm^{-1}.[44] These zero-field parameters are indicative of a high-spin heme with an axial histidine.

Neither the EPR signal from nitrosylferrocytochrome a_3 nor the high-spin cytochrome a_3 signal provide any information regarding the position of the histidine vis-a-vis Cu_B. A bridging imidazole has been proposed to provide the

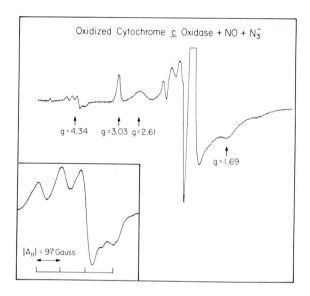

Fig. 6. X-band EPR spectrum of the nitrosylferrocytochrome a_3, Cu_B^{+2} species at 7 K. (Inset): Magnified view of the half-field transition region.

exchange coupling between the two metal centers.[45] However, the triplet EPR signal observed from the one-quarter-reduced NO-bound enzyme provides a measure of the interaction between cytochrome a_3 and Cu_B in the nitrosylferro-cytochrome a_3, Cu_B^{+2} complex and indicates that NO must bind between these two metal ions in this complex; in other words, the histidine imidazole of cyto-chrome a_3 is distal to Cu_B.

In order to obtain the one-quarter-reduced NO-bound enzyme, we exploited the high reduction potential of cytochrome a_3 to activate the disproportiona-tion of N_3^- in the presence of NO according to the overall reaction[10]

$$\text{ferricytochrome } a_3 + N_3^- + 2NO \longrightarrow \text{nitrosylferrocytochrome } a_3 + N_2 + N_2O.$$

The optical spectrum of the resultant nitrosylferrocytochrome a_3 suggests that the imidazole is still in place. The fact that a triplet EPR signal is observed for this complex indicates that NO binds to the ferrocytochrome a_3 and that the unpaired electron on NO is sufficiently close to interact magnetically with Cu_B^{+2}. The zero-field splitting D of the triplet can be estimated from the breadth of the $\Delta M_S = \pm 1$ transition. We obtained a value for $|D|$ of 0.07 cm^{-1}.

If we assume a purely dipolar interaction between the two spins, we estimate the distance between the two "spin centers" to be ~ 3.4 Å.

The triplet EPR signal from the nitrosylferrocytochrome a_3, Cu_B^{+2} species exhibits copper hyperfine splittings on the half-field transition at $g \sim 4$. The magnitude of the copper hyperfine interaction (~ 0.02 cm^{-1}) is characteristic of a tetragonal/rhombic geometry for Cu_B. A similar conclusion can be deduced from the results of Reinhammar et al.,[46] who have subsequently reported a new EPR signal from Cu_B^{+2}. This signal, which is obtained when the enzyme is reoxidized in the presence of bubbling O_2, exhibits a copper hyperfine coupling of 0.011 cm^{-1}. Karlsson and Andréasson[47] have assigned this signal to a $[Fe_{a_3}^{+3} - O_2^- \ HO^- - Cu_B^{+2}]$ species wherein Cu_B is in an unusual rhombic environment. If this interpretation is correct, then the ligation of O_2 to cytochrome a_3 must raise the reduction potential of this $[Fe_{a_3}^{+3} - O_2^-] \leftrightarrow [Fe_{a_3}^{+2} - O_2]$ complex above that of Cu_B, in a manner analogous to that observed in the nitrosylferrocytochrome a_3, Cu_B^{+2} species described above.

The ligands to Cu_B remain uncertain. However, preliminary ENDOR data (B. Hoffman and B. Reinhammar, private communication) suggest that there are three nitrogens associated with Cu_B. Comparative ENDOR studies of the native and the ^{15}N(His) yeast oxidase are in progress to ascertain whether these nitrogen ligands to Cu_B originate in histidine imidazoles. Recent EXAFS studies have suggested that there might be a cysteine ligand to Cu_B and that the cysteine sulfur might serve as a bridging ligand between cytochrome a_3 and Cu_B in some states of the oxidized enzyme.[13] Inasmuch as corroborative evidence for this cysteine from EPR and optical measurements is still lacking, the EXAFS result should be considered preliminary.

Structural Model

We present in Fig. 7 a model for oxidized cytochrome c oxidase. This model summarizes not only the structural results obtained to date, but also includes some preliminary information on the spatial disposition of the metal centers within the protein.

Generally speaking, the metal centers can be grouped into two pairs: 1) cytochrome a and Cu_A; and 2) the O_2 reduction site, which is composed of cytochrome a_3 and Cu_B. Recent EPR relaxation studies from our laboratory showed that cytochrome a and Cu_A are sufficiently close to one another that a magnetic dipolar interaction exists between them in the oxidized enzyme. This observation places the two metal centers within ~ 15 Å of each other.

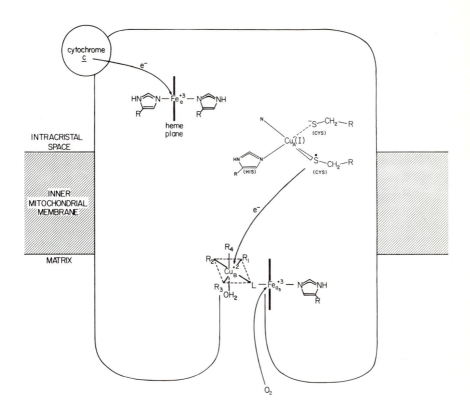

Fig. 7. A model for oxidized cytochrome *c* oxidase.

The close proximity of cytochrome a_3 and Cu_B in the O_2 reduction site is now well established. The distance between them seems to be about 5 Å,[10] but most likely varies with the state of the enzyme. It seems clear that this proximity is critical for efficient electron transfer and the stabilization of O_2 reduction intermediates during the O_2 reduction process. On the other hand, the two pairs of metal ions seem to be relatively far apart, as we have been unable to detect by EPR any effect of the spin state of the O_2 reduction site on the EPR relaxation behavior of cytochrome a or Cu_A.[48]

There now exists convincing evidence that cytochrome a and Cu_A are located near the cytosol side of the mitochondrial membrane,[2,14] where cytochrome c is known to bind. The location of the O_2 reduction site is, however, less certain. Recently, we obtained preliminary EPR results which place cytochrome a_3 and Cu_B closer to the matrix side than the cytosol side of the mitochondrial membrane. This distribution of the metal centers in the protein is consistent with the function normally attributed to cytochrome c oxidase, namely, the vectorial transfer of electrons to energize the inner mitochondrial membrane.

In terms of function, the general consensus is that cytochrome a and Cu_A serve to accept electrons from cytochrome c and transfer them to cytochrome a_3 and Cu_B at the O_2 reduction site. However, the exact sequence of this electron transfer remains unsettled, in part due to the rapid redox equilibrium which exists between these two centers.[49] The rate-determining step in the overall O_2 reduction process is electron transfer between the cytochrome a/Cu_A pair and the O_2 reduction site.[50] It has been known for some time that the turnover rate is sensitive to environmental factors, including the fluidity of the lipid medium in which the protein is embedded.[51] In light of this sensitivity and the apparently large distance over which the electrons must transfer between cytochrome a/Cu_A and the O_2 reduction site, we surmise that this electron transfer process may be coupled to some conformational change which not only serves to control directed electron transfer over the relatively large distance between the two metal ion pairs, but also functions to couple the redox energy to the pumping of protons across the mitochondrial membrane.

MECHANISM OF OXYGEN REDUCTION

Our structural results on cytochrome a_3 and Cu_B lend considerable insight into the role that these two metal centers play in the O_2 reduction process. The observation that the binding of O_2 raises the reduction potential of cytochrome a_3 above that of Cu_B has suggested to us that the role of cytochrome a_3 might be to anchor O_2 and its reduction intermediates, at least during the early stages of O_2 reduction. Cu_B would then serve to transfer electrons into the site. The close proximity of the two metal ions may serve both for efficient electron transfer and for the stabilization of reactive intermediates. On the basis of these considerations we proposed some years ago[39] a mechanism for dioxygen reduction which involves both a peroxo and a ferryl intermediate. Since that time, new evidence has been obtained in support of this mechanism, and additional results pertinent to the rest of the cycle have become available. An elaborated version of our earlier proposed scheme is presented in Fig. 8. For clarity, only the cytochrome a_3 - Cu_B site is shown.

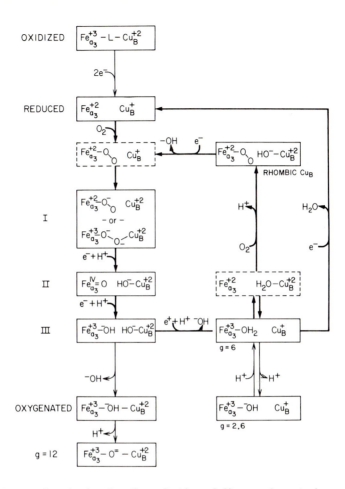

Fig. 8. Proposed mechanism for the reduction of dioxygen by cytochrome c oxidase. For clarity, only the metal centers of the oxygen binding site are shown.

The scheme begins with the binding of O_2 to reduced cytochrome a_3. This is followed by the rapid transfer of an electron from Cu_B to yield an intermediate which is, at least formally, equivalent to peroxide bound to the oxidized site. The proposal of such an intermediate has recently gained support from the observation[52] that oxidized cytochrome c oxidase binds hydrogen peroxide tightly ($K_d \sim 10\ \mu M$). It is likely that the bonding in this first intermediate bears resemblance to that in the oxidized enzyme-peroxide complex, although the exact nature of this intermediate deserves further study.

The next step in the sequence is the transfer of an electron from cytochrome a/Cu_A to the O_2 reduction site. In our earlier formulation, we proposed that this electron leads to rupture of the oxygen-oxygen bond and the formation of a ferryl species adjacent to Cu_B^{+2}. Indeed, evidence for such an intermediate has seen been reported by Karlsson et al.[53] In studies of the low-temperature reoxidation of the reduced enzyme, they reported the observation of an EPR signal from Cu_B^{+2} with unusual relaxation properties. In this intermediate, Cu_B^{+2} appears to be influenced by a nearby rapidly relaxing paramagnetic center such as Fe(IV), consistent with our previously proposed intermediate.

To complete the reduction of dioxygen, a second electron is transferred to the site from cytochrome a/Cu_A. This results in the formation of a species in which both metal ions of the O_2 reduction site are in the oxidized state. In single turnover experiments, Brudvig et al.[44] have shown that this species can relax slowly in succession to a number of "dead end" forms of the oxidized enzyme. In the scheme presented here, the first such species is denoted as "oxygenated." It is characterized by strong antiferromagnetic exchange coupling between cytochrome a_3 and Cu_B which is mediated by a readily exchangeable bridging ligand, possibly the hydroxyl anion. According to Brudvig et al.,[44] this form of the enzyme decays slowly ($t_{1/2} \sim 1$ hr) to another form which exhibits an X-band EPR signal near $g \sim 12$. The bridging ligand in this "g12" species was found to be almost substitutionally inert, and it has been proposed to be a μ-oxo.

The reduction of O_2 is a very rapid and efficient process. Accordingly, the direct observation of O_2 reduction intermediates is difficult. Chance et al.[54] have developed a low-temperature trapping technique capable of stabilizing several of the intermediate species, allowing them to be probed spectroscopically. With this technique, Clore et al.[55,56] recently completed a careful study of the intermediates involved in O_2 reduction and characterized them using optical and EPR spectroscopies. Starting from either the fully reduced enzyme or the mixed valence state in which only cytochrome a_3 and Cu_B are reduced, they have characterized three species which seem to correspond well spectroscopically and kinetically with the first three intermediates described in our mechanistic scheme. Accordingly, we have adopted the nomenclature of Clore et al. to describe these intermediates. Intermediate I of their study would correspond to our "peroxo-like" species; intermediate II to our ferryl species; and intermediate III to a form of the oxidized site immediately following the final electron transfer.

In the steady state cycling enzyme, we expect cytochrome a and/or Cu_A to be reduced. If this is so, the O_2 reduction site should be in a position to accept another electron from cytochrome a/Cu_A to form a partially reduced form of the site. Under normal conditions, we expect the resultant species formed to give the high-spin heme EPR signal at g ~ 6 and, at higher pH's, the low-spin heme EPR signal at g ~ 2.6. The fate of this intermediate depends on the relative availability of O_2 and reducing equivalents. Under low oxygen tension, the site can accept an electron from cytochrome a/Cu_A to re-form the fully reduced site and complete the cycle. In the presence of excess O_2, however, binding of O_2 can stabilize the cytochrome a_3^{+2}- O_2, Cu_B^{+2} species faster than an electron can transfer to the site from cytochrome a/Cu_A. The EPR spectrum of this site would then be expected to show only a normal Cu_B^{+2} EPR signal. It is precisely this species which we assign to the rhombic Cu_B^{+2} EPR signal observed by Reinhammar et al.[46] upon reoxidation of the reduced enzyme in the presence of bubbling O_2. Subsequent electron transfer from cytochrome a/Cu_A would lead quickly to intermediate I and a continuation of the O_2 reduction cycle.

SUMMARY

We have reviewed the progress that has been made toward elucidating the structures of the metal centers in cytochrome c oxidase. In our own work, we have employed NO and other ligands to probe the structure of the O_2 reduction site. We have also manipulated the yeast system *Saccharomyces cerevisiae* to enable direct incorporation of isotopically substituted amino acids into the enzyme. These approaches, coupled with low-temperature EPR and ENDOR spectroscopies have proven to be extremely powerful for obtaining unambiguous information on the ligands of the metal centers. From such structural studies a clear picture of the functioning enzyme is beginning to unfold.

ACKNOWLEDGEMENTS

This work was supported by Grant GM-22432 from the National Institute of General Medical Sciences, U. S. Public Health Service, and by BRSG Grant RR07003 awarded by the Biomedical Research Support Grant Program, Division of Research Resources, National Institutes of Health. C. T. Martin, D. F. Blair, G. W. Brudvig, and T. H. Stevens were recipients of National Research Service Awards (5T32GM-07616) from the National Institute of General Medical Sciences. J. Gelles was a recipient of a National Science Foundation Graduate Fellowship. This article is Contribution Number 6693 from the Division of Chemistry and Chemical Engineering, California Institute of Technology.

238

REFERENCES

1. Lehninger, A. L. (1975) Biochemistry, New York: Worth Publishers.
2. Wikström, M., Krab, K., and Saraste, M. (1981) in Cytochrome Oxidase, London: Academic Press.
3. Wikström, M. and Krab, K. (1979) Biochim. Biophys. Acta, 549, 177-222.
4. Lehninger, A. L., et al. (1981) in Mitochondria and Microsomes, Lee, C. P., Schatz, G., and Dallner, G., eds., Menlo Park, California: Addison-Wesley, pp. 459-479.
5. Anderson, S., et al. (1981) Nature, 290, 457-465.
6. Yasunobu, K. T., et al. (1979) in Cytochrome Oxidase, King, T. E., et al., eds., Amsterdam: Elsevier, pp. 91-101.
7. Steffens, G. J. and Buse, G. (1979) in Cytochrome Oxidase, King, T. E., et al., eds., Amsterdam: Elsevier, pp. 153-159.
8. Capaldi, R. A. (1979) in Membrane Proteins in Energy Transduction, Capaldi, R. A., ed., New York: Marcel Dekker, pp. 201-227.
9. Henderson, R., Capaldi, R. A., and Leigh, J. S. (1977) J. Mol. Biol., 112, 631-648.
10. Stevens, T. H., et al. (1979) Proc. Natl. Acad. Sci. USA, 76, 3320-3324. 3324.
11. Stevens, T. H. and Chan, S. I. (1981) J. Biol. Chem., 256, 1069-1071.
12. Stevens, T. H., et al. (1982) J. Biol. Chem., 257, 12106-12113.
13. Powers, L., et al. (1981) Biophys. J., 34, 465-498.
14. Blum, H., Leigh, J. S., and Ohnishi, T. (1980) Biochim. Biophys. Acta, 626, 31-40.
15. Andréasson, L.-E. (1975) Eur. J. Biochem., 53, 591-597.
16. Babcock, G. T., Vickery, L. E., and Palmer, G. (1976) J. Biol. Chem., 251, 7907-7919.
17. van Gelder, B. F. and Beinert, H. (1969) Biochim. Biophys. Acta, 189, 1-24.
18. Babcock, G. T., et al. (1981) Biochemistry, 20, 959-966.
19. Peisach, J. (1978) in Frontiers of Biological Energetics, Dutton, P. L., et al., eds., Vol. 2, New York: Academic Press, pp. 873-881.
20. Chan, S. I., et al. (1979) in Cytochrome Oxidase, King, T. E., et al., eds., Amsterdam: Elsevier, pp. 177-188.
21. Beinert, H. (1966) in The Biochemistry of Copper, Peisach, J., Aisen, P., and Blumberg, W. E., eds., New York: Academic Press, pp. 213-234.
22. Chan, S. I., et al. (1978) in Frontiers of Biological Energetics, Dutton, P. L., Leigh, J. S., Jr., and Scarpa, A., eds., Vol. 2, New York: Academic Press, pp. 883-888.
23. Hadley, J. H., Jr., and Gordy, W. (1977) Proc. Natl. Acad. Sci. USA, 74, 216-220.
24. Hemmerich, P. (1966) in The Biochemistry of Copper, Peisach, J., Aisen, P., and Blumberg, W. E., eds., New York: Academic Press, 15-34.
25. Peisach, J. and Blumberg, W. E. (1974) Arch. Biochem. Biophys., 165, 691-708.
26. Greenaway, F. T., Chan, S. H. P., and Vincow, G. (1977) Biochim. Biophys. Acta, 490, 62-78.
27. Hoffman, B. M., et al. (1980) Proc. Natl. Acad. Sci. USA, 77, 1452-1456.
28. Chan, S. I., et al. (1982) in Electron Transport and Oxygen Utilization, Ho, C., ed., Amsterdam: Elsevier, pp. 171-177.
29. Moore, C. E. (1952) in Atomic Energy Levels, Vol. II, United States Department of Commerce, National Bureau of Standards.
30. Powers, L., et al. (1979) Biochim. Biophys. Acta, 546, 520-538.
31. Hu, V. W., Chan, S. I., and Brown, G. S. (1977) Proc. Natl. Acad. Sci. USA, 74, 3821-3825.
32. Fee, J. A. (1975) Struc. Bonding (Berlin), 23, 1-60.
33. Van Camp, H. L., et al. (1978) Biochim. Biophys. Acta, 537, 238-246.
34. Wertz, J. E. and Bolton, J. R. (1972) Electron Spin Resonance, New York: McGraw-Hill.

35. Tweedle, M. F., et al. (1978) J. Biol. Chem., 253, 8065-8071.
36. Abragam, A. and Bleaney, B. (1970) Electronic Paramagnetic Resonance of Transition Ions, London: Oxford Press.
37. Hartzell, C. R. and Beinert, H. (1974) Biochim. Biophys. Acta, 368, 318-338.
38. Wilson, D. F. and Erecińska, M. (1978) Meth. Enzymol., 53, 191-201.
39. Brudvig, G. W., Stevens, T. H., and Chan, S. I. (1980) Biochemistry, 19, 5275-5285.
40. Blokzijl-Homan, M. F. J. and van Gelder, B. F. (1971) Biochim. Biophys. Acta, 234, 493-498.
41. Stevens, T. H., Bocian, D. F., and Chan, S. I. (1979) FEBS Lett., 97, 314-316.
42. Yonetani, T., et al. (1972) J. Biol. Chem., 247, 2447-2455.
43. Boelens, R., et al. (1982) Biochim. Biophys. Acta, 679, 84-94.
44. Brudvig, G. W., et al. (1981) Biochemistry, 20, 3912-3921.
45. Palmer, G., Babcock, G. T., and Vickery, L. E. (1973) Proc. Natl. Acad. Sci. USA, 73, 2206-2210.
46. Reinhammer, B., et al. (1980) J. Biol. Chem., 255, 5000-5003.
47. Karlsson, B. and Andréasson, L.-E. (1981) Biochim. Biophys. Acta, 635, 73-80.
48. Gelles, J., et al. (1982) in Frontiers in Biochemical and Biophysical Studies of Proteins and Membranes, Liu, T.-Y., et al., eds., New York: Elsevier-North Holland, in press.
49. Andréasson, L.-E., et al. (1972) FEBS Lett., 28, 297-301.
50. Gibson, Q. H. and Greenwood, C. (1965) J. Biol. Chem., 240, 2694-2698.
51. Vik, S. B. and Capaldi, R. A. (1977) Biochem., 16, 5755-5759.
52. Bickar, D., Bonaventura, J., and Bonaventura, C. (1982) Biochem., 21, 2661-2666.
53. Karlsson, B., et al. (1981) FEBS Lett., 131, 186-188.
54. Chance, B., Saronio, C., and Leigh, J. S. (1975) J. Biol. Chem., 250, 9226-9237.
55. Clore, G. M., et al. (1980) Biochem. J., 185, 139-154.
56. Clore, G. M., et al. (1980) Biochem. J., 185, 155-167.

Biochemical and Biophysical Studies of Proteins and Nucleic Acids,
Lo, Liu, and Li, eds.
241

THE INTRACELLULAR POLYMERIZATION OF SICKLE HEMOGLOBIN

A. N. SCHECHTER AND C. T. NOGUCHI
National Institutes of Health,
Bethesda, Maryland 20205

We have previously reported the detection, using natural abundance ^{13}C
nuclear magnetic resonance spectroscopy, of intracellular polymer (aggregates
of deoxyhemoglobin S) in sickle cell erythrocytes at high oxygen saturation
values (> 95%). In order to clarify the effect of cell heterogeneity on this
phenomenon, whole populations of SS erythrocytes were fractioned on discontinuous Stractan density gradients. For subpopulations with mean corpuscular
hemoglobin concentration (MCHC) values ranging from 29 g/dl to 42 g/dl, the
intracellular polymer fraction at complete deoxygenation varied from 0.45 to
0.75 of the total hemoglobin, in good agreement with the predicted values based on
deoxyhemoglobin S solubility. The values of measured intracellular polymer
at varying MCHC and oxygen saturations is in very good agreement with our
theoretical predictions based on solubility and the non-ideal behavior of water
and of concentrated hemoglobin solutions. The heterogeneity in intracellular
hemoglobin concentration of unfractionated sickle cell erythrocytes causes the
critical oxygen saturation for the appearance of polymer to shift to values
greater than 95%. These studies show that cell heterogeneity exaggerates further
the appearance, due to water and protein non-ideality, of polymer of deoxyhemoglobin S at high oxygen saturation. Further, we have recently found that
the predicted intracellular polymer fraction correlates well with hematological
severity (chronic anemia) in the various sickle syndromes. We believe that
intracellular polymer is a more important clinical parameter than "sickling."
The probable existence of polymer in arterial blood must be considered in
therapeutic approaches to sickle cell disease.

---

*
A full text of this contribution appears in T.-Y. Liu, S. Sakakibara, A. N.
Schechter, K. Yagi, H. Yajima, and K. T. Yasunobu (Eds.) Biophysical Studies
of Macromolecules: Frontiers in Biochemical and Biophysical Studies of
Macromolecules. Elsevier-North Holland, New York, 1983.

NUCLEAR MAGNETIC RESONANCE STUDIES OF COLLAGEN MOLECULAR DYNAMICS[*]

D. A. TORCHIA, L. S. BATCHELDER, S. K. SARKAR, AND C. E. SULLIVAN
Laboratory of Biochemistry, National Institute of Dental Research
Bethesda, Maryland 20205

Nuclear magnetic resonance spectroscopy of ^{2}H and ^{13}C labeled proteins is a powerful method for studying molecular dynamics of specific sites in fibrous proteins. Because ^{2}H and ^{13}C have low natural abundance the spectra of labeled proteins contain strong signals which are readily assignable to the labeled nuclei. The lineshape and spin lattice relaxation time of the labeled nucleus provides quantitative information about the rate and angular range of the motion of the labeled site. In the first part of the talk I will illustrate the application of this technique to elucidate molecular motion in crystalline amino acids and peptides. I will then describe our studies of reconstituted collagen fibrils labeled with $[3,3,3\text{-}d_3]$alanine, $[d_{10}]$leucine, $[2\text{-}d_1]$proline and $[d_7]$proline. Finally I will discuss experiments on intact rat tail tendon and rat calvaria which are labeled with $[1\text{-}^{13}C]$glycine. Taken together these experiments show that hydrated collagen possesses considerable dynamic activity at the molecular level, and that this dynamic behavior is greatly reduced when collagen is mineralized. The possible functional significance of these results will be discussed.

[*] A full text of this contribution appears in T.-Y. Liu, S. Sakakibara, A. N. Schechter, K. Yagi, H. Yajima, and K. T. Yasunobu (Eds.), Biophysical Studies of Macromolecules: Frontiers in Biochemical and Biophysical Studies of Macromolecules. Elsevier-North Holland, New York, 1983.

Biochemical and Biophysical Studies of Proteins and Nucleic Acids,
Lo, Liu, and Li, eds.

2-OXO ACID DEHYDROGENASE COMPLEXES AND THE PRINCIPLE OF PARSIMONY[*]

RICHARD N. PERHAM
Department of Biochemistry, University of Cambridge,
Tennis Court Road, Cambridge CB2 1QW, England, U.K.

The 2-oxo acid dehydrogenase multienzyme complexes consist of multiple copies
of each of three different enzymes which catalyze successive steps in the overall
reaction as shown:

$$\underset{RCCOOH+NAD^{+}+CoASH}{\overset{O}{\parallel}} \longrightarrow \underset{RC \sim SCoA+NADH+H^{+}+CO_2}{\overset{O}{\parallel}}$$

The three constituent enzymes of the pyruvate dehydrogenase complex are
pyruvate decarboxylase (E1, EC 1.2.4.1), lipoate acetyltransferase (E2, EC
2.3.1.12) and lipoamide dehydrogenase (E3, EC 1.6.4.3). For the 2-oxo-glutarate
dehydrogenase complex, the corresponding enzymes are 2-oxoglutarate decarboxylase (E1, EC 1.2.4.2), lipoate succinyltransferase (E2, EC 2.3.1.61) and
lipoamide dehydrogenase (E3, EC 1.6.4.3). In *Escherichia coli*, at least, all
the evidence indicates that the E3 component is identical for both complexes.

The E2 component forms a structural core of high M_r to which the E1 and E3
components are bound tightly but noncovalently. In the pyruvate and 2-oxo-
glutarate dehydrogenase complexes from *E. coli*, this core consists of 24 copies
of the E2 chain arranged with octahedral symmetry whereas in the pyruvate
dehydrogenase complexes of mammals, yeast, and *Bacillus stearothermophilus* the
core comprises 60 E2 chains arranged with icosahedral symmetry.

Despite the requirement for the participation of three enzymes in the overall
reaction, it is abundantly clear that the active sites are not necessarily
represented in equal numbers in the enzyme particle. For example, it is likely
that each E2 chain of the pyruvate dehydrogenase complex from *E. coli* can bind
up to 2 E1 chains and that the E1:E2 chain ratio in the complex is generally
greater than 1.0. On the other hand, the E3:E2 chain ratio is generally less
than 1.0 and has routinely been put as low as 0.5. The possibility of heterogeneity that this implies is borne out by ultracentrifugation analysis. Such

[*] A full text of this contribution appears in T.-Y. Liu, S. Sakakibara, A. N.
Schechter, K. Yagi, H. Yajima, and K. T. Yasunobu (Eds.) Biophysical Studies
of Macromolecules: Frontiers in Biochemical and Biophysical Studies of
Macromolecules. Elsevier-North Holland, New York, 1983.

heterogeneity may be due, at least in part, to competition between El and E3 subunits for space on the surface of the E2 core during the assembly process. Similar considerations are likely to apply to the 2-oxo acid dehydrogenase complexes from other sources.

During catalysis the substrate is carried in thioester linkage by lipoic acid residues which are in turn covalently linked to lysine side chains of the E2 core. It is supposed that these lipoyl-lysine "swinging arms" can rotate between the three different active sites as an essential feature of the mechanism and the physical mobility of the lipoyl groups inferred from spin-label experiments is consistent with this view. However, the distances between active sites in the *E. coli* pyruvate dehydrogenase complex have been estimated to be more than 5 nm by means of fluorescence energy transfer measurements, which is substantially greater than the span (about 2.8 nm) of a single swinging arm. Moreover, in all 2-oxo acid dehydrogenase complexes tested, it has been found that a given El subunit can bring about the reductive acylation of many of the lipoic acid residues in the E2 core and not just the lipoic acid residue(s) of the E2 subunit to which the El subunit is bound.

Thus, it is clear that these multienzyme complexes do not necessarily function by a simple direct transfer of substrate among three active sites organized in a strict geometrical arrangement on a 1:1:1 basis.

Biochemical and Biophysical Studies of Proteins and Nucleic Acids,
Lo, Liu, and Li, eds.

RESONANCE RAMAN STUDIES OF THE PRIMARY PHOTOCHEMISTRY IN
VISUAL PIGMENTS AND BACTERIORHODOPSIN[*]

RICHARD MATHIES
Department of Chemistry
University of California, Berkeley, California 94720

The lack of reliable vibrational assignments has often limited the utility
of resonance Raman spectroscopy of biological systems. In collaboration with
Prof. Johan Lugtenburg and coworkers at the University of Leiden, we have
addressed this problem for retinal-containing proteins by synthesizing a series
of isotopically labeled retinal derivatives. These derivatives have been
incorporated into bovine opsin and bacterioopsin to form visual pigment and
bacteriorhodopsin analogues. We have studied the primary photo-products of
these systems with low-temperature and time-resolved Raman techniques. Key
structural features have been identified in the vibrational spectra and models
for the structure of the retinal chromophore in the primary photoproducts have
been developed.

The Raman spectrum of bathorhodopsin, the primary photoproduct in vision, can
be obtained by trapping this intermediate in a photo-stationary steady-state
at $-196°C$. The most unusual feature of this bathorhodopsin spectrum is the
presence of three intense hydrogen out-of-plane (HOOP) wagging vibrations at
854, 875 and 922 cm^{-1}. We have assigned these modes to nearly isolated wagging
vibrations of the $C_{14}H$, $C_{10}H$ and $C_{11}H$ hydrogens, respectively.[1] The unusual
intensities (nearly as strong as the C=C stretches!) can be explained by conformational distortion of the retinal chromophore at the $C_{10}-C_{11}$, $C_{12}-C_{13}$ and
$C_{14}-C_{15}$ bonds. The vibrational isolation of the $C_{11}H$ wag from the $C_{12}H$ wag and
the unusually low frequency of the $C_{12}H$ wag (830-850 cm^{-1}) demonstrate that the
$C_{11}= C_{12}$ region of the chromophore in bathorhodopsin is strongly perturbed by
the protein. The most likely type of perturbation is a negatively-charged
residue (aspartate or glutamate) placed near carbon 12. Comparison of bathorhodopsin's skeletal "fingerprint" vibrations with those of all-_trans_ model
compounds indicates that the primary step in visual excitation does involve an

---

[*]
A full text of this contribution appears in T.-Y. Liu, S. Sakakibara, A. N.
Schechter, K. Yagi, H. Yajima, and K. T. Yasunobu (Eds.) Biophysical Studies
of Macromolecules: Frontiers in Biochemical and Biophysical Studies of
Macromolecules. Elsevier-North Holland, New York, 1983.

248

11-*cis* → *trans* isomerization. However, this "trans" chromophore is strongly
perturbed by steric and electrostatic interactions with the protein.

To obtain the Raman spectrum of the primary photo-product of bacteriorhodopsin
we have developed a liquid N_2-temperature spinning sample cell.[2] A rotating
bacteriorhodopsin sample is irradiated with spatially separated pump and probe
laser beams. The essential advantages of our successive pump-and-probe technique
are the circumvention of fluorescence inference and the reduction of deleterious
photochemistry. K, the primary photoproduct, exhibits intense HOOP modes at
957 and 811 cm^{-1}, indicative of a conformationally distorted chromophore. The
vibrations in the fingerprint region of K and its isotopic analogues are consistent with a 13-*cis* chromophore configuration. Therefore, the conversion of
bacteriorhodopsin to K involves a formal all-*trans* → 13-*cis* isomerization.
However, the intense HOOP's and the unusual Schiff base frequency (1610-1623
cm^{-1}) argue that the chromophore in K interacts strongly with its protein
environment. Thus, the primary steps in rhodopsin and bacteriorhodopsin are
very analogous. Photolysis in a restrictive protein binding site produces in
both cases an isomerized retinal chromophore which cannot relax conformationally
because it has "run into" the surrounding protein residues. The intensities
and frequencies of the HOOP vibrations provide a sensitive probe of these
strained "transition state" structures.

REFERENCES

1. Eyring, et al., (1982) Biochemistry, 21, 384.
2. Braiman and Mathies, (1982) Proc. Natl. Acad. Sci., 79, 403.

Published by Elsevier Science Publishing Co., Inc.
Biochemical and Biophysical Studies of Proteins and Nucleic Acids,
Lo, Liu, and Li, eds.

STUDIES ON THE *Limulus* AMEBOCYTE COAGULATION SYSTEM

TEH-YUNG LIU AND SHU-MEI LIANG
Division of Biochemistry and Biophysics, National Center for Drugs and
Biologics, Food and Drug Administration, 8800 Rockville Pike
Bethesda, Maryland 20205

INTRODUCTION

Limulus amebocyte lysate forms a gel when exposed to minute quantities of
bacterial lipopolysaccharides.[1] This unique property has allowed the development of a sensitive assay for the detection of contaminating pyrogen in
pharmaceuticals and drugs intended for human use.[2] Our laboratory has been
investigating the molecular mechanism of the lysate gelation. We have purified
a proclotting enzyme,[3,4] a coagulogen,[5] an active enzyme[6] and, more recently,
an endotoxin-binding protein from *Limulus* amebocyte membranes[7] that possesses
calmodulin-like activity.[8] Our study is directed toward understanding the
chemical and biological principles underlying the *Limulus* coagulation system.

Limulus amebocytes are normally discoid and highly granulated. However,
minute amounts of endotoxin cause cellular degranulation and lysis, followed
by the formation of an extracellular gel.[9] No enzymatic activity related to
the clotting occurs within the amebocytes. All factors which eventually
contribute to a clot must be expelled into the blood plasma when the cells
degranulate and lyse. Activation of a zymogen by Ca^{2+} and endotoxin is an
early stage in the clotting process,[3] in spite of the fact that endotoxin
cannot enter the intact cell. Consequently, *in vivo*, endotoxin must mediate
the degranulation of the amebocytes before activation of the zymogen occurs.

This paper summarizes our studies on the molecular mechanism involved in
the *Limulus* coagulation system and present new evidence in support of the
hypothesis that adenylate cyclase may play a role in the *Limulus* cascade system.

ENZYMATIC REACTIONS

Considerable progress has been made toward understanding the basic biochemical reactions essential to the formation of a fibrin-like gel in the *Limulus*
coagulation system. *Limulus* is characterized by having a single cell type in
the blood, the granular amebocyte.[10,11] The granules contain a proclotting
enzyme and a coagulogen, factors that eventually contribute to clot formation.

Since minute amounts of endotoxin cause the degranulation of the cell
followed by the activation of the proclotting enzyme, the isolation of the

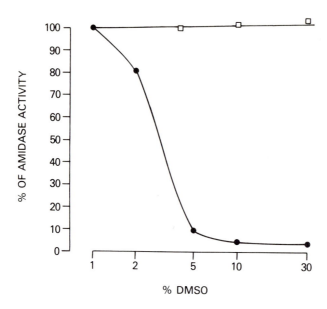

Fig. 1. Effect of dimethylsulfoxide on the endotoxin-preincubated *Limulus* amebocyte lysate. The amidase activity of the lysate was plotted versus the concentration of dimethylsulfoxide on a logarithmic scale. Endotoxin (2.5 ng/ml) was added to the lysate 1 hr before (□) or simultaneously with (●) the addition of dimethylsulfoxide.[4]

proclotting enzyme necessitates endotoxin-free conditions.[3] However, it has been found that activation of the proclotting enzyme occurs frequently during the purification procedure.[12] This endotoxin-induced activation of the pro-clotting enzyme has resulted in very poor yields or even failure in isolating the proclotting enzyme.[13] Dimethylsulfoxide was found to inhibit the endotoxin-induced clotting activity of *Limulus* amebocyte lysate.[4] The inhibition is dependent upon the concentration of dimethylsulfoxide and is competitive with endotoxin (Fig. 1). Further studies revealed that dimethylsulfoxide reversibly inhibited the activation of the proclotting enzyme to the active enzyme and that it had no effect on the active form of the clotting enzyme. Based on these findings, a procedure has been developed for the isolation of the pro-clotting enzyme from *Limulus* amebocyte in the presence of dimethylsulfoxide.

The purified proclotting enzyme has a molecular weight of about 150,000 and requires both calcium and endotoxin for inhibiting maximum activity.[3] The major active clotting enzyme has a molecular weight of about 80,000.[6] A minor

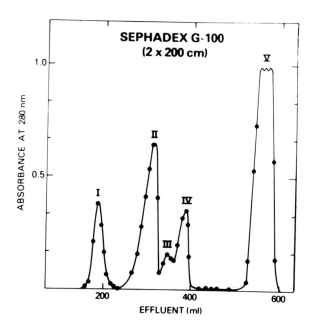

Fig. 2. Gel filtration of 30% acetic acid extract of *Limulus* lysate (100 mg) on Sephadex G-100. The column (2 x 150 cm) was equilibrated and eluted with 30% acetic acid. A_{280} absorbance of eluate at 280 nm.[5]

component with a molecular weight of about 40,000 can also be isolated with clotting activity. The amino acid compositions of the two proteins were similar indicating a "monomer-dimer" relationship but attempts to interconvert the 80,000 protein into the 40,000 protein have not been successful. Both enzymes are serine-proteases and were purified in the presence of Tween-20. Without the detergent, enzymatic activity decreased rapidly. Both enzymes are effective in causing the gelation of purified coagulogen in the presence of Ca^{2+} ion and exhibits similar specificity toward a synthetic substrate.[6]

For the isolation of the coagulogen, the amebocytes were lysed and extracted with 30% acetic acid. This step removes contaminating proteins as insoluble precipitates and preserves the coagulogen as an intact protein. The clear acid-extract was placed on a Sephadex G-100 column in 30% acetic acid to yield five fractions (Fig. 2). Only fraction II gave a positive clot when the purified *Limulus* enzyme was added. On sodium dodecylsulfate-acrylamide gel

252

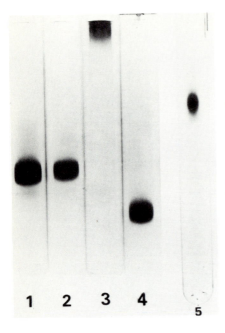

Fig. 3. Acrylamide gel (7%) electrophoresis in 0.1% sodium dodecyl sulfate,
0.1 M, pH 7.0 phosphate buffer of
coagulogen and the insoluble coagulin.
Gel 1, coagulogen from Peak II in
Fig. 2; Gel 2, reduced and carboxymethylated coagulogen; Gel 3, insoluble
coagulin; Gel 4, reduced and carboxymethylated insoluble coagulin; Gel 5,
6 M urea-acrylamide gel electrophoresis
of C-peptide in 0.1% sodium dodecyl
sulfate, 0.1 M, pH 7.2, phosphate
buffer. The acrylamide concentration
is 12.5% and the ratio of acrylamide
to methylenebisacrylamide is 10:1.[5]

electrophoresis (Fig. 3), both the native coagulogen and the reduced carboxymethylated coagulogen migrated to similar positions, suggesting that the
coagulogen is a single polypeptide chain, with a molecular weight of about
22,000. The protein is void of methionine and has a total of 18 half-cysteines,
all of which are involved in disulfide bond formation. Upon exposure to
purified clotting enzyme, the coagulogen is converted into insoluble coagulin
with a molecular weight of about 18,000 and a soluble peptide of molecular
weight of about 4,500. The two components can be separated by Sephadex G-50
gel-filtration in 30% acetic acid. The results of amino terminal analyses
and carboxyterminal analyses of the intact coagulogen and its cleavage products
are shown in Fig. 4 which summarizes our understanding of the *minimum mechanism*
involved in the *in vitro* gel formation resulting from mixing purified coagulogen

253

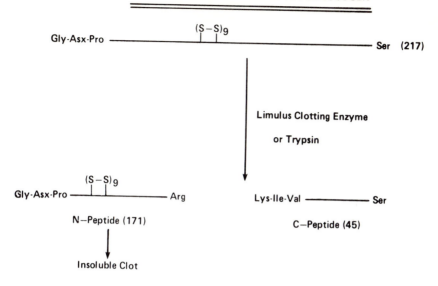

Fig. 4. Schematic representation of the clotting process of *Limulus* coagulogen.[5]

with purified clotting enzyme. The insoluble coagulin retains all of the disulfide bonds and the majority of the hydrophobic amino acid residues.[5]

DEGRANULATION PROCESSES

Limulus amebocytes aggregate and degranulate in response to bacterial endo-toxins. All of the clotting proteins released by the amebocytes are contained in numerous large granules filling the cytoplasma.[14] The cellular processes of degranulation, whereby the clotting proteins are secreted, are still obscure. Dumont et al.,[10] examined the ultrastructural changes of stimulated amebocytes and suggested that the granule membrane fused with the plasma membrane during release. Recent morphological studies of secretion in *Limulus* amebocytes by Ornberg and Reese[15] confirm this observation. They have quickly frozen secret-ing amebocytes to capture transient morphological events, showing that granule contents are released as a result of granule membrane fusion, i.e., release occurs by granule exocytosis. In addition, they find that more internal granules release by fusing with membranes of previously released peripheral granules.

TABLE 1

EFFECT OF *limulus* ENDOTOXIN BINDING PROTEIN (PROTEIN ACTIVATOR)
ON CYCLIC NUCLEOTIDE PHOSPHODIESTERASE[8]

A.	*Linulus* protein activator	Activation	Bovine brain calmodulin[a]	Activation
	μM	-fold	μM	-fold
	0	1.0	0	1.0
	7×10^{-3}	1.4	4×10^{-4}	1.6
	2×10^{-2}	2.2	1.3×10^{-3}	2.0
	7×10^{-2}	3.1	3.7×10^{-3}	2.8
	2×10^{-1}	3.8	2.4×10^{-1}	3.4
	7×10^{-1}	4.1	3.7	4.0
	2.0	4.5		

B.	Activator (0.2 μM)	EGTA (0.2 mM)	Ca^{2+} (1 mM)	Pheno-thiazine[b] (0.2 mM)	Activity	Activation
					pmol/ml/min	-fold
	−	−	−	−	97	1.0
	+	−	−	−	365	3.8
	+	−	−	+	100	1.0
	+	+	−	−	122	1.3
	+	+	+	−	347	3.6

The assay was done using 1 μM cyclic [G-^{3}H]AMP as substrate. Unless otherwise specified, the assay solution contained 50 μM $CaCl_2$. The detail of assay and the purification of protein activator from *Limulus* amoebocytes were described under "Experimental Procedures."

[a]**Sample** kindly provided by Dr. Claude B. Klee.
[b]2-Chloro-10-(3-aminopropyl)phenothiazine.

A protein that has been isolated from *Linulus* amebocyte membranes binds endotoxin.[7] The protein was purified by two independent methods, organic solvent extraction and affinity chromatography, both followed by gel filtration. Immunologic studies confirm that the protein is a component of amebocyte membranes. Although without enzymatic activity, the binding protein enhances *Limulus* lysate gelation. As a membrane associated endotoxin binding protein, it may act as an endotoxin receptor and mediate the characteristic degranulation of amebocytes exposed to endotoxin. The protein has an apparent molecular weight of 80,000 and contains trimethyllysine, a characteristic amino acid found in all calmodulin.

The *Limulus* protein is different from calmodulin isolated from other species, however, in its molecular weight (4 to 5 times greater), amino acid composition and antigenicity. Like calmodulin, the *Limulus* endotoxin-binding protein activates the hydrolysis of cyclic AMP by phosphodiesterase and this activation is Ca^{2+}-dependent[8] (Tables 1 A and B). The effect of the *Limulus* protein activator was inhibited by the addition of 2-chloro-10-(3-amino-propyl) phenothiazine (Table 1 B), a compound known to bind calmodulin.[1] Preincubation of phosphodiesterase with bovine brain calmodulin-modulating protein[16] completely abolished the activation of phosphodiesterase by the *Limulus* protein. The isolated *Limulus* protein, therefore, is mimics to calmodulin in its action on phosphodiesterase.

Calmodulin regulates many cellular events. It has also been shown to activate adenylate cyclase,[17,18] although the mode of its action has not been established; it has been difficult to relate cyclic AMP (cAMP) levels to physiological functions in intact cells, tissues, and organisms. Adenylate cyclases are activated by numerous effectors. Receptor-mediated activation of the enzymes by hormones are highly specific and are further dependent on interactions of an intracellular guanylnucleotide-binding subunit with the catalytic unit of adenylate cyclase.[19] Sodium fluoride,[20,21] guanylnucleotides[22] divalent cations such as manganese,[23] and calcium,[24] acitvate the enzyme independent of cell surface receptors, but fluoride is ineffective with intact cells, and guanyl nucleotides and divalent cations require access to intracellular sites. The hypotensive diterpene forskolin[25,26] is a potent and unique activator of adenylate cyclase in membranes from brain and other tissues. It elevates cAMP levels in intact cells of brain clices by interacting with the catalytic unit.[27]

We believe that adenylate cyclase and, hence, the level of cAMP may play a role in the *Limulus* cascade system for the following reasons:

A.   the *Limulus* endotoxin binding protein activates adenylate cyclase as effectively as rat testis calmodulin when incubated with pertussis adenylate cyclase (Table 2). A five-fold increase in enzymatic activity was observed when either protein was added to the pertussis adenylate cyclase. The *Limulus* endotoxin binding protein, by itself, had no adenylate cyclase activity.

TABLE 2

EFFECT OF *Limulus* ENDOTOXIN BINDING PROTEIN ON PERTUSIS ADENYLATE CYCLASE[28]

Components	Relative Activity (%)
Adenylate cyclase	100
Adenylate cyclase with rat testis calmodulin	519
Adenylate cyclase with endotoxin-binding protein	506
Rat testis calmodulin alone	1.89
Endotoxin-binding protein alone	0.43

TABLE 3

EFFECT OF FORSKOLIN, FLUORIDE, GUANYL NUCLEOTIDE AND *Limulus* ENDOTOXIN BINDING PROTEIN ON *Limulus* ADENYLATE CYCLASE ACTIVITY[28]

Activators	cAMP, pmol min^{-1} mg^{-1}
Barsal	0.1 $\pm$ 0.2
Forskolin	1.25 $\pm$ 0.3
NaF	1.90 $\pm$ 0.1
Guanyl nucleotide	2.75 $\pm$ 0.2
Limulus Endotoxin Binding Protein	1.65 $\pm$ 0.2

Limulus amebocytes were homogenized and adenylate cyclase was assayed with 0.2 mM ATP and a nucleotide regenerating system. Basal activities were obtained with the standard assay mixture. The concentration of forskolin, NaF, guanyl nucleotide were 100 µM, 10 mM, and 10 µM, respectively. The amount of *Limulus* Endotoxin Binding Protein present was 0.5 µM, or 4 µg per 100 λ of assay solution. Amebocyte homogenates (20 µg) were added to the reaction mixture at 0°C for 10 min, and the assay was initiated at 30° by the addition of the substrate. Incubations were carried out for 10 min. Values are means $\pm$ SEM for three determinations. When the amebocyte homogenates were preincubated with 0.1 mM of N-ethylmaleimide at 30° for 35 min, adenylate cyclase activity diminished to background level with or without subsequent treatment with activators.

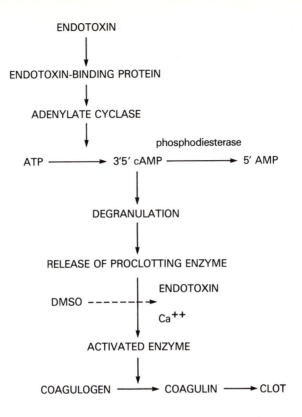

Fig. 5. Proposed mechanism of *Limulus* coagulation system.

B. When endotoxin is added to the *Limulus* amebocyte, the level of
cAMP in the cells increased 170%.

C. Forskilin, Na F, guanyl nucleotide and the *Limulus* endotoxin
binding protein enhances the adenylate cyclase activity in
Limulus amebocytes homogenates. Fifteen- to 30-fold increases
in the production of cAMP were observed (Table 3).

D. N-ethylmaleimide is a potent inhibitor of adenylate cyclase[29]
by abolishing the ability of effectors to cause the production
of cAMP in *Limulus* amebocyte homogenates. N-ethylmaleimide
prevents the degranulation of the *Limulus* amebocytes but does
not prevent the activation of the proclotting to enzyme
catalyzed by endotoxin in the *Limulus* lysate. This process
is inhibited by serine protease inhibitors such as
diisopropylfluorophosphate.

258

PROPOSED MECHANISM

The results of these experiments support the hypothesis that endotoxin is involved in two distinct processes (Fig. 5). First, it induces degranulation by binding to the receptor protein on the amebocyte membrane. The receptor protein possesses calmodulin-like activity and it, in turn, activates adenylate cyclase to increase the cAMP level in the amebocytes. The granules fused with the plasma membrane resulting in exocytosis and expulsion of clotting components. In the second step, the proclotting enzyme interacts with the receptor-bound endotoxin and converts itself into an active anzyme which acts on the coagulogen to form a clot. In Fig. 5 we summarize our proposal for a possible cellular and biochemical mechanism of the *Limulus* cascade system. The requirement of endotoxin in the second step of the reaction, namely, the activation of the proclotting enzyme to the active enzyme is the basis of *Limulus* lysate test for quantitating the amount of endotoxin.

The modulating effect of cAMP on degranulation and exocytosis in *Limulus* amebocytes is the subject of our further investigation.

REFERENCES

1. Levin, B. and Weiss, R. M. (1977) Mol. Pharmacol., 13, 690-697.
2. Cooper, J. F., Hochstein, H. D., and Seligmann, E. F., Jr. (1972) Bull. Paranter. Drug Assoc., 26, 153-162.
3. Tai, J. Y. and Liu, T.-Y. (1977) J. Biol. Chem., 252, 2178-2181.
4. Liang, S. M. and Liu, T.-Y. (1982a) Biochem. Biophys. Res. Commun., 105, 553-559.
5. Tai, J. Y., et al. (1977) J. Biol. Chem., 252, 4774-4776.
6. Seid, R. C., Jr. and Liu, T.-Y. (1980) in Frontiers in Protein Chemistry (Liu, T.-Y., Mamiya, G., and Yasunobu, K. T., eds., New York: Elsevier North-Holland Co., 1980.
7. Liang, S. M., Sakmar, T. P., and Liu, T.-Y. (1980) J. Biol. Chem., 255, 5586-5590.
8. Liang, S. M., Liang, C. M., and Liu, T.-Y. (1981) J. Biol. Chem., 256, 4968-4972.
9. Shirodar, M. W., Warwick, A., and Bang, F. B. (1960) Biol. Bull. (Woods Hole), 118, 324-337.
10. Dumont, J. N., Anderson, E., and Winner, G. (1966) J. Morphol., 119, 181-217.
11. Levin, J. and Bang, F. B. (1964) Bull. Johns Hopkins Hosp., 115, 265-274.
12. Schleef, R. R., Kenney, D. M., and Shepro, D. (1979) Thrombos. Haemostas (Stutty.), 41, 329-336.
13. Ohki, M., et al. (1980) FEBS Letters, 120, 217-220.
14. Murer, E. H., Levin, J., and Holme, R. (1975) J. Cellular Physiol., 86, 533-542.
15. Ornberg, R. L. and Reese, T. S. (1979) in Biomedical Applications of the Horshoe Crab (*Limulidae*), Cohen, E., et al., eds., New York: Alan R. Liss, Inc., pp. 125-130.
16. Cheung, W. Y. (1980) Science, 207, 19-27.
17. Cheung, W. Y., et al. (1975) Biochem. Biophys. Res. Commun., 66, 1055-1062.

18. Brostrom, C. O., et al. (1975) Proc. Natl. Acad. Sci. USA, 72, 64-68.
19. Rodbell, M. (1980) Nature (London), 284, 17-22.
20. Ralb, T. W. and Sutherland, E. W. (1958) J. Biol. Chem., 232, 1065-1076.
21. Perkins, J. P. (1973) Cyclic Nucleotide Res., 3, 1-64.
22. Londos, C., et al. (1974) Proc. Natl. Acad. Sci., USA, 71, 3087-3090.
23. Londos, C., et al. (1979) J. Supramol. Struct., 10, 31-37.
24. Bradham, L. S., Holt, D. A., and Sims, M. (1970) Biochem. Biophys. Acta, 201, 250-260.
25. Metzgar, H. and Lindner, E. (1981) Journal of Medical Science, 9, 99.
26. Daly, J. W., Padgett, W. L., and Seamon, K. B. (1981) Fed. Proc. Fed. Am. Soc. Expt. Biol., 40, 626.
27. Seamon, K. B., Padgett, W., and Daly, J. W. (1981) Proc. Natl. Acad. Sci. USA, 78, 3363-3367.
28. Liang, S. M. and Liu, T.-Y. (1982b) manuscript in preparation.
29. Ross, E. M., et al. (1978) J. Biol. Chem., 253, 6401-6412.

STRUCTURE AND FUNCTION RELATIONSHIPS OF ASPARTYL PROTEASES

JORDAN TANG
Laboratory of Protein Studies, Oklahoma Medical Research Foundation, and the
Department of Biochemistry and Molecular Biology, University of Oklahoma
Health Sciences Center, Oklahoma City, Oklahoma 73104

Aspartyl proteases are a group of pepsin-like enzymes which catalyze proteolysis predominantly in an acidic solution. Since pepsin was the first enzyme to be discovered and the second enzyme to be crystallized, it is no wonder that the study of the structure-function relationship of pepsin as an enzyme model has enjoyed a relatively long history in the development of enzymology. The term aspartyl proteases, which has substituted the name of acid proteases, originated from the knowledge in recent years that the active sites of all pepsin-like enzymes consist of two aspartyl residues. Therefore, the term aspartyl proteases is more logical since it conforms with the terminology of other groups of proteases, such as serine proteases, etc.

As compared to serine proteases, there are relatively few aspartyl proteases. Three major groups of aspartyl proteases are better studied. The gastric proteases, including pepsin, gastricsin and chymosin, have their main physiological function in the digestion of food. These proteases are secreted into stomachs as zymogens and activated in an acidic gastric juice. A large number of fungi secrete aspartyl proteases into growth media. The function of these enzymes is apparently the hydrolysis of proteins for nutrient requirements, as these microorganisms also secrete many other hydrolases such as cellulase and dextranases. The third group of aspartyl proteases are intracellular enzymes. These include lysosomal cathepsin D and E, granule-based enzyme such as proteinase A from yeast, and renin which converts angiotensionogen to angiotensin I. The enzyme properties and some structure-function relationships of these aspartyl proteases have been reviewed previously.[1-3] For the current discussion, emphasis is placed on more recent developments.

The structure of aspartyl proteases and their active center

A number of complete amino acid sequences of aspartyl proteases are now available. Pepsin from pig,[4,5] human,[6] and chicken,[7] and chymosin from cow have been sequenced.[8] The sequence of aspartyl proteases from *Penicillium janthinellum* (penicillopepsin),[9] and *Mucor miehei*[10] are also near completion. Recently, the sequence of renin has been obtained from both the protein

chemistry methods[11] and from c-DNA sequence.[12] From comparisons of these
sequences, it is clear that the primary structures of these aspartyl proteases
are homologous with one another. An example is illustrated in Fig. 1 in
which the structure of several aspartyl proteases are aligned.

There are four x-ray crystallographic structures of aspartyl proteases
which have been solved at high resolutions. Besides porcine pepsin,[13] the
crystal structures are available for three fungal enzymes: penicillopepsin,[9]
rhizopuspepsin (from *Rhizopus chinensis*),[14] and endothiapepsin (from *Endothia
parasitica*).[14] A wire model of alpha-carbon positions of penicillopepsin based
on the results of James and coworkers[9] is shown in Fig. 2A. The overall
folding of polypeptide chains in the tertiary structure of these enzymes are
very similar. The homology among these four crystal structures is much more
extensive than those observed among the primary structures of these enzymes.

The tertiary structures of aspartyl proteases contain two internally
homologous domains. The amino- and carboxyl-terminal halves of the molecules
are independently folded and highly homologous to each other.[15] This is
illustrated in Fig. 2B. Within each domain there are apparently two parts
with homology in the foldings of the polypeptide chains.[16,17] This internal
structural homology is likely to have derived from gene duplication and
fusion during the evolution of aspartyl proteases.

The substrate binding cleft is located essentially between the two lobes
in each of the four crystal structures. The cleft, which cuts across the
entire width of the molecule, is sufficient in size to bind substrate of about
seven amino acid residues. This is consistent with the known specificity of
aspartyl proteases which recognize at least seven residues around the scissile
peptide bond.[18-21] The catalytic residues, aspartyl residues 32 and 215
(pepsin numbers), are located deep in the center of the substrate binding
cleft. The catalytic function of these aspartyl residues had previously been
demonstrated by specific chemical modifications.[22-25] In the crystal structures, these aspartyl residues are pointing toward each other and are probably
hydrogen bonded. Although there are other residues located in the active
center near the aspartyl residues, such as serine-35 and tyrosine-75, their
functions are probably not catalytic.

Fig. 1. The amino acid sequence of several representative aspartyl proteases:
Porcine pepsin,[4] penicillopepsin,[9] mouse renin,[11] and porcine cathepsin D
light chain.[35] The numbering is based on the pepsin sequence. The overall
homology is apparent in the alignment. The similarity in sequences around
the active center residues D-32 and D-215 is particularly strong.

```
                                    1                 10                  20                  30                  40                  50
PORCINE PEPSIN                      I G D E P L E N Y L D T E Y F - - G T I G I G T P A Q D F T V I F D T G S S N L W V P S V Y C S - S - - L A C S D H N Q F
PENICILLOPEPSIN     A A S G V A T N T P T A N - D E E Y I T P V T I G - G T - T - - L N L N F D T G S S A D L W V F S T E L P A S - - Q Q S G H S V Y
MOUSE RENIN         T D L I S P V V L T N Y L N S Q Y Y - - G E I G I G T P P Q L F K V I F D T G S A N L W V P S T K C S - R L Y L A C G I H S L Y
PORCINE CATHEPSIN   G P I P E V L K N Y M D A Q Y Y - - G E I G I G T P P Q C F T V V F D T G S S N L W V P S I H C K - L L D I A C W I H H K Y
D LIGHT CHAIN

                    60                  70                  80                  90                  100                 110
PORCINE PEPSIN      N P D D S S T F E A T S Q E L S I T Y G T G S M - T G I L G Y D T V Q V G G I S D T N Q I F G L S E T E P G S F L Y Y A P F D
PENICILLOPEPSIN     N P S A T G K - E A S G Y T W S I S Y G D G S S A S G N V F T D S V T V G G V T A H G Q A V E A A Q Q I S A Q F Q Q D T N N D
MOUSE RENIN         E S S D S S S Y M E D G D D F T I H Y G S S G R V - K G F L S Q D S V T V G G I T V T - Q T F G E V T E L P L I P F M L A Q F D
PORCINE CATHEPSIN   N S G K S S T Y V S N G T T F A L H Y G S G S L - S G Y L S Q D T V S V P S N

                    120                 130                 140                 150                 160                 170
PORCINE PEPSIN      G I L G L A Y P S I S A S G A T P V F D N L W D Q G L V S Q D L F S V Y L S S N D D - - S G S V V L L G G I D S S Y Y T G S L
PENICILLOPEPSIN     G L L G L A F S S I N - T V Q P Q S Q T T F F D T V K S S L A Q P L F A V A L K H Q - - Q P G V Y D F G F I D S S K Y T G S L
MOUSE RENIN         G V L G M G F P A Q A V G G V T P V F D H I L S Q G V L K E K V F S V Y Y N R G P H L L G G E V V - L G G S D P E H Y Q G D F

                    180                 190                 200                 210                 220                 230            240
                                      S                                                                             N I Q            S E
PORCINE PEPSIN      N W V P V - S V E G Y W Q I T L D S I T M D G E T I A C S G G C Q A I V D T G T S L L T G P T S A I A - N I Q S D I G A - S E
PENICILLOPEPSIN     T Y T G V D N S Q G F W S F N V D S Y T A G S Q S G D G F - - S G I A D T G T T L L L L B D S V V S Q Y Y S G A Q Q D
MOUSE RENIN         H Y V S L - S K T D S W Q I T M K G V S V G S S T L L C E E G C E V V V D T G S S F I S A P T S S L K L I M Q A - L G A - K E

                    250                 260                 270                 280                 290
                                      S
PORCINE PEPSIN      N S D G E M V I S C S S I D S L P D I V F T I N G V Q Y P L S P S A Y I L Q - - D - - D D S - C T S G F E G M D V P T S S G E
PENICILLOPEPSIN     S N A G G Y V F X C S B V T B L P V S I S G Y - T A T V P G S L I N Y G P S - - G - - N G S T C L G G I Q S N - - - S G I G
MOUSE RENIN         K R L H E Y V V V C S Q V P T L P D I S F N L G G R A Y T L S S T D Y V L Q Y P N(RR) D K L - C T V A L H A M D I P P P T G P

                    300                 310                 320
PORCINE PEPSIN      L W I L G D V F I R Q Y Y T V F D R A N N K V G L A P V A
PENICILLOPEPSIN     F L I F G D I F L K S Q Y V V F D S D G P Q L G F A P Q A
MOUSE RENIN         V W V L G A T F I R K F Y T E F D R H N N R I G P A L A R
```

Fig. 1. The amino acid sequence of several representative aspartyl proteases: porcine pepsin,[4] penicillo-pepsin,[9] mouse renin,[11] and porcine cathepsin D light chain.[35] The numbering is based on the pepsin sequence. The similarity in sequences around the active center residues D-32 and D-215 is particularly strong.

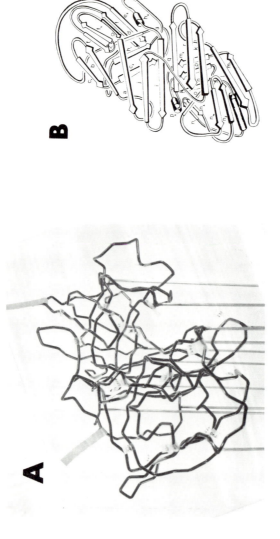

Fig. 2. (A) A wire model of penicillopepsin. The model was based on the results of Hsu, et al.[9] Only the asphacarbon positions are shown at the bend of the wire. Two attached tapes mark positions 67 and 183 which are the glycosylation sites in cathepsin D. (B) A schematic drawing of the tertiary structure of endothia-pepsin viewed along the two-fold axis of symmetry.[15]

Catalytic mechanism and substrate specificity

Although the structure of active centers of several aspartyl proteases have been determined, the catalytic mechanism is by no means clear at the present. A number of catalytic mechanisms have been proposed for pepsin so far based on kinetic,[20] chemical,[26] or crystallographic evidence.[9,27] There is nevertheless a general agreement that, like the catalytic mechanisms of serine and metallo proteases, the catalysis of aspartyl proteases should contain at least three components: a nucleophile for nucleophilic attack of the carbonyl carbon of the substrate (Fig. 3, N), a component for polarizing the carbonyl group (Fig. 3, E), and a proton donor for the leaving amide (Fig. 3, P). The nature of these components are, however, uncertain mainly due to a lack of detailed information on the orientation of substrates in the active center. Based on the fact that asparty-32 was a nucleophile in the pepsin reaction with an epoxide inactivator,[26] this residue was proposed to be the nucleophile in the peptic catalysis. However, from the model building based on the crystal structure of penicillopepsin, James and coworkers[9] concluded that aspartyl-32 was too deeply located in the substrate binding cleft to be engaged in the nucleophilic attack. It was suggested that a water molecule hydrogen bonded to one of the aspartyl residues may be the actual nucleophile. Both the aspartyl-215 and tyrosyl-75 residues have been suggested to be the proton donor, component P in Fig. 3.[26,9] From the more recent results on pepstatin binding in the active site, it appears that the primary function of tyrosyl-75 may be in the binding of substrates (see below). The elimination of tyrosyl-75 from a catalytic role would leave only two aspartyl residues in the active site to fulfill three functional components. On the other hand, the hydrogen ions are present at a relatively high concentration under the catalytic condition for aspartyl proteases, they may serve as component E or P (Fig. 3) as in a general acid catalyzed hydrolysis.

The most recent knowledge on the substrate orientation in aspartyl proteases has been derived from the analogy of the binding of pepstatin, a strong inhibitor of aspartyl proteases. It is known that the strong binding of pepstatin (with Ki = 10^{-11} for pepsin) is due to the presence in this peptide of statine residues which are transition-state analog inhibitors of aspartyl proteases.[28] The structure of statine as a transition-state analog also suggested that the carbonyl group of the substrate may be in tetrahedral conformation at the transition state of the aspartyl protease catalysis.[28] Rich and coworkers have carried out kinetic studies of inhibition of pepstatin and its synthetic analogs.[29] The pepstatin inhibition of pepsin was shown to contain at least

266

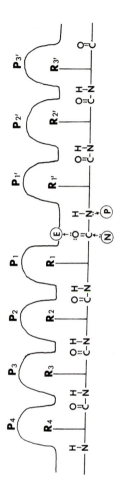

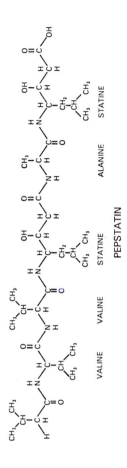

VALINE   VALINE   STATINE   VALINE   STATINE   ALANINE   STATINE

PEPSTATIN

Fig. 3. A schematic presentation of the active center of aspartyl proteases. The specificity is determined by at least seven binding loci for the side chains of the substrates: p1 to p4 on the N-terminal side and p1' to p3' on the C-terminal side of the scissiled bond. The catalysis of the peptide bond hydrolysis is carried out by at least three components at the active site: E, N, and P. The structure of pepstatin is shown below with its side chains positioned below the probable binding loci.

two steps: a fast initial step, and a slow final step.[30] It is interesting that all strong binding statine containing analogs of pepstatin exhibit this type of inhibition kinetics. But the analogs which are weaker inhibitors do not have the slow component in the inhibition kinetics. The reason for the two-step process is not clear. Among all six individual residues present in the structure of pepstatin, the statine at position 4 binds strongest to pepsin.[28] Therefore, the initial fast step could be the binding of statine at the P1 site of the enzyme (Fig. 3) which can be viewed as a "capture step." Since aspartyl proteases contain at least seven side chain-binding sites (Fig. 3), the possible combinations in the orientation of all other pepstatin side chains in these binding pockets must be large. After the initial "capture step," the time required in the second "fitting step" to reach a final lowest energy form may explain the slow step of the inhibition. However, recent results of James and coworkers[31] on the binding of pepstatin analog on penicillopepsin showed a distinct conformation change in the region of structure called "beta-flap," including residues 70-80. Schmidt, et al.[32] using proton nuclear magnetic resonance also observed proton signal changes as a result of conformation changes after the pepstatin binding to pepsin. Therefore, the slow step in pepstatin inhibition could be a consequence of combined factors.

The orientation of pepstatin in the crystal structure of rhizopuspepsin has been studied by Bott, Subramanian and Davies.[27] The binding of a pepstatin analog, a statine containing tetrapeptide isovaleryl-Val-Val-Sta-OEt, in the crystal structure of penicillopepsin has also been studied by James and coworkers.[31] There are some general similarities in the orientation of the critical statine residues in these two crystals. The hydroxyl group of statine is within hydrogen bond distance to the aspartyl-32 and -215. These results appear to support the hypothesis that statine binds as an analog of the transition state. The sidechain of statine residue is located near the equivalent of tyrosyl-75 in both crystals. This suggests that the role of tyrosyl-75 may be in the binding of the substrate in the P1 pocket (Fig. 3). Besides the statine residue, the binding of the rest of the molecules are apparently very different for the two inhibitor molecules. The reason for the difference is not clear at the present. However, it tends to argue for the hypothesis that the similarity in statine binding results from the identical "capture step" and the differences may be due to the "fitting step," which can be a consequence of the different inhibitor structures or different side chain binding pockets in the two crystals. Bott, et al.,[27] have proposed the substrate binding orientation and the hydrolytic mechanism derived from the analogy to the binding of statine.

Another convincing piece of evidence for the transition state analog
hypothesis of statine has come from the study by Rich, Bernatowicz and
Schmidt[33] who used a synthetic pepstatin analog in which the critical hydroxyl
group in statine-3 was substituted by a C^{13} labeled keto group. The C^{13} NMR
study showed that in the bound inhibitor the keto group had a tetrahedral
conformation.

The structure and function of renin

Renin is an aspartyl protease present in the plasma and intracellularly in
some organs such as kidney, submaxillary gland, and brain. The function of
renin is to produce angiotensin I from angiotensinogen by proteolysis. Renin
is inhibited by pepstatin and inactivated by active site-directed pepsin
reagents such as diazoacetyl and epoxide compounds. Therefore, renin is a
member of the aspartyl protease group even though the pH optimum of renin for
angiotensinogen is near neutrality. The substrate specificity of renin is
very stringent, it requires for its activity a strict sequence of eight amino
acids (although the species difference is known for this sequence). Therefore,
like other aspartyl proteases, renin must have also an extended substrate
binding site. The binding loci, on the other hand, must be much more strict in
the structural requirements of the substrate side chains.

The amino acid sequence of mouse submaxillary gland renin has recently been
determined by Inagami and co-workers using protein chemistry methods,[11] and
from c-DNA sequence by Panthier and co-workers.[12] The sequences are in good
agreement with minor differences which are apparently generated from the
structural microheterogeneity of the purified renin as a result of multiple
gene products. As expected, the amino acid sequence of renin is strongly
homologous to other aspartyl proteases throughout the length of the molecule
(Fig. 1). It is likely that the tertiary structure and catalytic mechanism
of renin are very similar to the other aspartyl proteases discussed above.

Lysosomal cathepsin D

Cathepsin D is a lysosomal aspartyl protease with a physiological role in
the catabolism of intracellular proteins. The structure and function relationships of this enzyme is interesting not only because it is one of the few
intracellular aspartyl proteases but also cathepsin D can be a model for the
lysosomal hydrolases, of which the biosynthesis, packaging and regulation are
active areas of current research. The best studied cathepsin D is the porcine
spleen enzyme which is about 50,000 daltons and consists of two polypeptide

chains.[34] The light chain sequence has been recently completed in our laboratory.[35] As shown in Fig. 1, this sequence is closely homologous to pepsin and even more so to renin. Within the 95-residue light chain sequence, there are 59% identical residues with renin and 49% identity with pepsin. Homology with other aspartyl proteases not shown in Fig. 1 are also extensive. These results would predict also closely homologous tertiary structures between cathepsin D and other aspartyl proteases. These enzymes are apparently derived from a divergent evolution in which the branching of cathepsin D and renin was a relatively late event.

A unique feature in the structure of cathepsin D is that it is a glycoprotein. The functions of the oligosaccharide units in lysosomal enzymes, including cathepsin D, are known to be in the receptor mediated endocytosis and in the targeting of these enzymes to be packaged in the lysosomes after their biosynthesis.[36,37] The critical structural signals for these biological functions are mannose-6-phosphate residues present in the asparagine-linked oligosaccharides.[38-44] This is a feature common to all lysosomal enzymes. Therefore, the oligosaccharide structures and their locations of attachment in the primary and tertiary structure of cathepsin D are interesting questions.

We have recently determined the amino acid sequence around the two oligosaccharide attachment sites in cathepsin D, one from each of the light and heavy chains.[45] As shown in Fig. 4, close homology between the glycopeptide sequence and the sequence of pepsin is apparent. Since cathepsin D and pepsin are likely to be closely related in their primary structures, these sequence alignments established the oligosaccharide attachment positions which correspond to pepsin residue numbers 67 and 183. The two glycopeptide sequences also show the normal glycosylation signals, in these cases Asn-X-Thr. In all four available crystal structures of aspartyl proteases, the glycosylation positions 67 and 183 are located on the surface of the molecules. An example is shown in Fig. 2A in which a wire model of penicillopepsin is shown with the corresponding glycosylation positions. It can be seen from this model, the oligosaccharide are attached in positions away from the substrate binding cleft of the enzyme.

There is considerable heterogeneity in oligosaccharide structure at each site. We found 5 oligosaccharide structures connected to position 67 in the light and 3 structures attached to position 183 in the heavy chain. As illustrated in Fig. 5, most of these oligosaccharides are derived from the high mannose-type which are known to be present on the lysosomal hydrolases.[46] However, oligosaccharide containing fucose was found to be one of the major

270

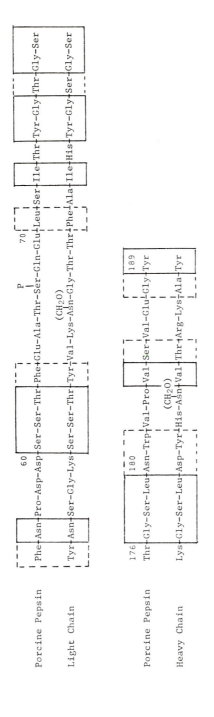

Fig. 4. The amino acid sequences near the two oligosaccharide attachment sites in cathepsin D and their homology to part of pepsin sequence. These alignments establish that the glycosylation positions in cathepsin D correspond to residue numbers 67 and 183.

271

OLIGOSACCHARIDE STRUCTURES FOUND IN THE CATHEPSIN D LIGHT CHAIN AT POSITION 67

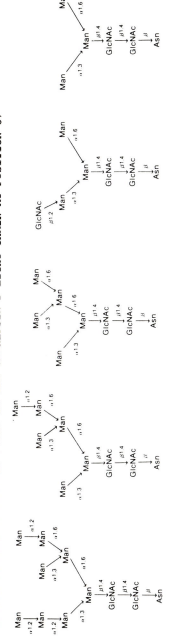

OLIGOSACCHARIDE STRUCTURES FOUND IN THE CATHEPSIN D HEAVY CHAIN AT POSITION 183

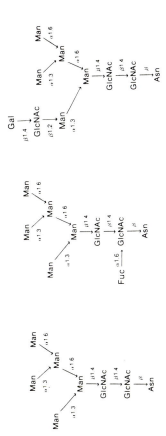

Fig. 5. Oligosaccharide structures found in cathepsin D. Five structures are attached to position 67 and three structures are connected to position 183 (see Fig. 4).

structures derived from the heavy chain.[45] Two other minor oligosaccharides from
the light chain were found to contain extra glucosamine and galactose (Fig. 5).
One possible source of the oligosaccharide heterogeneity is the difference in
oligosaccharide chain processing during the biosynthesis of cathepsin D. It is
well known that during the biosynthesis of N-linked oligosaccharides, glucoses
and mannoses are removed from the original oligosaccharide structure.[46] Different degrees of mannose removal and selective glycosylation of other sugars
during the subsequent processing could result in the structure heterogeneity.
Another possibility is that the observed oligosaccharide structures are
degradative products from the high mannose-type oligosaccharide within the
lysosomes. The functional reason for the presence of significant amounts of
the "hybrid" type of oligosaccharides is not clear at the present. It seems
probable that they serve as different recognition signals for the processing
of cathepsin D.

Aspartyl protease zymogens and their activation mechanisms

The zymogens for the aspartyl proteases from the stomach, such as pepsinogen,
progastricsin and prochymosin, are well known. All these zymogens are larger
than their respective enzymes. In the case of porcine pepsinogen, a 44-residue
activation peptide is cleaved off from the N-terminal side to produce pepsin.
The sizes and amino acid sequences of the activation peptides from all three
gastric zymogens are very similar.[47] The aspartyl proteases of the fungal
origin, such as penicillopepsin or rhizopuspepsin, are also secreted extracellular enzymes. However, no zymogen has been found for them. The amino acid
sequence of mouse prorenin has been obtained from the c-DNA sequence.[12] The
activation peptide of prorenin is also located at the N-terminal side with a
sequence highly homologous to the activation peptides from the gastric zymogens.
From a study of m-RNA directed biosynthesis of cathepsin D, Erickson, Conner
and Blobel[48] have shown that the *in vitro* synthesized protein had, in addition
to the "leader" sequence, an activation peptide region with a partial sequence
highly homologous to that of the other aspartyl protease zymogens. It is clear
that the zymogens of these aspartyl proteases are derived from a common
ancestral gene during evolution.

Although all the aspartyl protease zymogens are structurally homologous, at
least three different activation mechanisms are known for the zymogen to
protease conversions. The primary activation mechanism for the gastric zymogens
is an intramolecular process.[49-52] This mechanism has been studied in some
detail in the case of pepsinogen. After the acidification in the stomach the

273

conformation of the activation peptide region spontaneously changes to expose the active site of pepsin.[52] This is followed by the binding of the activation peptide in the active center which catalyzes the cleavage of a peptide bond at position 16-17 of pepsinogen.[53] The removal of the rest of the activation peptide up to position 44-45 is probably accomplished by the digestion of another pepsin molecule. This unique intramolecular mechanism is favorable under the physiological conditions.[50,52] A second activation mechanism for the gastric zymogens is the activation by the proteases they produce, i.e., the activation of pepsinogen by pepsin. This mechanism favors somewhat less acidic conditions and thus may not be physiologically very important.[50] A third activation mechanism is that of the activation of prorenin. The sequence of the prorenin shows that the cleavage site for zymogen activation and the cleavage site between the heavy and the light chains are in each case preceded by an Arg-Arg sequence.[12] Since this dipeptide sequence is the familiar recognition signal for the activation cleavage of many peptide hormone precursors, the cleavage of these sites in prorenin can be expected to be carried out *in vivo* by a specialized protease. The mechanism for the activation of procathepsin D is unknown.

ACKNOWLEDGEMENT

I would like to thank Dr. Robert Delaney for helpful suggestions in the preparation of this manuscript. The studies described from our laboratory were supported by Research Grants GM-20212 and AM-01107 from the National Institute of Health.

REFERENCES

1. Tang, J. (1976) Trends in Biochemical Sciences 1, 205-208.
2. Tang, J. (1977) Acid Proteases, Structure, Function and Biology, New York: Plenum Press.
3. Tang, J. (1979) Mol. Cell. Biochem., 26, 93-109.
4. Tang, J., et al. (1973) Proc. Nat. Acad. Sci. USA, 70, 3437-3439 (1973)
5. Morávek, L. and Kostka, V. (1974) FEBS Letters, 43, 207-211
6. Sepulveda, P. and Tang, J. Unpublished results.
7. Kostka, V., Keibvá, H., and Baudyś, M. (1981) in Proteinases and Their Inhibitors, Turk, V., and Vitale, L. J., eds., Oxford: Pergamon Press, pp. 125-130.
8. Foltmann, B., et al. (1977) Proc. Natl. Acad. Sci. USA, 74, 2321-2324.
9. Hsu, I.-N., et al. (1977) Nature, 266, 140-145.
10. Bech, A.-M. and Foltmann, B. (1981) Neth. Milk Dairy J., 35, 275-280; and and (1983) Foltmann, B., Personal communication.
11. Misono, K. S., Chang, J.-J., and Inagami, T. (1982) Proc. Natl. Acad. Sci. USA, 79, 4858-4862.
12. Panthier, J.-J., et al. (1982) Nature, 298, 90-92.
13. Andreeva, N., et al. (1978) J. Mol. Biol. (Russian), 12, 922-927.

14. Subramanian, E., et al. (1977) Proc. Nat. Acad. Sci. USA, 74, 556-559.
15. Tang, J., et al. (1978) Nature, 271, 618-621.
16. Andreeva, N. S. and Gustchina, A. E. (1979) Biochem. Biophys. Res. Commun., 87, 32-42.
17. Blundell, T. L., Sewell, B. T., and McLachlan, A. D. (1979) Biochim. Biophys. Acta, 580, 24-31.
18. Tang, J. (1963) Nature, 199, 1094-1095 (1963).
19. Powers, J. C., Harley, A. D., and Myers, D. V. (1977) in Acid Proteases, Structure, Function, and Biology, Tang, J., ed., New York: Plenum Press, pp. 141-157.
20. Fruton, J. S. (1976) Adv. Enzymol., 44, 1-36.
21. Morihara, K. (1974) Adv. Enzymol., 41, 179-243.
22. Rajagopolan, T. G., Stein, W. J., and Moore, S. (1966) J. Biol. Chem., 241, 4295-4297.
23. Bayliss, R. S., Knowles, V. R., and Wybrandt, G. B. (1966) Biochem. J., 113, 377-386.
24. Tang, J. (1971) J. Biol. Chem., 246, 4510-4517.
25. Chen, K. C. S. and Tang, J. (1972) J. Biol. Chem., 247, 2566-2574.
26. Hartsuck, J. A. and Tang, J. (1972) J. Biol. Chem., 242, 2575-2580.
27. Bott, R., Subramanian, E., and Davies, D. R. (1982) Biochemistry, 21, 6956.
28. Marciniszyn, J., Jr., Hartsuck, J. A., and Tang, J. (1976) J. Biol. Chem., 251, 7088-7094.
29. Rich, D. H. and Sun, E. T. (1977) in Peptides - Proceedings of the Fifth American Peptide Symposium, Goodman, M. and Meinhofer, J., eds., John Wiley, pp. 209-212.
30. Rich, D. H. and Sun, E. T. O. (1980) Biochem. Pharmacol., 129, 2205-2212.
31. James, M. N. G., et al. (1982) Proc. Natl. Acad. Sci. USA, 79, 6137-6141.
32. Schmidt, P. G., Bernatowicz, M. S., and Rich, D. H. (1982) Biochemistry, 21, 1830-1835.
33. Rich, D. H., Bernatowicz, M. S., and Schmidt, P. G. (1982) J. Amer. Chem. Soc., 104, 3535-3536.
34. Huang, J. S., Huang, S. S., and Tang, J. (1979) J. Biol. Chem., 254, 11405-11417.
35. Takahashi, T. and Tang, J. (1983) J. Biol. Chem., in press.
36. Neufeld, E. F. and Ashwell, G. (1980) in The Biochemistry of Glycoproteins and Proteoglycans, Lennarz, W. J., ed., New York: Plenum Press, pp. 252-257.
37. Sly, W. S. and Fischer, H. D. (1982) J. Cell. Biochem., 18, 67-85.
38. Hickman, S., Shapiro, L. J., and Neufeld, E. F. (1974) Biochem. Biophys. Res. Commun., 57, 55-61.
39. Kaplan, A., Achord, D. T., and Sly, W. S. (1977) Proc. Natl. Acad. Sci. USA, 74, 2026-2030.
40. Distler, J., et al. (1979) Proc. Natl. Acad. Sci. USA, 76, 4235-4239.
41. Natowicz, M. R., et al. (1979) Proc. Natl. Acad. Sci. USA, 76, 4322-4326.
42. Tabas, I. and Kornfeld, S. (1980) J. Biol. Chem., 255, 6633-6639.
43. Varki, A. and Kornfeld, S. (1980) J. Biol. Chem., 255, 10847-10858.
44. Hasilik, A., et al. (1980) Proc. Natl. Acad. Sci. USA, 77, 7074-7078.
45. Takahashi, T. and Tang, J. (1983) J. Biol. Chem., in press.
46. Kornfeld, R. and Kornfeld, S. (1980) in The Biochemistry of Glycoproteins and Proteoglycans, Lennarz, W. J., ed., New York: Plenum Press, pp. 1-34.
47. Foltzmann, B. and Jensen, A. L. (1982) Eur. J. Biochem., 128, 63-70.
48. Erickson, A. H., Conner, G. E., and Blobel, G. (1981) J. Biol. Chem., 256, 11224-11231.
49. Bustin, M. and Conway-Jacobs, A. (1971) J. Biol. Chem., 246, 615-620.
50. Al-Janabi, J., Hartsuck, J. A., and Tang, J. (1972) J. Biol. Chem., 247, 4628-4632.
51. McPhie, P. (1972) J. Biol. Chem., 247, 4277-4281.
52. Marciniszym, J., Jr., et al. (1976) J. Biol. Chem., 251, 7095-7102.
53. Kay, J. and Dykes, C. W. (1976) Biochem. J., 157, 499-502.

STRUCTURE AND FUNCTION OF PROTEINASE INHIBITORS FROM SNAKE VENOM

CHEN-SHENG LIU AND TUNG-BIN LO
Institute of Biological Chemistry, Academia Sinica and
Institute of Biochemical Sciences, National Taiwan University,
Taipei, Taiwan, R.O.C.

ABSTRACT

Many varieties of protein inhibitors of proteinases are present extensively in nature. From snake venoms, nine proteinase inhibitors or related peptides have been chemically characterized. They all belong to single-headed, Kunitz-type inhibitors and consist of 57 to 60 amino acid residues with invariant 6 half-cystines. The so-called reactive site residue, P_1, is occupied by either lysine (5 cases), arginine (one case), tyrosine (two cases) or methionine (one case). The inhibition towards bovine trypsin and/or chymotrypsin ranges from none to strong actions.

Recently we have isolated two analogous proteinase inhibitors, VIIIb and IX from the venom of *Bungarus fasciatus*. The complete amino acid sequences have been determined by microsequencing technique. Fraction IX consists of 65 amino acid residues and VIIIb consists of 62 residues which is identical to the N-terminal 62 amino acid sequence of IX. About 60% homology could be observed between IX and inhibitor II of *Vipera russelli*. For the determination of the location of three disulfide bridges, the intact molecule of IX was subjected to thermolysin digestion, and the peptides were separated by thin layer peptide mapping technique. The cystine-containing peptides (nitroprusside positive) were isolated and directly sequenced by DABITC/PITC double coupling method. The location of disulfide bridges were found to be: Cys^7-Cys^{57}, Cys^{16}-Cys^{40} and Cys^{32}-Cys^{53}. The results coincide with those present in inhibitor II of *Vipera russelli*.

These two analogous proteinase inhibitors were shown to be chymotrypsin inhibitor and not trypsin inhibitor. The P_1 residue is Asn. The preliminary study indicated the dissociation constant of enzyme-inhibitor complex, Ki, is 3.0×10^{-7} and 1.9×10^{-7} M, respectively, for VIIIb and IX.

INTRODUCTION

Many varieties of protein inhibitors of proteinases are present extensively in nature as illustrated in Table 1 and Table 2.[1] Noteworthy in Table 1 are two facts that bromelain originally present in pineapple stem is inhibited by a protein also in pineapple stem, and *Ascaris*, a large intestine parasite,

TABLE 1

SOME INHIBITORS OF THIOL, CARBOXYL, AND METALLO PROTEINASES

Enzyme inhibited	Source of inhibitor
Thiol proteinase	
Papain, ficin	Bauhinia seeds
Bromelain	Pineapple stem
Papain	Rabbit skin
Carboxyl proteinase	
Pepsin	*Ascaris lumbriocoides*
Cathepsin D	Potato
Pepsin	Bauhinia seeds
Metallo proteinase	
Carboxypeptidase A	*Ascaris lumbriocoides*
Collagenase (human gastric mucosal)	β_1 serum protein
Aminopeptidase	*Neurospora crassa*

produces at least two different inhibitors respectively against pepsin and
carboxypeptidase A. The first case is indicative of an axiom, where there is
a proteolytic enzyme, there is an inhibitor.[2] The second case suggests one of
the physiological functions of proteinase inhibitor acts as a defense mechanism
and may signify the main role on the defensive side of the known proteinase
inhibitors.[1] As serine proteinases are easily available in most laboratories,
it is no wonder that the most studied group encompasses at least seven families
as shown in Table 2. Some of the typical protein structures of the serine
proteinase inhibitors are illustrated in topological way in Fig. 1. They
manifest clearly the multiple varieties in the primary structure.

Snake venom is one of the rich sources of proteinase inhibitors and nine
proteinase inhibitors or related peptides have been chemically characterized,
i.e., inhibitor II of *Vipera russelli*, inhibitor II of *Naja nivea*, inhibitor II
of *Hemachatus haemachatus*, toxins B, E, I and K of *Dendroaspis polylepsis polylepsis*, and toxins of $C_{13}S_1C_3$ and $C_{13}S_2C_3$ of *Dendroaspis angusticeps*.[3-5] In
this paper we report the primary structure as well as the preliminary kinetic
study on the two analogous proteinase inhibitors obtained from the venom of
Bungarus fasciatus.

TABLE 2

SOME INHIBITORS OF SERINE PROTEINASES

Families of protein inhibitors	Source of inhibitor
A. Pancreatic trypsin inhibitor (Kunitz) family	1. bovine pancreas
	2. red sea turtle egg white
	3. snake venom
B. Pancreatic secretory trypsin inhibitor (Kazal) family	1. bovine pancreas
	2. human pancreas
	3. chicken ovoinhibitor
C. *Streptomyces* subtilisin inhibitor family	1. *Streptomyces albogriseolus*
	2. *Streptomyces antifibrinolyticus*
D. Bowman-Birk inhibitor family	1. lima bean
	2. soy bean
	3. adzuki bean
E. Soybean trypsin inhibitor (Kunitz) family	1. soybeans
	2. winged bean
F. Potato inhibitor family	1. var. Russet Burbank
	2. var. Dan Shaku Imo
G. Alpha-I-proteinase inhibitor family	1. human plasma

MATERIALS AND METHODS

Isolation and purification of the two proteinase inhibitors, VIIIb and IX, from crude venom of *Bungarus fasciatus* (Lot No. BF 10 STL, Miami Serpentarium Laboratories, U.S.A.) were carried out as described earlier.[6] The amino acid sequences have recently been determined in our laboratory as follows.[7] Reduced and carboxymethylated inhibitors were digested with trypsin, and tryptic peptides were isolated by peptide mapping techniques on cellulose thin-layer plate (Polygram Cel 300 of Macherey Nagel, West Germany).[8] Amino acid sequences were determined by DABITC/PITC double coupling method.[9] Finally, alignment of all tryptic peptides were established by the analyses of chymotryptic peptides and *Staphylococcus aureus* V8 protease digested peptides.

For the location of the three disulfide bonds in the proteinase inhibitor, intact molecule of IX was subjected to thermolysin digestion at pH 6.5. Cystine-containing peptides on the peptide map were revealed by nitroprusside test.[10] After amino acid determination of the corresponding peptides, the three Cys- peptides were subjected to the direct sequence analyses using double coupling method as mentioned above.

Topological structures of proteinase inhibitors

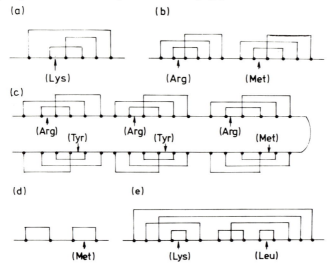

Fig. 1. Topological structures of proteinase inhibitors, (a) bovine pancrease; (b) canine submandibular glands; (c) avian egg white ovoinhibitor; (d) *Streptomyces albogriseolus*; (e) lima bean. Solid circle and arrow indicate, respectively, a half cystine residue and a reactive site. The amino acid in the parenthesis corresponds to P_1 residue.

As for the preliminary kinetic study, the inhibition parameter, Ki, was determined by the Lineweaver-Burk plotting. Inhibitor assay method according to Schwert and Takenaka[11] was essentially followed in the way described by Strydom.[12] Briefly, ATEE (N-acetyl-L-tyrosine ethyl ester) was prepared at the concentration of 0.5 mM to 3 mM in 0.02 M $CaCl_2$, 0.05 M tris/HCl buffer, pH 8.4. α-Chymotrypsin was first dissolved in 0.001 N HCl to make a 0.5 mg/ml solution. A solution containing 3 ml of substrate solution, 0.15 ml buffer and 0.05 ml enzyme solution was used for the enzyme activity, while a mixture of 3 ml of substrate solution and the preincubated solution containing 0.1 ml inhibitor solution, 0.05 ml buffer and 0.05 ml enzyme solution was used for the inhibitor assay. The absorbance was measured at 240 nm with 10 sec interval. The reciprocal of initial velocity (Vo) versus reciprocal of the substrate concentration (So) was plotted in the usual way.

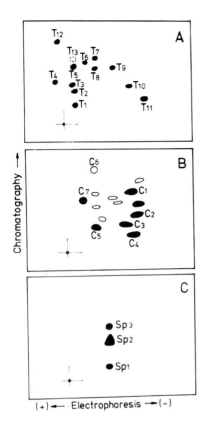

Fig. 2. Peptide maps of RCM-IX on cellulose t.l.c. plates:
(A) tryptic peptides; (B) chymotryptic peptides; (C) peptides obtained by S. aureus V$_8$ protease digest.

RESULTS AND DISCUSSION

The peptide maps of tryptic, chymotryptic and *Staphylococcus aureus* V8 protease of RCM IX are shown in Fig. 2, respectively, as A, B, and C. RCM VIIIb gave almost identical tryptic peptide map to RCM IX except spots T_3 and T_9 were missing and a new T_{13} (dotted spot in Fig. 2A) appeared. The complete amino acid sequence of inhibitor IX is shown in Fig. 3. IX consists of 65 amino acid residues while VIIIb is identical to the N-terminal 62 amino acid sequence of IX.

When the amino acid sequence of present proteinase inhibitor is compared with other known proteinase inhibitors from snake venoms, it is readily seen that 6 half-cystine positions are kept invariable, and IX is most similar to *Vipera russelli* inhibitor, R VV-II.[13] About 60% homology (36/60) could be observed and most of the amino acid substitutions (18 out of 24), can be

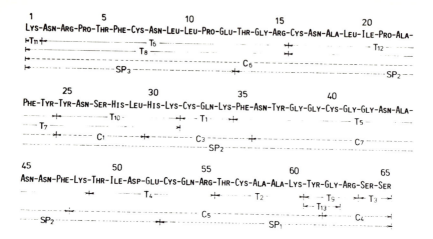

Fig. 3. Amino acid sequence of proteinase inhibitor IX.

attributed to one base replacement in their codons as shown in Fig. 4.[7] The
peptide map of thermolysin digested IX is shown in Fig. 5. Nitroprusside
positive peptides 1, 2, and 3 were recovered and sequenced directly after
amino acid analysis. For peptide 1 (Arg_1, Asx_1, Gly_4, $1/2Cys_2$), the first step
showed up only DABTH-Gly, the second, DABTH-Gly and DABTH-Arg and the third,
DABTH-Cys-Cys-DABTH. For peptide 2 (Asx_1, Thr_1, $1/2Cys_2$, Phe_1), the first
step revealed the presence of DABTH-Phe and DABTH-Thr, the second, DABTH-Cys-
Cys-DABTH and the third, DABTH-Asn. For peptide 3 (Lys_1, His_1, Asx_1, Glx_1,
$1/2Cys_2$, Ile_1, Leu_1), the first step resulted in DABTH-Leu/Ile, the second
DABTH-His and DABTH-Asp and the third, DABTH-Lys and DABTH-Glu. From those
results and the sequence shown in Fig. 3 the following linkages were substan-
tiated: Cys^7-Cys^{57}, Cys^{16}-Cys^{40} and Cys^{32}-Cys^{53} as shown in Fig. 6. They
were identical to those obtained in Russel's viper inhibitor.[13]

Our previous study showed that proteinase inhibitors of VIIIb and IX acted
against chymotrypsin but not to trypsin.[6] In proteinase inhibitors the reactive
site residue (P_1) generally corresponds to the specificity of the cognate
enzyme, i.e., inhibitors with P_1 Lys and Arg tend to inhibit trypsin and those
with P_1 Tyr, Phe, Trp, Leu and Met inhibit chymotrypsin. Table 3 shows the
amino acid residues in the sequences surrounding the reactive site of some
inhibitors from five snake venoms. It should be noted, however, that the
inhibitor with the reactive site residue corresponding to the enzyme specificity
does not necessarily act as a real inhibitor. Thus, toxin K (P_1Lys) and toxin I

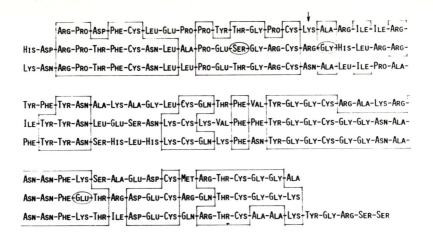

Fig. 4. Comparison of amino acid sequence of (from the top line) bovine pancreatic trypsin inhibitor, trypsin inhibitor of *Vipera russelli* and chymotrypsin inhibitor of *Bungarus fasciatus*. The sequences are aligned with respect to the position of the half cystine residues and the position of the invariant amino acids are boxed. The circles in boxed regions indicate variant amino acid residues. An arrow indicates the reactive site residue (P$_1$) proposed by Laskowski and Kato.[1]

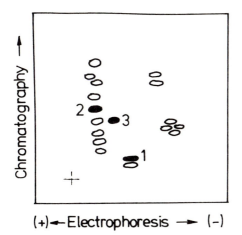

Fig. 5. Peptide map of thermolysin digest of IX. Three cystine-containing spots are marked as solid spots. Electrophoresis was carried out in pyridine:acetic acid:acetone:water (2:4:15:73, by volume), pH 4.5 first, and then ascending chromatography in pyridine:butanol:acetic acid:water (10:15:3:12, by volume).

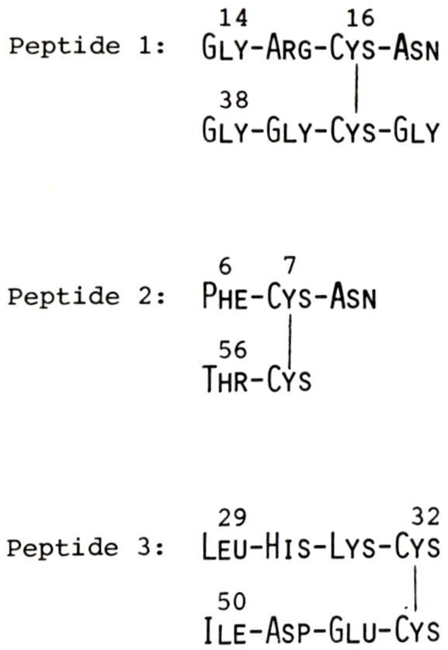

Peptide 1:
$$
\begin{array}{cc}
14 & 16 \\
\text{GLY-ARG-CYS-ASN} \\
\end{array}
$$

38
GLY-GLY-CYS-GLY

Peptide 2:
6 7
PHE-CYS-ASN

56
THR-CYS

Fig. 6. Amino acid sequences of cystine containing peptides.

Peptide 3:
29 32
LEU-HIS-LYS-CYS

50
ILE-ASP-GLU-CYS

(P_1Tyr) do not inhibit trypsin and chymotrypsin, respectively, but toxin I (P_1Lys) is a strong inhibitor of trypsin.[1] P_1 residue of the present inhibitors is asparagine (Asn[17]). It is rather unusual, however, Joubert and Strydom[14] reported that chymotrypsin slowly split Asn[45]-Arg[46] bond in trypsin inhibitor E of *D. polylepsis polylepsis* (Black mamba) venom, so it may suggest Asn residue possibly becomes targets of chymotrypsin in some substrates.

Lineweaver-Burk plots of proteinase inhibitors, VIIIb and IX are shown in Fig. 7 and Fig. 8, respectively. Clearly, both of them belong to the competitive

TABLE 3

AMINO ACID RESIDUES IN THE SEQUENCES SURROUNDING THE REACTIVE SITE OF INHIBITORS

	P_4	P_3	P_2	P_1	P_1'	P_2'	P_3'	P_4'	P_5'	P_6'
Snake venoms										
H. haemachatus	G	L	C	K	A	Y	I	R	S	F
Naja nivea	G	L	C	K	A	R	I	R	S	F
Black mamba (E)	G	P	C	K	A	S	I	P	A	F
Black mamba (K)	G	P	C	K	R	K	I	P	S	F
Black mamba (I)	G	R	C	Y	Q	K	I	P	A	F
Vipera russelli	G	R	C	R	G	H	L	R	R	I
B. fasciatus	G	R	C	N	A	L	I	P	A	F

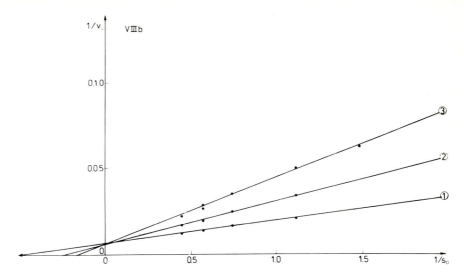

Fig. 7. Lineweaver-Burk plot of inhibitory action of VIIIb. (1) control;
(2) enzyme/inhibitor molar ratio is 1:1;   (3) enzyme/inhibitor molar ratio is
1:2.

inhibitors. The dissociation constant of enzyme inhibitor complex, Ki, is
3.0×10^{-7} M and 1.9×10^{-7} M, respectively, for VIIIb and IX. In contrast, Ki
of Russel's viper inhibitor is in the order of 10^{-10} M which signifies quite
strong inhibitory action.[15] Therefore, the present proteinase inhibitors are
rather weak. It may be due to the presence of rather nonspecific amino acid
residue, Asn, at the reactive site.

   The physiological function of proteinase inhibitors of snake venoms still
remains obscure. The relatively frequent occurrence may indicate that they
play some role. It is interesting that Strydom[12] observed a synergistic
effect in toxicity by mixing test of *gusticep*-type protein FS$_2$ and proteinase
inhibitor I from *D. polylepsis polylepsis* venom.

ACKNOWLEDGEMENTS

   Thanks are due to Messrs. Tse-Chong Wu and Wen-Juin Hsieh for their technical
assistances in structure studies and also to Mr. Chi-Ching Chen for his assistance in preliminary kinetic studies. This work was supported in part by
grant (NSC71-0203-B001-03) from the National Science Council, Taipei, ROC.

284

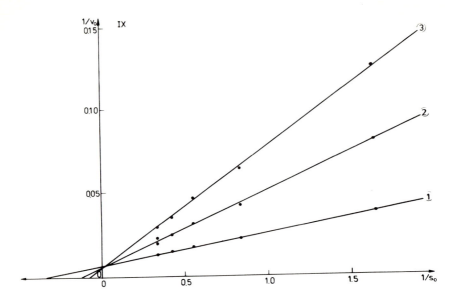

Fig. 8. Lineweaver-Burk plot of inhibitory action of IX. (1) control;
(2) enzyme/inhibitor molar ratio is 1:1; (3) enzyme/inhibitor ratio is 1:2.

REFERENCES

1. Laskowski, Jr., M and Kato, I. (1980) Ann. Rev. Biochem., 49, 593-626.
2. Kato, I. (1979) Protein, Nucleic Acid and Enzyme (in Japanese), 24, 667-679.
3. Tu, A. (1977) in Venoms: Chemistry and Molecular Biology, New York: John
   Wiley, pp. 139-147.
4. Joubert, F. J. and Taljaard, N. (1980) Hoppe-Seyler's Z. Physiol. Chem.,
   361, 661-674.
5. Strydom, D. J. and Joubert, F. J. (1981) Hoppe-Seyler's Z. Physiol. Chem.,
   362, 1377-1384.
6. Liu, C. S., et al. (1981) Proc. Natl. Sci. Counc. B. ROC, 5, 355-360.
7. Liu, C. S., Wu, T. C., and Lo, T. B. (1983) Int. J. Peptide Protein Res.,
   (in press).
8. Gracy, R. W. (1977) Methods Enzymol., 47, 195-204.
9. Chang, J. Y., Brauer, D., and Wittmann-Liebold, B. (1978) FEBS Lett., 93,
   205-214.
10. Easley, C. W. (1965) Biochim. Biophys. Acta, 107, 386-388.
11. Schwert, G. W. and Takenaka, Y. (1955) Biochim. Biophys. Acta, 16, 570-575.
12. Strydom, D. J. (1976) Eur. J. Biochem., 69, 169-176.
13. Takahashi, H., et al. (1974) J. Biochem., 76, 721-733.
14. Joubert, F. J. and Strydom, D. J. (1978) Eur. J. Biochem., 87, 191-198.
15. Takahashi, H., Iwanaga, S., and Suzuki, T. (1974) J. Biochem., 76, 709-719.

Biochemical and Biophysical Studies of Proteins and Nucleic Acids,
Lo, Liu, and Li, eds.

THE EDMAN DEGRADATION IN PROTEIN SEQUENCE ANALYSIS

ROBERT L. HEINRIKSON
Department of Biochemistry, The University of Chicago,
Chicago, Illinois 60637

"While careful colleagues fear the rational study of proteins will
encounter at the present time insuperable difficulties because of
the complex composition and physical properties of these substances,
other optimistically disposed observers (among whom I like to call
myself) incline to the view that one should at least attempt with
all current means available to lay seige on that impregnable fortress;
it is only through bold measures that the limit of capability of our
methods can be assessed."

E. Fischer, 1906

In this little gem taken from a talk delivered 77 years ago to the German
Chemical Society,[1] the great chemist and father of biochemistry identified with
prophetic insight two crucial conditions for furthering our understanding of
biology at the molecular level. One was that more physical and organic
chemists get involved in studies of biological systems. After all, one of the
definitions of organic chemistry, i.e., that it is "the chemistry of carbon
compounds," certainly applies to biochemistry as well. Nevertheless, students
of the "purer" forms of chemical inquiry were reluctant to engage in such
work and this attitude, though tempered somewhat in recent years, still
prevails today to a large degree. Fortunately, an "immortal" few have taken
on the challenge of chemical complexity afforded by living forms and it is
their contributions which have brought us to this new age of molecular biology.

The second observation of Fischer, really related indirectly to the first,
had to do with the importance of developing new methodologies to cope with
these seemingly unassailable problems. Of course, the history of science is
replete with examples of new breakthroughs brought about by the application
of new methods. Such occasions often opened whole new fields of inquiry.
Certainly from Fischer's perspective in 1906, our understanding of biology in
chemical terms would progress little without the development of methods equal
to the new challenge, and this development would be, in turn, dependent upon
the participation of organic and physical chemists.

Fischer's prophecy, of course, turned be right on the mark; our contemporary understanding of the molecular architectures of biological macromolecules is a consequence of the application of physical and chemical tools and

the design of new methods. Because the subject of this Chapter begins in
1950, no mention will be made of the important advances in physical biochemistry,
the purification and characterization of biopolymers, enzyme kinetics, and the
study of metabolism which, taken together, comprised the major efforts of
biochemistry in the first half of this century. One development, however, was
of such fundamental significance as to place in a class by itself, for without
it we would never have been able to attain our present level of chemical
sophistication in molecular biology. This was, of course, the widespread
application from the 1940s to the present of chromatographic methods for purifying and characterizing compounds of biochemical interest. Column chromatography was first described in 1906 by the Russian chemist Mikhail Tswett who
resolved the pigments in a petroleum ether extract of spinach leaves on calcium
carbonate. Ironically this date coincides with that of Fischer's talk! This
method, which would provide the arsenal for attack upon the "impregnable
fortress" of protein structure was even then at hand! However, it was nearly
40 years before the idea was resurrected by two English chemists, Martin and
Synge[2] who pioneered development of separations by partition chromatography on
paper. Thereafter followed an explosion in all aspects of separation chemistry
resulting in the development of partition, ion-exchange, and adsorption procedures for purifying proteins, amino acids, peptides, nucleic acids and their
building blocks, lipids, and carbohydrates. The stage was now set for the new
age of structural biochemistry.

In 1950, a Swedish chemist, Pehr Edman, published a series of reactions by
means of which amino acids may be removed one at a time from the amino terminus
of a protein or peptide.[3] These reactions, which have come to be known as the
Edman degradation,[4] constitute a truly remarkable contribution to protein
chemistry. Almost from the beginnings of this discipline in the 1950s right
up to the present time, the Edman procedure has been, and continues to be,
the cornerstone of protein sequence analysis. In fact, the strategies for
determining the covalent structures of proteins have been dictated, by and
large, by the prevailing state-of-the-art as regards application of the Edman
degradation. The present chapter will consider within an historical framework
this dependence of sequencing strategies on the Edman degradative technology.
Further, the current state of primary structural analysis in the 1980s will
be evaluated in light of recent developments in automation of the procedure,
high performance liquid chromatography, and nucleic acid sequencing.

The author was privileged to have been associated as a post-doctoral fellow
with two of the founders of the field of protein chemistry, Stanford Moore and

William H. Stein. These organic chemists pioneered development of chromatographic methods for protein purification and separation of amino acids which
established a quantitative foundation in protein analysis for all who followed.
In the course of their now classic structural analysis of the first enzyme,
ribonuclease, they became the first to prescribe a general strategy that could
be followed for sequencing large proteins. This strategy had at its heart the
Edman degradation. The greatness of these men in science was matched, if not
exceeded, by their humility, honesty and unswerving devotion and commitment to
their field and to their colleagues and young coworkers. Dr. Stein died in
1980 and Dr. Moore passed away this year only a few days after this Conference
in Taipei. It is to their fond memory that the present chapter is humbly
dedicated.

THE EDMAN DEGRADATION

Before discussing the interdependence of sequencing strategies and the
Edman procedure, it may be worthwhile to say a few words about the reaction
sequence *per se*. So much has been written about the Edman degradation that it
would be almost impossible (and for the purpose of this article, inappropriate)
to cite all of the important contributions in this regard. Rather, I will
attempt to summarize the wide body of relevant literature as it pertains to
each aspect of the degradative cycle. The "3 Cs" of the Edman degradation,
coupling, cleavage, and conversion, are outlined in Fig. 1.

Coupling. The alpha- and epsilon-amino groups of proteins react efficiently
with isothiocyanates at about pH 10. Edman's choice of phenylisothiocyanate[3,4]
was predicated on a similar series of reactions, first devised by Bergmann and
Meckeley[5] and developed further by Abderhalden and Brockmann[6] in which phenylisocyanate was the coupling reagent. Problems with reagent and by-product
insolubility and the vigorous acidic conditions required for cyclization and
cleavage of the phenylcarbamyl peptide which led, in turn, to undesirable
splitting of peptide bonds prevented any widespread practical application of
this method prior to Edman's innovation. Edman showed that when the oxygen is
replaced by the more nucleophilic sulfur atom, attack on the carbonyl carbon in
the second step leading to cyclization and cleavage is promoted under conditions
in which peptide bonds are highly stable.

A number of isothiocyanates have been employed for a variety of reasons.
Use of sulfonated derivatives[7] provides insertion of a strong negative charge
at all epsilon-amino groups which increases the polarity of the peptide and
thereby minimizes losses during extraction with organic solvents. A new

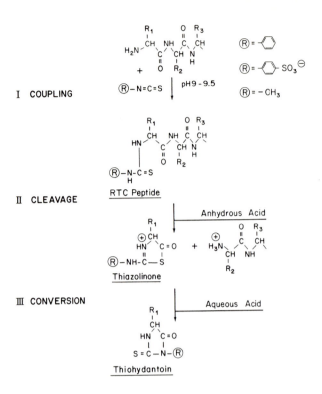

Fig. 1. The reactions of the Edman degradation. Although the isothiocyanate may vary considerably in structure, the most commonly employed reagent is that originally described by Edman, phenylisothiocyanate. Trifluoroacetic acid and heptafluorobutyric acid have been used extensively for the cleavage step of the procedure.

reagent, 4-N,N-dimethylaminoazobenzene-4-isothiocyanate (DABITC) provides a colorimetric method of high sensitivity for manual sequencing.[8] Methyliso-thiocyanate has been used in conjunction with mass spectrometric methods for analysis of the resulting methylthiohydanotin derivatives[9] in Step 3 (Fig. 1). Peptides terminating in lysine residues have been coupled at both ends to insoluble supports bearing phenylisothiocyanate groups. Subsequent exposure to anhydrous acid in Step 2 (Fig. 1) leads to cleavage of the peptide from the support at the N-terminus only, so that solid phase Edman degradation can now be carried out efficiently.[10] Clearly, there are as many possible variations in the coupling chemistry as the imagination can conceive. Similarly, a wide variety of solvents have been employed for the coupling reaction;[11-13] the conditions of choice often depend upon whether the procedure is being performed manually or in an automated fashion, and what methods are used for monitoring the sequence analysis (*vide infra*). It is safe to say that Edman's original choice, phenylisothiocyanate (PITC), is still the reagent most widely employed in the degradation. At a temperature of 50° (and with excess PITC), reaction is complete in a few minutes.[13] The first, or coupling, step thus yields a phenylthiocarbamyl (PTC) peptide derivative from which excess PITC, reaction by-products, and, in some case, nonvolatile buffers such as Quadrol must be removed by extraction with organic solvents before proceeding further.

Cleavage. Exposure of the PTC-peptide derivative to highly volatile anhydrous acids such as trifluoracetic acid or heptafluorobutytric acid leads to attack by the sulfur atom of the substituent on the carbonyl carbon of the first peptide bond, followed by cleavage of the first amino acid from the residual peptide chain (Step 2, Fig. 1). The resulting anilinothiazolinone (ATZ) amino acid derivative is then removed from the residual peptide by extraction with some organic solvent such as ethylacetate, butyl chloride, etc., and the peptide is ready for another cycle of degradation.

Conversion. Subjecting the ATZ-amino acid to $1M$ HCL for 10 minutes at 80° leads to its conversion to a more stable phenylthiohydantoin (PTH) derivative (Step 3, Fig. 1). The PTH amino acids are the derivatives most commonly identified in the direct Edman procedure. A final word should be said here about how to monitor the course of sequence analysis. Again, many choices may be available depending upon whether manual or automated procedures are employed. As shown in Fig. 2, these methods may be divided roughly into two categories, Sampling (or indirect) and Direct methods. The latter, which involves identification of the PTH derivative at each cycle is required for automated procedures where sampling of the reaction mixture after each cycle is neither feasible nor desirable. Direct analysis may be used as well for manual sequencing, but with the use of dansyl-Edman[14] strategies in conjunction with the manual Edman degradation, higher sensitivities may be achieved with sampling of the residual peptide at each cycle. As mentioned above, the use of DABITC[8] with PITC in the coupling step has also provided high sensitivity capabilities for manual sequencing.

THE EDMAN DEGRADATION AND PROTEIN SEQUENCING STRATEGIES

As might be inferred from the foregoing account of developments in chromatographic separation methods, Edman's publications[3,4] came on the scene at just the right time. Sanger had already begun his work on the insulin structure.[20-22]

290

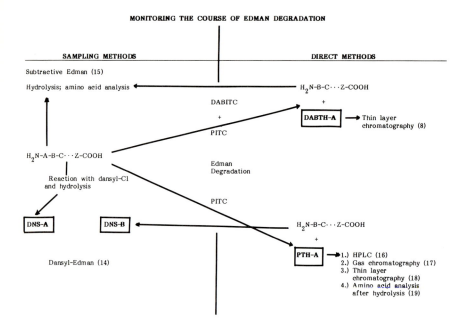

Fig. 2. Schematic representation of the various methods available for following the course of Edman degradation. These may be classified under two general
headings: 1) Direct--analysis of the PTH amino acid, and 2) Sampling (or
indirect)--analysis of the residual peptide.

His goal was to prove that proteins had discrete sequences and he chose insulin
as a subject of analysis because the hormone was relatively small and abundant.
His strategy was to use random cleavage of the purified A and B chains by
partial acid hydrolysis and to use a combination of N-terminal and compositional
analyses on the purified peptides. Paper chromatography was his primary tool
for separating amino acids and the host of peptides generated in this study,
and it is a tribute to his technical virtuosity that he was able to elucidate
correctly the insulin sequence[22] with these methods.

Meanwhile, Moore and Stein had turned their attention in the late 40s and
early 50s to devising column chromatographic procedures for amino acid analysis
and papers describing an automated ion-exchange procedure appeared from their
laboratory in 1958.[23,24] Having developed a sound quantitative basis for

pursuing sequence studies of their own, they decided to study an enzyme and pancreatic ribonuclease was their choice. Although relatively abundant and small in size, this enzyme, nevertheless, presented problems more typical of the vast majority of proteins and required the formulation of a new strategy that could find general application in protein chemistry. Two aspects of that strategy were 1) the use of specific proteolytic enzymes to generage fragments, and 2) the use of the manual Edman procedure for sequencing the fragments. The order of application of the proteases was dictated by their level of specificity and hence trypsin was chosen as a primary cleavage agent followed by chymotrypsin, pepsin, papain, etc. Protein disulfide bonds were cleaved by performic acid oxidation[25] a method that later gave way to the milder reduction and alkylation procedure.[26] This cleavage strategy will be referred to as "extensive specific" as opposed to the "random" method of Sanger (Table 1). Because the manual Edman procedure was useful for up to 10 cycles of information it was best applied to smaller peptides. Very often further cleavages were necessary and the task of fragmenting and reconstructing the ribonuclease chain was one which, in this pioneering effort, required about four years and the participation of many young scholars.[25,27-30]

Workers who followed in the 1960s continued to use the "extensive-specific" cleavage strategy and by the end of the decade, Dayhoff[31] had published four volumes of the "Atlas of Protein Sequence and Structure." Hundreds of proteins and protein fragments had been sequenced by that time using the manual Edman procedure and the field of molecular evolution came into its own. Edman recognized, however, the several disadvantages of his scheme as applied manually. In addition to being exceedingly tedious and time consuming, it was not possible with this system to obtain high repetitive yields from cycle to cycle. Oxygen is notoriously difficult to keep out of the reaction, and it is easy to oxidize PITC to phenylisocyanate, thus blocking any chains so modified from subsequent degradation. In 1967, Edman and Begg[32] published a paper describing the construction of a protein sequencer which performed automatically the coupling and cyclization steps of the degradation (Fig. 1). This liquid phase instrument had at its heart a spinning glass cup into which protein samples were applied in solution and then dried as a film on the wall of the cup. The cup itself was housed in a chamber that could be evacuated or flushed with N_2. Reagents and extraction solvents were added to the cup in a controlled fashion and drying at various stages of the degradation was done under vacuum or under N_2. Extraction of materials was effected by allowing the stream of solvent to pass over the dried protein and up the walls of the cup until it

TABLE 1

STRATEGIES FOR PROTEIN SEQUENCING

Dates	Fragmentation	Peptides	Rationale
1950s	Random: H^+, nonspecific proteases	Small	Compositional and N-terminal analysis of many overlapping peptides: F. Sanger, insulin (mole quantities required)
1960s	Extensive, Specific: Trypsin, chymotrypsin etc.	Small	Moore and Stein; ribonuclease: Development of ion exchange methods for amino acid analysis, peptide and protein purification. Use of substractive and dansyl Edman method (μmole quantities)
1970s	Limited Specific: CNBR, trypsin at Arg etc.	Large	Edman and Begg development of automated Edman method. Fewer fragments simplify purification; methodology geared to large peptides increased efficiency (nmole).
	1) Random	Small	a) In conjunction with DNA sequencing, isolate only a few. b) Possible adaptation to a completely automated procedure using GC or HPLC- mass spectrometry.
	2) Extensive Specific	Small	Development of reverse phase HPLC technology has placed new emphasis on this approach since nearly all peptides in a mixture can be resolved in one step.
1980s		Large	Hunkapiller and Hood development of gas- liquid solid phase sequenator with picomole sensitivity. Purification of fragments by PAGE or HPLC.

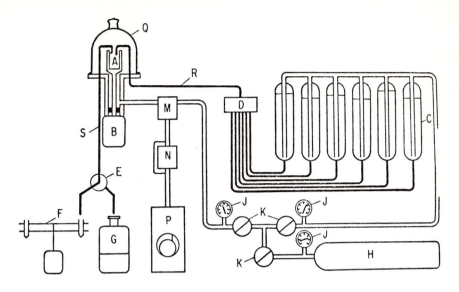

Fig. 3. Schematic description of the prototype spinning cup liquid phase
sequenator developed by Edman and Begg.[32] A, spinning cup; B, electric motor;
C, reagent/solvent reservoir; D, valve assembly; E, outlet stopcock assembly;
F, fraction collector; G, waste container; H, nitrogen cylinder; J, pressure
gauges; K, pressure regulators; M, three-way valve; N, two-way valve with
bypass; P, rotary vacuum pump; Q, bell jar; R, feed line; S, effluent line.

reached the top where it was collected in a groove and removed either to
waste or, when collecting the desired ATZ derivative, to a fraction collection.
A schematic representation of the system taken from the original paper of Edman
and Begg is shown in Fig. 3.

The liquid phase spinning cup sequencer created a sensation in protein
structural analysis. Automation not only minimized the tedium associated with
the manual procedure, but, more importantly, it set the stage for a whole new
approach to sequencing in that the focus was changed from small to large
fragments. The higher repetitive yields (95-98%) made possible by the more
efficient exclusion of oxygen from the system permitted analysis further into
the polypeptide chain and runs of 60 cycles were not uncommon. In short, more
information could be generated more easily and efficiently from large peptides.
Thus, "limited-specific" cleavage methods became of paramount importance in
the new automated strategies (Table 1). Because of their relative low abundance in proteins, methionyl, arginyl, and tryptophyl residues became prime

targets for cleavage. Cyanogen bromide had long been known to be an excellent method for cleavage at methionines,[33] and tryptic cleavage at arginines was possible following amino group modification by any of a host of reagents. Several procedures for cleavage at tryptophan were devised and methods were also described for scission of the rare Asn-Gly and Asp-Pro bonds (see reference 34 for a review and references). These various tactics provided the needed arsenal for generating a few large fragments which were eminently well suited for the liquid phase instruments. The impact here was in relation to the savings in time and effort gained in the purification of fragments. It is much easier to purify a few large pieces than many small ones. Often, most, if not all, cyanogen bromide fragments may be isolated in a single, high yield gel permeation step.[35,36] The higher yields and greater efficiency of sequencing in turn led to a reduction in the amount of protein necessary for analysis. Further refinements in the delivery system, valves and cup design[8,37] led to optimization of the sensitivity of the liquid phase instrumentation so that one to ten nmole levels of sample could be sequenced. Widespread application of the spinning cup instruments caused an exponential increase in the generation of protein sequence information and essentially abolished the "size" barrier so that investigators no longer shunned giant molecules like beta-galactosidase.[38]

At about the same time as the liquid phase sequencers were becoming popular, Laursen[39] described the construction of an automated solid phase instrument in which peptide samples were coupled covalently to an insoluble support. On the positive side, the solid phase method offered advantages in the lower cost of reagents and of construction of the instrument as compared to its liquid phase rival. Moreover, covalent attachment of the sample to an insoluble support prevented extractive losses. The latter feature may be crucial for sequencing of highly hydrophobic and organic solvent-soluble peptides which would soon disappear from the spinning cup machine. The major obstacle to the widespread use of the solid phase instrument is the problem associated with the coupling chemistry. Much work has been done in this regard to develop specific chemistry for sets of peptides having select C-terminal residues. The problem, however, has still not been resolved satisfactorily. Moreover, liquid phase instrumentation can be employed in the sequencing of large and small peptides and is often perfectly suitable for sequencing of hydrophobic peptides in the presence of carrier substances so that in terms of its general applicability it continues to be the most popular method for automated Edman degradation.

The narrative thus far describes the state-of-the-art of protein sequencing in the mid 1970s. As described in Table 1, the prevailingly popular cleavage strategy changed from "random" to "extensive-specific" to "limited-specific" over the course of 25 years of this development. However, the 1970s ushered in another new age of sequence analysis in which any one of these strategies could well be a reasonable approach depending upon the technology involved.

One of the most important of the new developments of the 1970s which has led to the recent revolution in protein sequence analysis has been the description of methods for nucleic acid sequencing.[40,41] The explosion of technology in genetic engineering and recombinant DNA research has made feasible the cloning of the structural gene for any protein. This, coupled with the fact that nucleic acids may be sequenced much more readily than their protein translation products, makes it possible to generate, in principle, the sequence of any protein in a most expeditious fashion. Indeed, serious consideration must be given to the question of whether to undertake a two to five year protein sequence analysis when the nucleic acid chemist might be able to accomplish the task in a few months' time. Some have even remarked facetiously, that the new technology spells doom for protein chemistry and protein sequence analysis.[42] This statement is a bit premature for a number of reasons.

1) It is not always easy and straightforward to isolate the desired structural gene or the mRNA; in many cases, this phase of the operation has required as much or more time than sequencing the protein itself.

2) Given the nucleic acid, it is fairly routine to generate its sequence and, hence, the structure of the protein corresponding to it. But the nucleic acid chemist feels much more confident in his assignments given a framework of information regarding the molecular weight and partial sequences from terminal and internal regions of the protein derived from analysis of the protein itself. This has led to several fruitful collaborations[43] in a new strategy of protein sequencing where the brunt of the effort is born by the nucleic acid sequencer, and the protein chemist provides a few definitive stretches of sequence as sign-posts along the way (Table 1). This minimizes errors which would shift the reading frame and give a nonsense sequence, and it helps to recognize introns.

3) The "nascent" protein product is seldom the same as that studied in its functional form. Nucleic acid sequencing gives a structure

which, although highly informative, does not necessarily relate
to the biologically interesting protein. Protein chemistry is
still required to identify aspects of post-translational modification and their important functional consequences.

4) Chemical modification studies which often provide crucial
information as to the catalytic or regulatory importance of
particular amino acid side chains in proteins will always require
the full repertoire of protein chemistry.

5) The protein chemist will be closely involved with the discovery
of new proteins, present perhaps in only minute quantities, with
new or highly refined functions. This input will be a critical
factor in initiating the productive two-pronged approach to
elucidating the structures of these as yet uncharacterized
proteins.

Clearly, the new technology in nucleic acid sequencing has changed the rules
of the game but it has by no means put the protein chemist out of business.
These new developments can only be seen as a positive influence on the science
of the 1980s. There remains an enormous number of proteins to be sequenced
and such information always yields important new ideas as regards biological
activity and relationships among proteins of similar or diverse function. This
decade will see more and more collaborative efforts between nucleic acid and
protein chemists in solving these important structural problems.

Having dealt with the new dimension given to protein sequencing from the
world of nucleic acids, consideration will now be given to two further developments originating in the 1970s which have enhanced the ability of the protein
chemistry to do his job. These developments have been in the area of high
performance liquid chromatography (HPLC) and microsequencing. Protein sequencing as a field has grown along two major lines; increasing automation and
increasing sensitivity. The new technologies in separation chemistry and
microsequencing work hand in hand to provide a new capability in the structural
analysis of proteins.

Just as chromatography paved the way for molecular biology as we know it
today, recent refinements of that process, exemplified by micro-particle gel
permeation, ion-exchange and reverse phase column chromatography (HPLC) have
impacted profoundly on every phase of separation chemistry. As regards the
field of protein chemistry, the greatest influence has been in the separation
by HPLC techniques of proteins, peptides and PTH derivatives. Although only
the smaller, more stable proteins are readily purified by reverse-phase
systems,[4] almost any protein is amenable to the conditions of HPLC on silacious

gel permeation columns. The latter separations based upon size can be accomplished with great resolution in a few minutes time and have proven effective both for analytical and preparative purposes.[45] Reverse-phase HPLC technology with octyl- and octadecyl-silica columns has found important application in peptide separations. Peptides ranging in size from a few to more than 100 residues have been purified from complex mixtures in only a few hours. Separations based upon this principle involve gradients of decreasing polarity since hydrophobic interactions are of major importance in binding to the stationary phase. Hermodson and co-workers[46] have explored the use of a tri-fluoroacetic acid solvent system with increasing amounts of acetonitrile in the gradient. This system has many advantages because it is a good solvent for most peptides, it is totally volatile and it can be monitored for peptide content at 220 nm. Fears that it might be corrosive to parts on pumps and fittings appear to be unwarranted. HPLC of peptides provides a new way of "mapping" protein digests which offers greater speed and resolution compared to paper or thin-layer methods and a nearly quantitative means for isolating and analyzing the peptides. Peptides detected by their A_{220} can be collected as they emerge from the column in quantities sufficient for compositional and sequence analysis, so the advantage of a volatile solvent is obvious. Not all peptides, however, are soluble at low pH and we have used a 0.05 M solution of ammonium bicarbonate in acetonitrile for separations at pH 7.5. This has provided a useful complement to the acid solvent system.

What this all means, in effect, is that the major hurdle in protein sequencing, i.e., peptide purification, has been diminished to the point where almost any peptide mixture can be resolved in two steps; gel filtration and HPLC. This, in turn, influences the strategy employed. If the sequencing is to be done in conjunction with nucleic acid analysis and the total structural study is not the major objective, only a few relatively short peptides need be analyzed so that a "random" or "extensive-specific" cleavage strategy would be indicated. In those few cases where "limited" cleavage procedures are not possible, the HPLC methods facilitate application of strategies that lead to larger numbers of fragments. In any case, the days of week-long separations of peptides on ion-exchange columns run in pyridine buffers and monitored with ninhydrin would appear, thankfully, to be over.

The other major aspect of protein sequencing that has been revolutionized by HPLC is in the area of PTH separations. Prior to application of HPLC methods, a battery of separation systems was used to identify and quantitate PTH derivatives. Gas chromatography[17] provided a good method for the stable derivatives, but Arg, Lys, His, Gln, Asn, and Ser were often difficult, if not impossible, to identify. Then, one would have to resort to thin layer[18] methods or to amino acid analysis following hydrolytic conversion of the PTH derivative to its parent amino acid.[19] All of this was very time consuming. For example, in order to keep pace with one automated sequencer producing PTH derivatives at the rate of one per hour, required roughly two days of analysis time per day of operation. Application of reverse-phase HPLC methods to this task (e.g., Rose and Schwartz[16]) has been refined over the past ten years to the point where *all* of the usual PTH amino acids can be separated and quantitated in a single six minute isocratic run.[47] Thus, a few hours' analysis time will suffice to keep pace with *two* sequencers in continuous operation, and the reliability of the results is greater. Continued exploration of columns containing three micron beads will, no doubt, further reduce the separation time. In this field one seems to have the feeling that anything is possible.

The stage is now set for consideration of the final development of the 1970s, protein microsequencing. In the never-ending quest for greater and greater sensitivity a number of approaches have been tried. One is to use radioactive PITC in the Edman procedure. This extrinsic labeling device suffers from most of the problems associated with the chemistry itself so that sensitivity beyond 10 nmoles is seldom realized. The same is true for the dansyl-Edman[14] and DABITC[8] methods, and these suffer as well from their limitation to manual procedures. Intrinsically labeled proteins can be sequenced with carriers at much higher sensitivity, extending down to the picomole range. The problem here, of course, is introduction of the labeled amino acids.

After careful consideration of the problem of how to extend the sensitivity of the Edman degradation to the picomole range, Lee Hood and his co-workers at the California Institute of Technology constructed a microsequencer[48] based upon a totally new design that would perform all of the steps of the procedure (Fig. 1). Several problems associated with the design of the spinning-cup liquid phase sequencer had been recognized both by the group at CIT[37] and by Wittman-Liebold[8] at the Max Plank Institute in Berlin. This led to major improvements in the cup design, vacuum and valving systems, the programming device, and solvent delivery system of the commercial Beckman 890 Sequencer.[8,37] However, the major problem left unsolved was fundamental to the design itself, the spinning cup. It is essential that delivery of solvents and reagents to the cup be controlled precisely so that the film and/or solution of protein always be at the same height in the cup from cycle to cycle. If not, incomplete reaction will result, or material will be washed from the cup. In

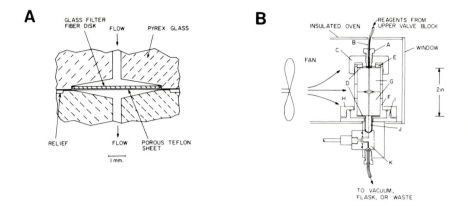

Fig. 4. Diagrams of the cartridge assembly of the gas-liquid solid phase microsequenator developed in the laboratory of Dr. Lee Hood at the California Institute of Technology. A, Enlarged detail of reaction chamber; B, Cartridge assembly. A) aluminum top fitting; B) Teflon tubing; C) aluminum cap; D) Teflon washer; E) keyed aluminum washer; F) 304 stainless steel cartridge body; G, Pyrex glass rod; H) aluminum locking ring; I) aluminum mounting base; J) Kel-F bottom fitting; K) Kel-F valve block. Taken from Hewick, et al.[48]

practice such precise control is difficult to achieve and this leads to a lower repetitive yield. Further problems of contamination arise from the design of the solvent delivery and valve system.

Hood's group decided to use the basic Edman degradation but to make it work at the 10 pmole level by 1) miniaturizing the sequencer, 2) redesigning the vacuum valving and solvent delivery systems, 3) careful purification of all solvents, and 4) use of computer-assisted HPLC for PTH identification. The important innovation was replacement of the cup and drive assembly with an immobile 50 µl reaction chamber into which the protein or peptide is placed as a Polybrene[49,50] film on a glass filter disc. Solvents and reagents flow in and out of the chamber (Fig. 4). Those which are more polar and hence more likely to dissolve the sample are introduced as a vapor thus minimizing losses. Nonpolar solvents such as benzene are washed through the sample in the liquid phase. Thus, the microsequencer is really a gas/liquid/solid phase instrument. In this sequencer there are many fewer moving parts and the delivery and vacuum systems are much more efficient. The micro-computer controlled

TABLE 2

POLYPEPTIDES ANALYZED WITH THE GAS-LIQUID SOLID PHASE SEQUENATOR[a]

Sample	Residues identified/ total residues	Amount	
		pmoles	micrograms
Angiotensin	8/8	500	0.5
Angiotensin	6/8	50	0.05
Somatostatin	14/14	1,400	2.0
Insulin, B chain	30/30	300	1.0
Neuropeptide B from Aplysia	31/34	500	2.0
Dynorphin	14/17	20	0.04
Myoglobin	90/153	10,000	165.0
Myoglobin	22/153	5	0.08
Larval cuticle protein 1 from Drosophila	55/166	850	15.0
Larval cuticle protein 3 from Drosophila	36/96	900	9.0
22,000-dalton membrane phospho-protein from Aplysia	23/200	15	0.3
Human histocompatibility antigen HLA-DR, α-chain	49/300	700	23.0
Human histocompatibility antigen HLA-DR, β-chain	39/240	500	13.0
Human erythropoietin	28/150	100	1.6
Human melanoma cell surface antigen	13/850	60	5.5

[a]Taken from Hewick, et al.[48]

programmer is virtually foolproof, a considerable advance over the cumbersome and unreliable punched-tape step counting system.

The paper by Hewick et al[48] which describes the microsequencer provides details in regard to its construction and operation. Figure 4 and Table 2 taken from that article give information about the sample chamber and results obtained from various amounts of a number of proteins and peptides, respectively. At very low quantities of sample in the 5 to 50 pmole range, computer-assisted baseline correction and peak enhancement were used in the HPLC analysis of PTH derivatives. At the 1 nmole level one is working near maximal capacity and with this quantity of sample, many cycles of degradation are possible. Thus, the microsequencer provides an instrument of more reliable design which is one or two orders of magnitude more sensitive than the most improved spinning cup instruments available.

CONCLUSIONS

The foregoing brings us to our contemporary view of protein sequencing in 1983. As outlined in Table 1, the vast majority of sequencing strategies have, over the years, been linked inextricably with the most advanced methods for performing the Edman degradation. Now we have reached the dawn of a new day when the burden of complete sequence analysis may tend to rest less and less with the protein chemist and more and more with the nucleic acid sequencer. Nevertheless, protein chemistry will always be important both as an adjunct to nucleic acid sequencing and in its own right as the only means for assessing aspects of protein processing.

Clearly, the strategies for protein structural analysis in the 1980s will vary considerably depending upon the technologies employed. The amazing fact is that nearly all of these strategies will continue to be based upon the timeless Edman degradation. The new protein microsequencing methods will undoubtedly, play an important role in future studies. These methods are ideally suited for use in conjunction with the most highly resolving systems available for protein and peptide purification. Thus, proteins purified by polyacrylamide gel electrophoresis may be obtained in quantities sufficient for microsequencing and this opens up a whole new realm of hitherto unapproachable problems of great biological interest. The HPLC systems offer a complementary means of obtaining peptides from these rare proteins so that partial, if not complete sequence analysis is a present reality. Deciphering of unique stretches of sequence in this way will provide the blueprint for synthesis of oligonucleotide probes to facilitate isolation of related nucleic acids for sequencing.[51] Microsequencing capabilities employing the Edman degradation, the new technology in the manipulation of genetic materials, and the seemingly endless refinement in speed and resolution of chromatographic methods will provide the tools necessary for opening an exciting new era of molecular biology in the 1980s.

REFERENCES

1. Fischer, E. (1906) Berichte, 39, 530.
2. Martin, A. J. P. and Synge, R. L. M. (1941) Biochem. J., 35, 1358.
3. Edman, P. (1950) Acta. Chem. Scand., 4, 283.
4. Edman, P. (1956) Acta. Chem. Scand., 10, 761 (1956)
5. Bergmann, M. and Miekeley, A. (1927) Ann. Chem., 458, 40.
6. Abderhalden, E. and Brockmann, H. (1930) Biochem. Z., 225, 386.
7. Braunitzer, G., Schrank, B., and Ruhfus, A. (1970) Hoppe-Seyler's Z. Physiol. Chem., 351, 1589.
8. Wittman-Liebold, B. (1980) in Polypeptide Hormones, Beers, R. F., Jr., and Bassett, E. G., eds., New York: Raven Press, p. 87.
9. Richards, F. F. and Lovins, R. E. (1972) Methods Enzymol., 25B, 314.

302

10. Stark, G. R. (1971) Biochemical Aspects of Reactions on Solid Supports, New York: Acad. Press.
11. Heinrikson, R. L. and Kramer, K. J. (1974) in Prog. in Bioorg. Chem., Kaiser, E. T. and Kezdy, F. J., eds., New York: Wiley, p. 141.
12. Tarr, G. E. (1975) Anal. Biochem., 63, 361.
13. Tarr, G. E. (1977) Methods Enzymol., 47E, 335.
14. Gray, W. R. (1972) Methods Enzymol., 25B, 333.
15. Konigberg, W. (1967) Methods Enzymol., 11, 461.
16. Rose, S. M. and Schwartz, B. D. (1980) Anal. Biochem., 107, 206.
17. Pisano, J. J. and Bronzert, T. J. (1969) J. Biol. Chem., 244, 5597.
18. Jeppsson, J.-O. and Sjöquist, J. (1967) Anal. Biochem., 18, 264.
19. Smithies, O., et al. (1971) Biochemistry, 10, 4912.
20. Sanger, F. and Tuppy, H. (1951) Biochem. J., 49, 463.
21. Sanger, F. and Thompson, E. O. P. (1953) Biochem. J., 53, 353.
22. Sanger, F. (1959) Science, 129, 1340.
23. Moore, S., Spackman, D. H., and Stein, W. H. (1958) Anal. Chem., 30, 1185.
24. Spackman, D. H., Stein, W. H., and Moore, S. (1958) Anal. Chem., 30, 1190
25. Hirs, C. H. W., Moore, S., and Stein, W. H. (1956) J. Biol. Chem., 219, 623.
26. Hirs, C. H. W. (1967) Methods Enzymol., 11, 199.
27. Hirs, C. H. W. (1960) J. Biol. Chem.
28. Hirs, C. H. W., Moore, S., and Stein, W. J. (1960) J. Biol. Chem., 235, 633.
29. Spackman, D. H., Stein, W. H., and Moore, S. (1960) J. Biol. Chem., 235, 648.
30. Smyth, D. G., Stein, W. H., and Moore, S. (1963) J. Biol. Chem., 238, 227.
31. Dayhoff, M. O. (1969) Atlas of Protein Sequence and Structure, National Biomedical Research Foundation, Vol. 4.
32. Edman, P. and Begg, G. (1967) Eur. J. of Biochem., 1, 80.
33. Gross, E. (1967) Methods Enzymol., 11, 238.
34. Walsh, K. A., et al. (1981) Ann. Rev. Biochem., 50, 261 (1981).
35. Sterner, R., Noyes, C., and Heinrikson, R. L. (1974) Biochemistry, 13, 91.
36. Marcus, F., et al. (1982) Proc. Natl. Acad. Sci. USA, 79, 7161.
37. Hunkapiller, M. W. and Hood, L. E. (1981) in Chemical Synthesis and Sequencing of Peptides and Proteins, Liu, T.-Y., et al., eds., New York: Elsevier-North Holland, p. 111.
38. Fowler, A. V. and Zabin, I. (1978) J. Biol. Chem., 253, 5521.
39. Laursen, R. A. (1972) Methods Enzymol., 25B, 344.
40. Sanger, F., et al. (1977) Nature, 265, 687.
41. Maxam, A. M. and Gilbert, W. (1977) Proc. Natl. Acad. Sci. USA, 74, 560.
42. Malcolm, A. D. B. (1978) Nature, 275, 90.
43. Biemann, K. (1981) in Chemical Synthesis and Sequencing of Peptides and Proteins, Liu, T.-Y., et al., eds., New York: Elsevier-North Holland, p. 31.
44. Henderson, L. E., Sowder, R., and Oroszlan, S. (1981) in Chemical Synthesis and Sequencing of Peptides and Proteins, Liu, T.-Y., et al., eds., New York: Elsevier-North Holland, p. 251.
45. Regnier, F. E. and Goodling, K. M. (1980) Anal. Biochem., 103, 1.
46. Mahoney, W. C. and Hermodson, M. A. (1980) J. Biol. Chem., 255, 11199.
47. Heinrikson, R. L., unpublished results.
48. Hewick, R. M., et al. (1981) J. Biol. Chem., 256, 7990.
49. Tarr, G. E., et al. (1978) Anal. Biochem., 84, 622.
50. Klapper, D. G., Wilde, C. E., III, and Capra, J. D. (1978) Anal. Biochem., 85, 126.
51. Noyes, B., et al. (1979) Proc. Natl. Acad. Sci. USA, 76, 1770.

CYTOCHROME P-450-CAM: THE ROLE OF SOME SPECIFIC CYSTEINE RESIDUES

KERRY T. YASUNOBU, MITSURU HANIU, AND IRWIN C. GUNSALUS[*]
Department of Biochemistry-Biophysics, John A. Burns Medical School,
University of Hawaii, Honolulu, Hawaii; [*]Department of Biochemistry
School of Chemical Sciences, University of Illinois, Urbana, Illinois.

SUMMARY

The status of our chemical studies in progress in our laboratory on cytochrome P-450 are discussed. The amino acid sequence of cytochrome P-450-CAM
from *Pseudomonas putida* has been completed. Chemical modification of the
cysteine residues are being investigated. Cytochrome P-450-CAM contains cysteine
residues at positions 56, 83, 134, 146, 240, 283, 332, and 355 from the NH_2-
terminus of the enzyme. Using the probe 2-bromoacetamido-4-nitrophenol, a
yellow-colored peptide was isolated which demonstrated cysteine residue 355 as
the surface situated cysteine residue which is probably involved in enzyme
dimerization which occurs during freeze-thaw cycle of the enzyme during physicochemical studies or during enzyme purification in the absence of reducing
agents. In the enzyme-substrate complex, the reactivity of the cysteine residues were residue 355>56>332>134 or 146 and a small fraction of residues 240.
In order to get reaction of the other cysteine residues of P-450-CAM with
iodoacetic acid, it was necessary to deheme the enzyme. Evidence from comparative sequence investigations suggested that cysteine 134 may possibly be
the residue chelated to the heme iron in cytochrome P-450-CAM. Single crystal
X-ray diffraction of the P-450-CAM is in progress and the validity of our
tentative conclusion that cysteine-134 is the axial heme ligand in P-450-CAM
has been shown to be correct. Cysteine residues 56, 332, and 146 are possible
-SH groups involved in the substrate-induced shift of iron to high spin and in
putidaredoxin ferric-ferrous reduction.

INTRODUCTION

There is extreme interest in the structure and function of cytochrome P-450
monooxygenases at the present time. The cytochrome P-450 enzymes have been
isolated from bacteria,[1-4] yeast,[5,6] plants,[7,8] insects,[9] and mammals[10-14] in
very high purified forms. Cytochrome P-450 monooxygenases are required for a
number of important functions in animals. They are required for the hydroxylation and oxidation of a variety of drugs, for the hydroxylation of biologically
important steroids, for cholesterol side chain cleavage, and for metabolism of

xenobiotics.[15] The bacterial cytochrome P-450 monooxygenases are being used as a model for the animal cytochrome P-450 enzymes since they are water soluble, crystallizable, and are lipid free.

One of the goals at the present time is to determine the structure of these monooxygenases. The amino acid sequence of the *Pseudomonas putida* cytochrome P-450 has recently been completed by classical chemical sequencing procedure.[16-18] The rat liver cytochrome P-450 sequence has been inferred from the c-DNA sequence of the cytochrome P-450 gene and the rabbit liver cytochrome P-450-LM-2 sequence has been established from the combined gene sequence and chemical sequence data. This is possible since the sequence of the rat and rabbit cytochrome P-450 enzymes show great sequence homology. The complete structure of the *Pseudomonas putida* cytochrome P-450 is under investigation by the single crystal X-ray diffraction technique by Poulos et al.[19] at the University of California, San Diego. The availability of the complete structure of the *Pseudomonas putida* cytochrome P-450-CAM will make it possible to carry out well planned structure-function investigations.

In several previous reports, preliminary structure-function investigations on the role of the cysteine residues in P-450-CAM have been reported.[20,21] In the present report, the results of our continuing investigation on the role of specific cysteine residues in P-450-CAM are presented.

EXPERIMENTAL PROCEDURES

Materials. Cytochrome P-450-CAM was isolated from *Pseudomonas putida* PpG 786 (ATCC 29,607) according to established procedures.[22] 2-Bromoacetamido-4-nitrophenol was purchased from Sigma Chemical Co. Iodoacetic acid 1-^{14}C labeled, 22 mCi/nmol, was purchased from ICN Chemical and Radioisotope Division.

Methods. Sulfhydryl group modification with 2-bromoacetamido-4-nitrophenol. About 0.3 ml (0.2 μmole) of native P-450-CAM was added to 0.7 ml of 0.1 M phosphate buffer, pH 7.0 which has been flushed with N_2 at 0°C. About 1 μmole of mercaptoethanol was added and then the sample was dialyzed for 6 hours against 1 ℓ of 0.1 M phosphate buffer, pH 7.0. The sample was transferred to a test tube and 0.6 mg (2 μmoles) of 2-bromoacetamido-4-nitrophenol in 0.5 ml of 50% dimethyl formamide was added at 0°C. The total volume was 2.0 ml and was separated into five, 0.4 ml portions. The reaction was stopped at different time intervals by the addition of excess mercaptoethanol. Three percent HCl-acetone precipitation was then carried out. The precipitate was collected, washed and then dissolved in 1.0 ml of 0.1 N NH_4OH and the absorbance determined at 410 nm.

For the modification only of the surface located cysteine residue, the conditions were slightly altered as follows. About 0.22 μmole of P-450-CAM in 3.0 ml of nitrogen flushed, ice cold 0.1 M phosphate buffer, pH 7.0, was reacted for 10 minutes at 0°C with 0.6 mg of 2-bromoacetamido-4-nitrophenol in 3 ml of 50% dimethylformamide. Then 100 mmoles of dithiothreitol were added and then dialysis against water was continued for 48 hours. The derivative was then precipitated by the addition of 3% HCl-acetone.

Trypsin digestion of the derivative was performed after the addition of 2.0 ml of water and sufficient 33% N-ethylmorpholine to adjust the pH to 8.0. Then 0.5 mg of TPCK-trypsin was added and digestion was allowed to proceed for 12 hours at 25°C. The sample was dried with a stream of nitrogen. The sample was then dissolved in 0.1 M NH_4OH and chromatographed on a column of Sephadex G-75 (1.7 x 70 cm) equilibrated with 0.1 M NH_4OH as shown in Fig. 1.

The yellow-colored peptide containing eluate was pooled and rechromatographed. The yellow-colored peptide fraction was then pooled and dried. Further purification was achieved by peptide mapping. The peptide was applied at the origin of a thin layer silica get plate (20 x 20 cm). Electrophoresis (Fig. 2) in the first dimension was performed in a DeSaga-Heidelberg apparatus using 150V for 2 hours and pH 3.70 pyridine-HOAc buffer. In the second dimension, 1-butanol: pyridine: HOAc: H_2O (15:10:3:12, v/v) was used after which the yellow peptide was scrapped off the plate and then extracted with 2 ml of 0.1

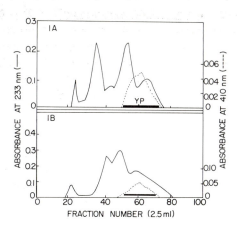

Fig. 1. Sephadex G-75 chromatography and rechromatography of tryptic peptide from S-β-carboxamido-4-nitrophenol-P-450-CAM. YP and bar indicate regions where yellow peptide were pooled.

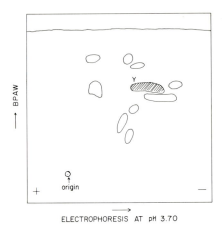

Fig. 2. Fingerprint of Sephadex G-75 chromatographed yellow peptide. Electro-phoresis was followed by chromatography in n-butanol:pyridine:HOAc"H$_2$O.

N NH$_4$OH. After centrifugation, the sample was dried with a nitrogen stream. The sample was dissolved in 5% pyridine solution and aliquots were taken for amino acid analysis and for Edman degradation.

<u>Modification of the -SH groups of P-450-CAM with ^{14}C-iodoacetic acid</u>. About 0.21 μmole of P-450-CAM in 0.3 ml of buffer was reacted with 3.76 μmole of ^{14}C-ICH$_2$COOH at 0°C. At 1, 5, 30, 60, 660, 1440 and 3600 minutes, 50 μl aliquots were removed and excess mercaptoethanol were added to stop the reaction. Three percent HCl-acetone was added to the individual samples and the precipi-tate collected. The precipitates were then dissolved in 3.0 ml of 8 M urea - pH 8.6, 0.1 M Tris buffer, flushed with N$_2$, and treated with 10 μl of mercapto-ethanol. It was then reacted with 50 mg of iodoacetic acid adjusted to pH 8.6 with NaOH for 20 minutes at room temperature. Each sample was then passed individually through a Sephadex G-25F column (1.7 x 70 cm) with 0.1 M NH$_4$OH as solvent. The resulting S-β-carboxymethyl-cysteinyl-P-450-CAM was lyophilized.

Tryptic digestion of the CysCm-P-450-CAM (1.2 mg) were performed as follows. The samples were suspended in 2.0 ml of water, the pH adjusted to 8.0 with 33% N-ethylmorpholine, and 0.1 mg of TPCK-trypsin in 0.1 ml of water was added. Digestion of the derivatized enzyme was allowed to proceed for 48 hours at 40°C. About 0.1 ml aliquots were taken for radioactivity determination, 0.2 ml for amino acid analysis and 2.0 ml for fingerprinting the samples.

AMINO ACID SEQUENCE OF THE P. PUTIDA CYTOCHROME P-450 cam

```
                               20                                          40
T T E T I Q S N A N L A P L P P H V P E H L V F D F D M Y N P S N L S A G V Q E

                           60                                          80
A W A V L Q E S N V P D L V R Q N G G H W I A T R G Q L I R E A Y E D Y R H F S

                       100                                          120
S E Q P F I P R E A G E A Y D F I P T S M D P P E Q R Q F R A L A N Q V V G M P

                           140                                          160
V V D K L E N R I Q E L A Q S L I E S L R P Q G G Q N F T E D Y A E P F P I R I

                       180                                          200
F M L L A G L P E E D I P H L K Y L T D Q M T R P D G S M T F A E A K E A L Y D

                           220                                          240
Y L I P I I E Q R R Q K P G T D A I S I V A N G Q V N G R P I T S D E A K R M Q

                   260                                          280
G L L L V G G L D T V V N F L S F S M E F L A K S P E H R Q E L I Q R P E R I P

               300                                          320
A A Q E E L L R R F S L V A D G R I L T S D Y E F H G V Q L K K G D Q I L L P Q

                   340                                          360
M L S G L D E R E N A Q P M H V D F S R Q K V S H T T F G H G S H L Q L G Q S L

                   380                                          400
A R R E I I V T L K E W L T R I P D F S I A P G A Q I Q H K S G I V S G V Q A L

           412
P L V W N P A T T K A V
```

Fig. 3. Amino acid sequence of cytochrome P-450-CAM. Circled residues show the positions of the various cysteine residues.

For peptide mapping, Eastman cellulose chromatographic TLC sheets were used. The electrophoresis apparatus used was a Desaga-Heidelberg instrument. In the first dimension, electrophoresis with 15% AcOH - 10% HCOOH solvent mixture was used. In the second dimension electrophoresis using AcOH:pyridine:H_2O (pH 3.6) buffer was used. Radioactive peptide localization was by radioautography using Kodak AR X-ray film. Exposure time was 1 week. After the position of the radioactive peptide was determined, the spot was cut out, the silica gel mixed with 10 ml of Aquasol (New England Nuclear), shaken and the eluate counted in a Beckman LS-100 liquid scintillation counter.

Amino acid analysis. Acid hydrolyzates of the samples were analyzed in the Beckman Model 121MB automatic amino acid analyzer. Manual Edman degradation was used to determine the NH_2-terminal sequence of the isolated peptides as described previously.[17,18]

RESULTS

Location of cysteine residues in enzyme. The amino acid sequence of the cytochrome P-450-CAM is shown in Fig. 3. The 8 cysteine residues present are located at residues 56, 83, 134, 146, 240, 283, 332, and 355 from the NH_2-

TABLE 1

IDENTIFICATION OF CYSTEINE RESIDUES LABELLED BY ^{14}C-IODOACETIC ACID

Tryptic Peptide Number	Sequence	Cysteine Residue Modified
1	Arg-Met-Cys-Gly-Leu......Lys	240
2	Ile-Gln-Glu.....Cys...Cys...Arg	134/146[*]
3	Cys-Asn-Gly.............Arg	56
4	Lys-Gly-Asp.....Arg...Cys...Arg	332
5	Val-Ser-His..........Cys...Arg	355

[*] One of the cysteine residues reacted.

terminus of the enzyme. It is assumed that there are no disulfide bonds present in the monooxygenase.[20] Summarized in Table 1 are the various tryptic peptides that have been isolated from the CysCm-P-450-CAM. The possibility existed that these peptides could be isolated using the fingerprinting technique to determine which sulfhydryl group(s) of P-450-CAM are being modified by reaction with iodo-acetic acid under different experimental conditions.

Kinetics of the reaction of the cysteine residues with 2-bromoacetamido-4-Nitrophenol. Prior to the use of iodoacetic acid as a probe of the cysteine residues, the fairly specific sulfhydryl reagent 2-bromoacetamido-4-nitrophenol was tested, as shown in Fig. 4. If the maximum number of cysteine residues which reacted in 60 hours was taken to be 4, about one -SH group reacted with the probe in 10 minutes. In order to determine which of the cysteine residues reacted rapidly, the experiment was repeated at 4°C and for 10 minutes. The labeled enzyme was precipitated with 0.3% HCl-acetone solution, purified by chromatography twice on a column of Sephadex G-75 followed by purification using fingerprinting conditions. The amino acid composition plus three steps of Edman degradation of the purified yellow peptide showed that the highly reactive, surface-oriented -SH group was cysteine residue 355.

Sulfhydryl group modification of P-450-CAM with ^{14}C-labeled iodoacetic acid. The kinetics of labelling of the cysteine residues of P-450-CAM with radio-active iodoacetic acid as described in the experimental section led to the curve shown in Fig. 5. The curve resembled the one observed with 2-bromo-acetamido-4-nitrophenol. After localization of the radioactive peptides by autoradiography, the spots were cut out and extracted with 0.2 N NH_4OH. Ali-quots were used to measure the radioactivity, the amino acid composition and

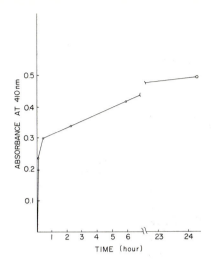

Fig. 4. Kinetics of reaction of cytochrome P-450-CAM with 2-bromoacetamido-4-nitrophenol.

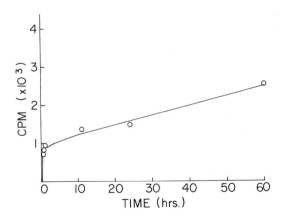

Fig. 5. Kinetics of reaction of P-450-CAM with 1-^{14}C-iodoacetic acid.

310

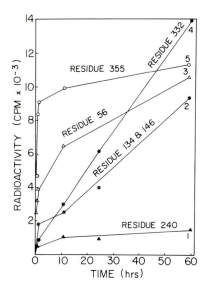

Fig. 6. Reaction of type I cysteine residues of P-450-CAM with 1-^{14}C-iodoacetate.

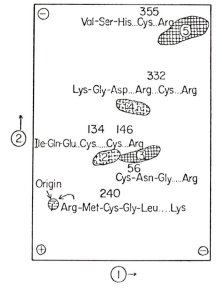

Fig. 7. Fingerprint of tryptic digest of the 1 hour reaction product. Radio-autography was used to detect the peptides.

the NH_2-terminal sequences of the radioactive peptides. The kinetics and the reactivity of the various cysteine residues are summarized in Fig. 6. It is clear that at least four different cysteine residues reacted. These were cysteine residues 355, 56, 332, and 134 or 146 which reacted strongly with iodoacetic acid. Residue 240 reacted only slightly. A fingerprint of the labeled tryptic peptides of cytochrome P-450 obtained for the enzyme which was reacted with ^{14}C iodoacetic acid for 60 minutes is shown in Fig. 7 to illustrate the type of separation obtained in the experiments.

Comparative amino acid sequences of related cytochrome P-450 enzymes. Akhrem et al.[23] have shown that trypsin cleaves the bovine adrenal cortical cytochrome P-450-SCC into two fragments FP-1 and FP-2 with molecular weights of 27,000 and 22,000, respectively, as shown in Fig. 8. FP-1 is the NH_2-terminal fragment and contains only one cysteine residue which is a heme ligand. FP-2 is the COOH-terminal fragment and contains three cysteine residues. The important point is that the heme is liganded to a cysteinyl sulfur which is situated in the NH_2-terminal half of the cytochrome P-450. Since the spectral properties of the class of enzyme designated at cytochrome P-450 are all very similar, the heme binding would also be expected to be similar for this class of heme proteins, i.e., the cysteine residue chelated to the heme iron might possibly be located in the NH_2-terminal portions of all the various cytochrome P-450 enzymes.

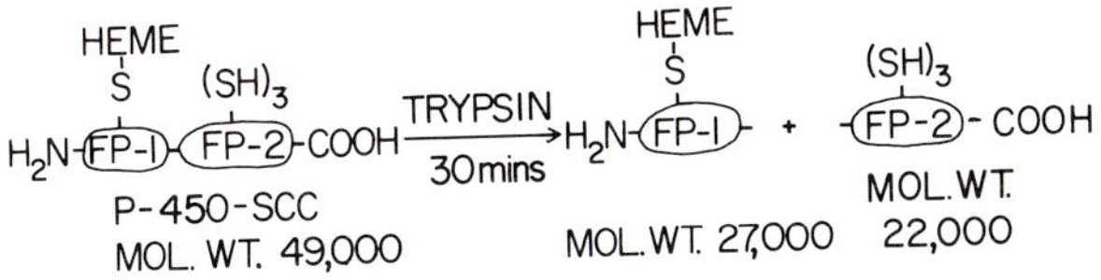

Fig. 8. Trypsin derived large fragments of bovine adrenal cytochrome P-450-SSC illustrating heme binding portion.

Comparison of the amino acid sequences of cytochrome P-450-CAM and rat liver cytochrome P-450-LM-2. As mentioned earlier the amino acid sequence of the cytochrome P-450-CAM has been completed and the amino acid sequence of the rat liver enzyme can be inferred from the c-DNA sequence.[24] If the assumption is made that the sulfhydryl residue which is ligated to the heme iron group is first of all present in the NH_2-terminal half of the enzyme and if one makes a further assumption that the sequence about the heme binding regions in the

different P-450 monooxygenases are alike, a comparison of the sequences of the
bacterial and the rat enzymes may possibly indicate which cysteine residue is
chelated to the heme iron in the two enzymes. Figure 9 shows the sequence
alignment which gives the best sequence homology. According to this alignment,
the suggestion can be made that cysteine residue 134 is the one which is
chelated to the heme iron. Also, if this prediction is correct, cysteine
residue 146 and not residue 134 reacts with iodoacetic acid.

Single crystal X-ray diffraction study of cytochrome P-450-CAM. Poulos et
al.[19] have succeeded in obtaining crystals of P-450-CAM with suitable dimensions

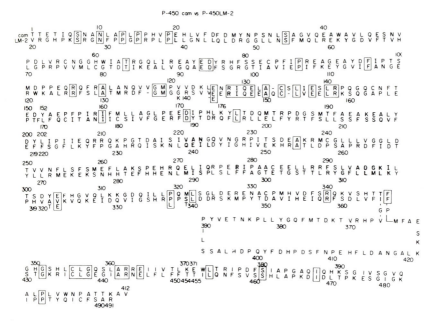

Fig. 9. Sequence comparison of rat liver cytochrome P-450b and P-450-CAM.
Residues squared off are identical sequences and - indicate possible deletions.

for X-ray crystallographic study of the enzyme. Isomorphous crystalline preparations have also been prepared. The data to 4 Å has been completed and currently
the data to 3 Å is being analyzed. The heme in the enzyme, contrary to other
heme-proteins, is located near the surface with the plane of the porphyrin ring
facing the surface of the enzyme. In other heme proteins, only the edge of the
porphyrin ring is facing the surface of the protein. The cysteine residue
chelated to the heme iron is situated between the surface of the enzyme and the
heme. Most importantly, cysteine-134 is the axial heme ligand from the X-ray

data, a conclusion which fits nicely with the chemical data.[19] The data to 3 Å
will provide much greater detailed structure soon.

DISCUSSION

There are eight cysteine residues in P-450-CAM and possible roles of these
residues include: 1) dimerization; 2) heme chelation; 3) H-bonding of substrate; 4) possible involvement in electron transfer from putidaredoxin; and
5) modulation of the spin state in the E-S complex. Prior studies as well as
the present investigation show that there are two classes of cysteine residues
in cytochrome P-450-CAM. Type I -SH reacts regardless whether heme is
present or not. Type II -SH reacts only after the heme is removed from the
enzyme. Among the type I -SH groups the order to reactivity of the -SH groups
were residue 355>56>332>134 or 146 and >240. Residue 240 only reacted to a
small extent.

In the present investigation, use of both 2-bromoacetamido-4-nitrophenol and
$1-^{14}C$-labeled iodoacetic acid to modify the P-450-CAM -SH residues showed that
the most reactive, surface situated and the -SH group involved in enzyme
dimerization was cysteine 355. It has been reported that dimer is fully
active[20] and therefore, this cysteine residue is not catalytically important.
However, it is interesting to note that despite the general lack of sequence
homology between cytochrome P-450-CAM and the rat liver microsomal cytochrome
P-450b, the region about cysteine 355 in the bacterial enzyme and cysteine
residue 436 in the mammalian enzyme shows sequence homology.

P-450-CAM | Phe-Gly-His-Gly-Ser-His-Leu-Cys-Leu-Gly-Gln-Ser-Leu-Ala-Arg-Arg-Glu

P-450-b | Phe-Ser-Thr-Gly-Lys-Arg-Ile-Cys-Leu-Gly-Glu-Gly-Ile-Ala-Arg-Asn-Glu

Cysteine residue 355 of P-450-CAM exists in a domain which is positively charged
with Chou-Fasman calculations suggest that this -SH is part of a β-pleated
sheet region.

Various indirect evidences point to cysteine residue 134 as a -SH group
ligand. The evidences are: 1) the cysteine residue involved in heme iron ligation present in cytochrome P-450-SCC is present in the NH_2-terminal part of the
enzyme; 2) sequence comparison of P-450-CAM and rat liver cytochrome P-450b
suggest that when cysteine 134 (NH_2-terminal portion of P-450-CAM) and cysteine
residue 154 in the rat liver enzyme are aligned, the best sequence homology is
observed as shown below.

P-450-CAM | E N R I Q E L A - C S L I E S L R P - Q C -

P-450-b | E E R I Q E A Q C - L V E E L R K S Q G -

Now it is clear from single crystal X-ray data that cysteine 134 is the axial heme ligand in cytochrome P-450-CAM.[19] Thus of the two cysteine residues (cysteine residues 134 and 146) present in tryptic peptide T-9, cysteine residue 146 was labeled while cysteine residue 134 was not since it was liganded to the heme iron. Chou-Fasman calculations predict -SH-134 is in an α-helical region. Among the other type I sulfhydryl groups, cysteine residues 56, 332 and 146 are possible -SH groups involved in substrate induced shift of heme iron to high spin and in putidaredoxin ferric-ferrous reduction.

ACKNOWLEDGEMENTS

The research described here was supported in part by Grants GM 22556 from the National Institutes of Health and RIAS-SER 77-06923 from the National Science Foundation to Kerry T. Yasunobu. Grants AM562, GM 21161 and GM 16406 from the National Institutes of Health to Irwin C. Gunsalus.

REFERENCES

1. Katagiri, M., Ganguli, B., and Gunsalus, I. C. (1968) J. Biol. Chem., 243, 3543-3546.
2. Berg, A., Gustafsson, J-A., and Ingelman-Sundberg, M. (1976) J. Biol. Chem., 251, 2831-2838.
3. Appleby, C. A. and Daniel, R. M. (1973) in Oxidases and Related Systems, King, T. E., et al., eds., Vol. 2, Baltimore: University Press, pp. 515-528.
4. Cardini, G. and Jurtshuk, P. (1970) J. Biol. Chem., 245, 2789-2796.
5. Lindenmayer, A. and Smith, L. (1964) Biochim. Biophys. Acta., 93, 445-461.
6. Lebcault, J. M., Lode, E. T., and Coon, M. J. (1968) Biochem. Biophys. Res. Commun., 42, 413-419.
7. Potts, J. R. M., Weklych, R., and Conn, E. E. (1974) J. Biol. Chem., 249, 5019-5026.
8. Russell, D. W. (1971) J. Biol. Chem., 246, 3870-3878.
9. Ray, J. W. (1967) Biochem. Pharmacol., 16, 99-107.
10. Cooper, D. Y., et al., eds. (1975) Cytochromes P-450 and b_5. New York: Plenum Press, pp. 1-554.
11. Coon, M. J., et al., eds. (1980) Microsomes, Drug Oxidations, and Chemical Carcinogenesis, Vols. I and II, Academic Press.
12. Gustafsson, J. A., et al., eds., (1980) Biochemistry, Biophysics and Regulation of Cytochrome P-450, Amsterdam: Elsevier/North Holland Biomedical Press, pp. 1-621.
13. Kato, R., et al., eds., 5th International Symposium on Microsomes and Drug Oxidations, July 1981, Tokyo: Japan Scientific Press, in press.
14. Coon, M. J., Chiang, Y. L., and French, J. S. (1979) in The Induction of Drug Metabolism, Estabrook, R. W. and Lindenlaub, E., eds., F. K. Schattauer Verlag Stuttgart, pp. 201-211.
15. Gunsalus, I. C. and Sligar, S. G. (1978) Advanc. Enzymol., 47, 1-44.
16. Haniu, M., et al. (1982) Biochem. Biophys. Res. Commun., 105, 889-894.
17. Haniu, M., Tanaka, M., et al. (1982) J. Biol. Chem., in press.
18. Haniu, M., Armes, L. G., et al. (1982) J. Biol. Chem., in press.
19. Poulos, T. (1982) Private communication.

315

20. Lipscomb, J. D., et al. (1978) Biochem. Biophys. Res. Commun., 83, 771-778.
21. Armes, L. G., et al. (1981) Fed. Proceed., 40, Abstract 704, p. 1662.
22. Gunsalus, I. C. and Wagner, J. C. (1978) Methods Enzymol., 52, 166-188.
23. Adhrem, A. A., et al. (1980) in Biochemistry, Biophysics and Regulation of Cytochrome P-450, Gustafsson, J. A., et al., eds., Elsevier/North Holland, 13, 57-64.
24. Fujii-Kuriyama, Y., et al. (1982) Proc. Natl. Acad. Sci., 79, 2793-2797.

BOVINE PLASMA AMINE OXIDASE: QUANTITATIVE KINETICS AND SPECTRAL STUDIES
OF THE REACTION WITH CARBONYL REAGENTS AND POSSIBLE EVIDENCE FOR THE
PRESENCE OF A COPPER ORGANIC COFACTOR CHELATE

HIROYUKI ISHIZAKI AND KERRY T. YASUNOBU
Department of Biochemistry-Biophysics, John A. Burns Medical School
University of Hawaii, Honolulu, Hawaii 96822

ABSTRACT

   Bovine plasma amine oxidase (a Cu-amine oxidase), is a pink type II copper
protein. The reaction of the enzyme with phenylhydrazine, hydralazine and
isoniazid at pH 7.0 and 25°C were investigated by kinetic and spectral methods.
Kinetic studies disclosed the presence of one type of carbonyl containing
organic cofactor. Pseudo first order rate constants at 0.08 min^{-1} were noted
in the reaction of enzyme with hydralazine and isoniazid while the kinetics
for the reaction with phenylhydrazine was more complicated. The kinetics of
inhibition led to the conclusion that the organic cofactor was present at the
active site and the irreversible nature of the reaction suggested that Schiff's
bases were formed with the inhibitor (or substrate). Phenylhydrazine, hydralazine and isoniazid inactivated enzymes showed maxima at 447, 490, and 350 nm
with molar absorbancy indeces of 37,000, 21,000, and 34,000 $M^{-1}cm^{-1}$, respectively. Difference spectra of the native versus the enzyme hydrazone form of
the enzyme suggested perturbation of the aromatic amino acid(s) in the enzyme.
ESR spectra of the enzyme-carbonyl reagent complexes were slightly different
from the spectrum of the native enzyme but indicated the carbonyl reagents did
not form copper chelates nor did they reduce the copper in the enzyme significantly. The CD spectrum of the enzyme-phenylhydrazine complex showed a new
maximum at 447 nm (ellipticity of 43,400 deg-cm^2/0.1 mole) which corresponded
to the maxima observed in the visible wavelength spectrum. Thus, the phenyl-
hydrazine-enzyme complex is optically active and is due possibly to the
presence of an optically active copper-organic cofactor-phenylhydrazone complex
at the active site of the enzyme. Titration of the organic cofactor with the
carbonyl reagents by three different methods showed there is one mole of co-
pactor per mole of enzyme.

   Cu-amine oxidases (diamine oxidases) (E. C. 1.4.3) have been isolated in
highly purified or homogeneous states from plasma,[1-3] pig kidney,[4] placenta,[5]
pea seedling,[6] and *Aspergillus niger*.[7] The enzyme catalyzes the following
reaction.[8]

$$RCH_2NH_2 + O_2 + H_2O = RCHO + NH_3 + H_2O_2$$

The molecular weights of these enzymes have been reported to be about 170,000-195,000[9,2,4] and there is present one g atom of copper per 85,000-90,000 g of protein.[10,6,11] The one exception is the reported molecular weight of 252,000 reported for the *A. niger* oxidase[7] but in view of the ease with which amine oxidase polymerizes,[9] association of the enzyme could account for the high molecular weight. The enzyme contains copper and a carbonyl-containing organic cofactor which was reported to by pyridoxal phosphate.[12,2,10,7] Recent studies have forced us to conclude that the organic cofactor is not pyridoxal phosphate[13,14] in agreement with Mann[6] and Hill and Mann[15] who concluded that the cofactor is an aromatic carbonyl compound of unknown structure. Phenyl-hydrazine has been used as a carbonyl reagent by many investigators[12,16,11,15] but in view of the objections to the use of phenylhydrazine because of its instability in aerobic aqueous solution,[17] we have also used hydralazine and isoniazid to probe the active site and to determine the moles of organic co-factor present. The results of these investigations are presented in this report.

MATERIALS AND METHODS

Materials

Crystalline preparations of bovine plasma amine oxidase B were isolated by a slight modification of the procedure previously developed in our laboratory[1] and the enzyme was homogeneous when the purity was checked by disc electro-phoresis. The specific activity of the enzyme was about 1,000. Phenylhydra-zine, hydralazine, isoniazid and benzylamine were purchased from Sigma Chemical Co. The other common reagents used were purchased from the usual standard chemical companies.

Methods

Enzyme assay. The enzyme was assayed by the procedure of Tabor et al.[8] in which the oxidation of benzylamine to benzaldehyde is monitored at 250 nm. One unit of enzyme was defined as the amount of enzyme which caused a 0.001 absorbance increase per minute at 25° when benzylamine (3.3 mM final concentra-tion) was dissolved in a final volume of 3.0 ml in 0.1 M potassium phosphate buffer, pH 7.0. Protein concentrations were determined by use of an $E_{1cm}^{1\%}$ at 280 nm of 20.8. All spectrophotometric measurements were made in the Cary Model 14 automatic recording spectrophotometer. All pH measurements were made in the Corning Digital 112 Research pH meter.

ESR spectra were recorded in the Varian E-4 spectrometer at 77°K and at 9.10 GHz. The concentration of the enzyme-inhibitor complexes were about 10.8

mg per ml and were frozen in liquid nitrogen before the spectra were recorded. Freezing and thawing of the sample had no significant effect on the spectra. The g values were obtained from the equation hv = gβH and the hyperfine splitting constants were obtained by measuring the magnetic field differences between the two peaks in g region.[18]

CD spectrum of the enzyme-phenylhydrazine complex was recorded in the Cary Model 61 spectropolarimeter and cylindrical quartz cells of 1.0 cm path was used. The full scale setting was 0.02° and the slit width was programmed so that a constant 20 Å bandwidth was maintained. Ellipticity was defined as (θ) = (θ°) M/10 cl, where θ° is the observed ellipticity in degrees at wavelength λ; M is the gram molecular weight (170,000); and 1, is the path length of the cell in cm; and c, is the concentration of the enzyme in g/cm^3.

RESULTS

Type of inhibition of carbonyl reagents. Standard assay conditions were used and the reaction mixture contained 102 units of enzyme, 0.2-3.33 mM of benzylamine and 0.0014, 7.28 and 273 mM phenylhydrazine, hydralazine, and isoniazid, respectively. When the substrate and inhibitors were added simultaneously to the enzyme, competitive inhibition was observed and K_I values of 1.4 x 10^{-6}, 1.4 x 10^{-5}, and 8.1 x 10^{-4} M, respectively, noted. When the inhibitors were first preincubated for 10 minutes with the enzyme and tested, noncompetitive inhibition was observed. In a separate experiment, when 1.0 mg of enzyme was reacted with excess carbonyl reagents for 10 minutes and the mixture passed through a 1 x 15 cm column of Sephadex G-25, the eluted enzyme was inactive which verified that an irreversible complex was formed.

Rates of reaction of enzyme with the carbonyl reagents. The rate of reaction of the enzyme with phenylhydrazine was complicated as shown in Fig. 1 even when the enzyme/inhibitor ratios were 1 and 0.5. Since it is reported that phenylhydrazine is unstable, it was allowed to stand for 0, 1 and 2 hours in 0.1 M potassium phosphate buffer, pH 7.0 and 25°C. As shown in Fig. 2, as phenylhydrazine ages, there is a decrease in the inhibition or potency.

When 73 units (0.14 μM) of enzyme were reacted with 5.6 μM of hydralazine or 145 μM isoniazid in 3.0 ml of 0.1 M potassium phosphate buffer, pH 7.0, the Ray and Koshland[19] plot (Fig. 3) showed that the reaction obeyed pseudo first order kinetics. Preincubation of these carbonyl reagents in buffer for up to 4 hours in 0.1 M potassium phosphate buffer, pH 7.0 had no effect on the inhibitor potency nor the kinetics of inhibition. For hydralazine and isoniazid, the pseudo first order rate constants were identical, i.e., 0.08 min^{-1}.

320

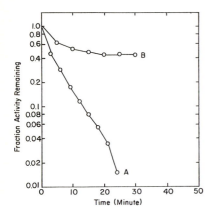

Fig. 1. Time-dependent inhibition of plasma amine oxidase B by phenylhydrazine. Preincubation mixtures contained 88 units (0.17 µM; 1129 units/mg) enzyme and 0.34 (A), and 0.17 µM (B) phenylhydrazine in 2.9 ml of 0.1 M potassium phosphate buffer, pH 7.0 at 25°C. Following preincubation for times indicated 0.1 ml of 0.1 M benzylamine was added to the reaction mixture and absorbance change at 250 nm was measured.

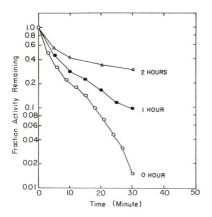

Fig. 2. Time-dependent inhibition of plasma amine oxidase B by oxidized phenylhydrazine. Preincubation mixtures contained 73 units (0.14 µM; 1014 units/mg) enzyme and 0.28 µM phenylhydrazine in 2.9 ml of 0.1 M potassium phosphate buffer, pH 7.0. Phenylhydrazine used in the experiments was incubated alone for 0 (o-o-o), 1 (●-●-●), and 2 hours (Δ Δ Δ) in the buffer prior to reaction. Following preincubation, 0.1 ml of 0.1 M benzylamine was added to reaction mixture and absorbance change at 250 nm was measured.

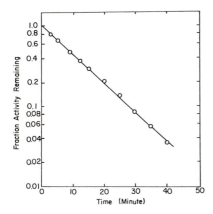

Fig. 3. Time-dependent inhibition of plasma amine oxidase B by hydralazine and isoniazid. Preincubation mixtures contained 73 units (0.14 μM; 1014 units/mg) enzyme and 5.6 μM hydralazine in 2.9 ml of 0.1 M potassium phosphate buffer, pH 7.0 at 25°C. Following pre-incubation, 1.0 ml of 1.0 M benzylamine was added to the reaction mixture and absorbance change at 250 nm was measured.

Ultraviolet and visible absorption spectra of the enzyme derivatives. Bovine plasma amine oxidase is pink and shows a broad absorption band centered at 480 nm.[1] When phenylhydrazine, hydralazine and isoniazid were added to the enzyme, the color of the enzyme changed to bright yellow, bright orange and faint yellow, respectively. New visible absorption maxima were observed at 447, 490, and 350 nm, respectively, as shown in Fig. 4 and the molar absorbancy indeces were calculated to be 37,000, 21,000, and 34,000 M^{-1} cm^{-1}, respectively, for these derivatives. No intermediates were detected during these reactions. The spectra of the products in each case did not change, at least for several hours, once the reaction was complete.

Changes observed in the ultraviolet region are partly due to the addition of an aromatic ring of the hydrazine inhibitors to the enzyme, but, a most interesting change seen, was the drop in the absorbance around 270-290 nm which is indicative of a perturbation of the aromatic amino acid residue(s) in the enzyme derivatives (Fig. 5). The increase in absorbance from 290 nm upwards will be discussed in greater detail in the section to follow.

Visible CD spectrum of enzyme-phenylhydrazine complex. The CD spectrum of the bovine amine oxidase has been previously reported by Ishizaki and Yasunobu.[21] When the CD spectrum of the enzyme-phenylhydrazine complex was determined in the visible wavelength region, a CD maximum corresponding to the maximum observed by visible absorption spectroscopy was noted at 447 nm as

322

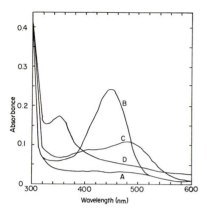

Fig. 4. Visible absorption spectra of plasma amine oxidase B in the presence of hydrazine-derivatives. Spectra were recorded of enzyme (6.64 μM; 912 units/mg) in 3.0 ml of 0.1 M potassium phosphate buffer, pH 7.0, without inhibitor (A), with 13.2 μM phenyl-hydrazine (B), 265 μM hydralazine (C), and 6.64 μM isoniazid (D).

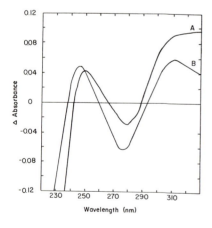

Fig. 5. Ultraviolet difference spectra of plasma amine oxidase B in the presence of hydrazine derivatives. Spectra were recorded of enzyme (1.76 μM; 1059 units/mg) in 3.0 ml of 0.1 M potassium phosphate buffer, 7.05 μM with hydralazine (A), and 1.76 μM isoniazid (B). Spectra were recorded with the inhibited enzyme as the sample and the native enzyme as the reference.

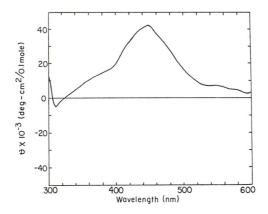

Fig. 6. Visible CD spectra of plasma amine oxidase B in the presence of phenylhydrazine. Spectra were recorded of enzyme (10.45 mg/ml; 825 units/mg) in
219 ml of 0.1 M potassium phosphate buffer, pH 7.0, with 610 μM of phenylhydrazine at 28°C after 15 minutes preincubation. No further significant change of
ellipticity was observed at phenylhydrazine concentration higher than 610 μM.

shown in Fig. 6. The result indicated that the enzyme-phenylhydrazine was
optically active and the ellipticity was calculated to be 43,400 deg-cm^{-2}/0.1
mole. Heat denaturation of the complex caused this CD maximum to disappear.

ESR spectra of the amine oxidase-inhibitor complexes. The bovine plasma
amine oxidase is a type II copper protein[18] and it was possible that the
hydrazine derivatives of the enzyme were either forming copper chelate complexes
or were reducing the cupric copper in the enzyme and were inhibiting the enzyme
by these reactions. The ESR spectra of the three derivatives of the enzyme are
shown in Figs. 7A-C. For the experiments, about 10 mgs of enzyme were reacted
with an excess of inhibitor (4-fold, 40-fold, and 500-fold molar excess of
phenylhydrazine, hydralazine and isoniazid) until completely inactivated. ESR
spectra were then recorded at 77°K in 0.1 M potassium phosphate buffer, pH 7.0.
The various ESR spectral parameters calculated are summarized in Table 1.

Titration of the number of carbonyl cofactor. It has been reported that Cu-
amine oxidases contain 1-4 moles of organic cofactor per mole of enzyme.[2,10-12,16]
Thus, accurate measurements are desired to ascertain the moles of cofactor in
the enzyme. This was checked by three different methods. The methods included:
1) spectral titrations of the enzyme utilizing the 447 nm absorption maximum
of the phenylhydrazine complex (Fig. 8); 2) CD titrations of the enzyme under
similar conditions as shown in Fig. 8; and 3) by activity measurements under

324

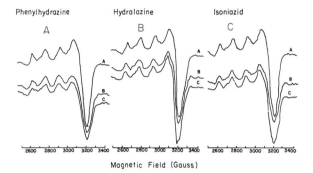

Phenylhydrazine   Hydralazine   Isoniazid

Magnetic Field (Gauss)

Fig. 7. Effect of phenylhydrazine on EPR spectrum of plasma amine oxidase B (10.8 mg/ml; 983 units/mg) in 0.1 M potassium phosphate buffer, pH 7.0: with enzyme alone (A); enzyme in the presence of 0.127 (B) and 0.254 μM (C) phenylhydrazine. EPR experimental conditions were: microwave power, 20 mW microwave frequency, 9.10 GHz; modulation amplitude, 10 gaus, receiving gain, 1000; time constant, 3 sec; scanning rate, 250 gauss/min; and temperature, 77°K. The same settings were used to obtain Figs. 7B and 7C. Fig. 7B. Effect of hydralazine on EPR spectrum of plasma amine oxidase B (10.8 mg/ml; 983 units/mg) in 0.1 M potassium phosphate buffer, pH 7.0: with enzyme alone (A); enzyme in the presence of 1.23 (B) and 2.46 μM (C) hydralazine. Fig. 7C. Effect of isoniazid on EPR spectrum of plasma amine oxidase B (10.8 mg/ml; 983 units/mg) in 0.1 M potassium phosphate buffer, pH 7.0: with enzyme alone (A); enzyme in the presence of 28.87 (B) and 57.75 μM (C) isoniazid.

similar conditions as shown in Fig. 9. These experiments indicated that there was 0.8, 1.0, and 1.2 moles of organic cofactor per mole of enzyme.

DISCUSSION

There is much controversy and confusion about the number and the identify of the organic cofactor present in the Cu-amine oxidases. The proposals made to date include the following: 1) the cofactor is pyridoxal phosphate;[12,2,16,10] 2) the cofactor is not pyridoxal phosphate but a new aromatic carbonyl-containing organic cofactor;[15] 3) there is no organic cofactor but instead an essential cysteine residue which reacts with phenyldiimide present in the phenylhydrazine;[22] and 4) there is no organic cofactor but a cysteine sulfonic acid which is catalytically essential.[we] It has also been reported that the

TABLE 1

EPR SPECTRAL PARAMETERS OF NATIVE AND INHIBITED ENZYME

Inhibitor	Enzyme/Inhibitor molar ratio	g_{II}	g_I	A(gauss)
1. Enzyme alone	0	2.28	2.06	155
2. Enzyme + ØNHNH$_2$	2	2.29	2.07	145
3. Enzyme + ØNHNH$_2$	4	2.29	2.07	145
4. Enzyme + Hydralazine	20	2.29	2.07	145
5. Enzyme + Hydralazine	40	2.29	2.07	145
6. Enzyme + Isoniazid	500	2.29	2.07	145
7. Enzyme + Isoniazid	1000	2.29	2.07	145

The enzyme concentration was 10.8 mg per ml and was dissolved in 0.1 M potassium phosphate buffer, pH 7.0

enzyme contains 1-4 moles of organic cofactor which contains a carbonyl group. In order to clarify the above-mentioned points, the reaction of the bovine plasma amine oxidase with the carbonyl reagents phenylhydrazine, hydralazine and isoniazid were investigated in detail. However, before proceeding with a discussion of the results, a critique of the above-mentioned proposals is warranted at this point. The view that the organic cofactor is pyridoxal phosphate must be abandoned simply because the reports to date have not established the cofactor to be pyridoxal phosphate.[13-15] Probably the conclusion of Hill and Mann[15] which is that the organic cofactor is an aromatic compound with a carbonyl group best summarizes our present knowledge of the nature of the cofactor. Cysteine residues can be ruled out since complete reaction of all of the cysteine residues with CH$_3$Hg Cl yields a nearly fuuly active enzyme.[24] Also, the Cu-free apoenzyme contains the organic cofactor which still reacts with carbonyl reagents[12,15] and this point has been neglected by those investigators who claim that Cu-amine oxidases do not contain an organic cofactor.

In the present study, our goal was to obtain evidence that phenylhydrazine, hydralazine and isoniazid are forming hydrazones with the carbonyl containing organic cofactor. It is necessary to rule out that the carbonyl reagents were forming chelate complexes with the copper in the enzyme; that the carbonyl reagents were reducing the copper in the enzyme, or were inactivating the enzyme by reacting with carbohydrate moiety present in the enzyme.[25]

Since phenylhydrazine was used in previous investigations to probe the nature of the organic cofactor in the enzyme, the initial experiments utilized this

326

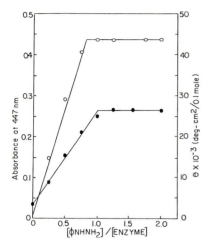

Fig. 8. Spectrophotometric titration of plasma amine oxidase B with phenylhydrazine. Enzyme (1.13 mg/ml; 912 units/mg) in 1.0 ml of 0.1 M potassium phosphate buffer, pH 7.0 was titrated with 5 µl portions of 0.664 µM phenylhydrazine and absorbance at 447 nm were read (o-o-o); enzyme 10.45 mg/ml; 825 units/mg) in 2.9 ml of 0.1 M potassium phosphate buffer, pH 7.0 was titrated with 5 µl-portions of 8.91 µM phenylhydrazine and ellipticity at 447 nm were read (●-●-●). Dilution effects on absorbance and ellipticity at 447 nm were corrected to 1.0 and 2.9 ml, respectively.

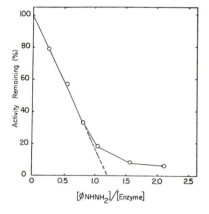

Fig. 9. Titration of plasma amine oxidase B with phenylhydrazine. Reaction mixtures contained 75 units (0.69 µM; 632 units/mg) of enzyme and various concentrations of inhibitor in 0.1 M potassium phosphate buffer, pH 7.0. The enzyme activity was measured after 10 minutes preincubation.

carbonyl reagent. However, due to the instability of this reagent, many of the experiments were repeated with hydralazine and isoniazid which were stable. In addition, the enzyme used in the present study was purer than previous preparations of the enzyme and the peak at 410 nm observed previously, which appears to be due to slight contamination of a heme protein, is absent.

Our first experiments involved confirming the role of the organic cofactor. Phenylhydrazine, hydralazine and isoniazid were all competitive inhibitors when the substrate inhibitor was added simultaneously with the substrate and non-competitive inhibitors when the inhibitor was first preincubated with the enzyme. This is the exact type of inhibition to be expected if the carbonyl reagents are forming Schiff's bases with the organic cofactor according to the formal mechanism derived for the oxidation of benzylamine.[26] Crabbe and Bardsley[27] have also shown that these inhibitors are active site directed reagents of kidney diamine oxidase.

Ray and Koshland[19] have proposed a different type of kinetic analysis to characterize the groups involved in enzyme action. When the inhibition of amine oxidase with either hydralazine or isoniazid was studied by their method, the inhibition was shown to follow first order kinetics and showed that one kind of cofactor in the enzyme reacted. However, the results with phenylhydrazine were complicated and the cause was shown to be due to instability of it in aerobic aqueous solutions.[17]

Spectral studies of the reaction showed that the phenylhydrazine, hydralazine and isoniazid complexes absorbed at 447, 490, and 350 nm and the molar absorbancy indeces were 37,000, 21,000, and 34,000 $M^{-1}cm^{-1}$. Lindstrom and Petterson[11] have reported the molar absorbancy index of the phenylhydrazine complex to be 36,000 while the high value of 62,000 $M^{-1}cm^{-1}$ was reported by Yamada et al.[16] for the A. niger Cu-amine oxidase. No intermediates were detected and once the product was formed, it was stable for at least several hours. Changes were also observed in the ultraviolet wavelength region when the carbonyl reagents reacted with the enzyme, the most interesting of which was the decrease in absorbance at about 280 nm which is indicative of a change in the environment of the tryptophan and tyrosine residues.[20]

ESR studies of the copper in the inhibitor-enzyme complexes were examined and it was concluded that the slight changes observed were not due to the formation of copper-chelate complexes nor due to the reduction of the cupric copper in the enzyme as the major factors responsible for enzyme inhibition. Cu-nitrogen chelates show fine structure in the g_1 region[28,29] due to nuclear interactions of the copper and nitrogen atoms and this was not observed. There was only a slight reduction of the cupric copper in the enzyme, much less than could account for the inhibition.

As stated earlier, it is important to quantitate the number of organic co-factor present in the enzyme since there are reports of widely differing values. The value of one mole of organic cofactor per mole of enzyme was determined by three different methods and this value agrees with the value reported by Lindstrom and Petterson[11] but differs from the values of 2 and 4 reported by others.

We would like to suggest that based on the present and previously published reports, bovine plasma amine oxidase contain 2 g atoms of copper and one mole of a carbonyl containing organic cofactor. The copper in the enzyme and the organic cofactor appear to be chelated. This suggestion is based on the obser-vation that native enzyme shows a maximum at about 480 nm while the Cu-free apoenzyme shows a maximum at 390 nm.[30] When copper is added to the apoenzyme, the maximum at 390 disappears and the band at 480 reappears. CD studies now show that the native enzyme exhibits a number of optically active bands in the visible wavelength region[21] which are destroyed when the enzyme is heat denatured. These CD extrema are most likely due to the assymetric binding of the copper to the protein which ESR studies indicate is of rhombic symmetry.[31] Since the CD spectrum of the phenylhydrazine complex showed that it was optically active too, one interpretation is that the phenylhydrazine is react-ing to give rise to the optically active Cu-organic cofactor chelate complex. Recent NMR studies of Kluetz and Schmidt[32] indicate close proximity of copper to the amine binding site of the enzyme.

ACKNOWLEDGEMENTS

This research was supported in part by the National Institutes of Health Grant MH 21539 and Grant RIAS-SER 77 06923 from the National Science Foundation.

REFERENCES

1. Yamada, H. and Yasunobu, K. T. (1962a) J. Biol. Chem., 237, 1511-1516.
2. Buffoni, F. and Blaschko, H. (1964) Proc. Roy. Soc., B., 161, 153-167.
3. McEwen, C. J., Jr., Cullen, K. T., and Sober, A. J. (1966) J. Biol. Chem., 241, 4544-4556.
4. Yamada, H., et al. (1967a) Biochem. Biophys. Res. Commun., 29, 723-727.
5. Bardsley, W. G., Crabbe, M. J. C., and Scott, I. V. (1974) Biochem. J., 139, 169-181.
6. Mann, P. J. G. (1961) Biochem. J., 79, 623-631.
7. Yamada, H., Adachi, O., and Ogata, K. (1965) Agricult. Biol. Chem. (Tokyo), 29, 912-917.
8. Tabor, C. W., Tabor, H., and Rosenthal, S. M. (1954) J. Biol. Chem., 208, 645-661.

9. Achee, F. M., et al. (1968) Biochemistry, 7, 4329-4335.
10. Mondovi, B., et al. (1967) Arch. Biochem. Biophys., 119, 373-381.
11. Lindstrom, A. and Petterson, G. (1973) Eur. J. Biochem., 34, 565-568.
12. Yamada, H. and Yasunobu, K. T. (1963a) J. Biol. Chem., 238, 2669-2675.
13. Watanabe, K., et al. (1972) Adv. in Biochem. Psychopharmacol., 5, 107-116.
14. Inamasu, M., Konig, W. A., and Yasunobu, K. T. (1973) J. Biol. Chem., 249, 5265-5268.
15. Hill, J. M. and Mann, P. J. G. (1964) Biochem. J., 91, 171-182.
16. Yamada, H., Suzuki, H., and Ogura, Y. (1972) Adv. in Biochem. Psychopharmacol., 5, 185-201.
17. Misra, H. P. and Fridovitch, I. (1976) Biochemistry, 15, 681-687.
18. Vanngard, T. (1972) in Biological Applications of Electron Spin Resonance, Schwartz, H. M., Bolton, J. R., and Borg, D. C., eds., New York: Wiley Interscience, pp. 411-447.
19. Ray, W. J. and Koshland, Jr., D. E. (1961) J. Biol. Chem., 236, 1973-1979.
20. Wetlaufer, D. G. (1962) Adv. Prot. Chem., 17, 304-404.
21. Ishizaki, H. and Yasunobu, K. T. (1974) Adv. in Exptl. Med. and Biol., New York: Plenum Press, 74, 575-588.
22. Hidaka, H. and Udenfriend, S. (1970) Arch. Biochem. Biophys., 140, 174-180.
23. Beneitez, L. V. and Allison, W. S. (1974) J. Biol. Chem., 249, 6234-6243.
24. Wang, T-M., Achee, F. M., and Yasunobu, K. T. (1968) Arch. Biochem. Biophys., 128, 106-112.
25. Watanabe, K. and Yasunobu, K. T. (1972) J. Biol. Chem., 245, 4612-4617.
26. Oi, S., Inamasu, M., and Yasunobu, K. T. (1970) Biochemistry, 9, 3378-3383.
27. Crabbe, M. J. C. and Bardsley, W. G. (1974) Biochem. J., 139, 183-189.
28. Gould, D. and Mason, H. S. (1966) in The Biochemistry of Copper, Peisach, J., Aisen, P., and Blumberg, W., eds., New York: Academic Press, pp. 35-47.
29. Lindstrom, A., Ollson, B., and Petterson, G. (1974) Eur. J. Biochem., 48, 237-243.
30. Yamada, H. and Yasunobu, K. T. (1962b) J. Biol. Chem., 237, 3077-3092.
31. Yamada, H., et al. (1963b) Nature, 198, 1092-1093.
32. Kluetz, M. and Schmidt, P. G. (1977) Biochemistry, 16, 5191-5199.

PROTEIN SYNTHESIS BY SOLUTION METHOD

HARUAKI YAJIMA, NOBUTAKA FUJII AND KENICHI AKAJI
Faculty of Pharmaceutical Sciences, Kyoto University, Kyoto, 606, Japan

ABSTRACT

In order to find a more feasible procedure for protein synthesis, several
new devices have been made. Of these, thioanisole-mediated deprotection and
the sulfoxide of cysteine were discussed and syntheses of gastrin-releasing
peptide and epidermal growth factor were presented in this symposium.

INTRODUCTION

Recently, we introduced a new deprotecting procedure with MSA (methanesul-
fonic acid) or trifluoromethanesulfonic acid (TFMSA)/TFA (trifluoroacetic acid)[1]
as a feasible one of procedures for protein synthesis by solution method.
These acids have enough acidities to cleave a large number of protecting groups,
as well as the MBS (p-methoxybenzenesulfonyl) group from arginine.[2] General
synthetic scheme for proteins by this deprotecting procedure is illustrated in
Fig. 1.

We applied this principle to the synthesis of bovine pancreatic ribonuclease (RNase) A and succeeded in obtaining a crystalline protein with the
full enzymatic activity.[3] However, the final yield in the deprotection, air-
oxidation and subsequent purification steps remained in approximately 6%.
Accumulation of various side reactions, that is a minor side reaction at each
residue, seems to be responsible for lowering the yield. Synthesis of proteins
by solution method is still far from a routine work.

We are still keeping up our efforts to improve these situations, such as
suppression of side reactions and exploration of new procedures. Currently,
a new arginine derivative, arg- (Mts) (mesitylene-2-sulfonyl) was introduced.[4]
This protecting group can be removed more easily than the previous ones, MBS
(p-methoxybenzenesulfonyl) and Tos (tosyl). A new carboxyl-activating procedure with thiazolidine-2-thione was introduced.[5] A new type of side reaction
at Tyr residue, i.e., O-sulfonylation from Arg-derivative, was elucidated.[6]
In this symposium, we wish to discuss mainly our observations on sulfur-
containing amino acids. One is the role of sulfur compounds during the
acidolytic deprotection and the other is the sulfoxide of cysteine derivative.
Then, we wish to present our recent synthetic works on gastrin-releasing
polypeptide (GRP),[7] and epidermal growth factor (EGF).[8]

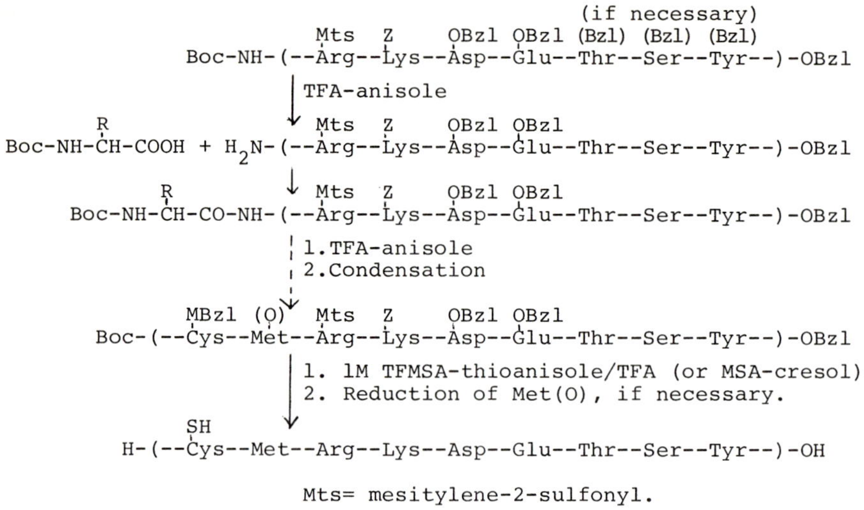

$$\text{(if necessary)}$$

$$\overset{\text{Mts}}{\underset{|}{}}\quad \overset{\text{Z}}{\underset{|}{}}\quad \overset{\text{OBzl}}{\underset{|}{}}\overset{\text{OBzl}}{\underset{|}{}}\text{(Bzl)}\;\text{(Bzl)}\;\text{(Bzl)}$$

$$\text{Boc-NH-(--Arg--Lys--Asp--Glu--Thr--Ser--Tyr--)-OBzl}$$

$$\downarrow \text{TFA-anisole}$$

$$\underset{\text{Boc-NH-CH-COOH}}{\overset{R}{\underset{|}{}}} + \text{H}_2\text{N-(--}\overset{\text{Mts}}{\underset{|}{\text{Arg}}}\text{--}\overset{\text{Z}}{\underset{|}{\text{Lys}}}\text{--}\overset{\text{OBzl}}{\underset{|}{\text{Asp}}}\overset{\text{OBzl}}{\underset{|}{\text{Glu}}}\text{--Thr--Ser--Tyr--)-OBzl}$$

$$\downarrow$$

$$\underset{\text{Boc-NH-CH-CO-NH-(--}\overset{\text{Mts}}{\underset{|}{\text{Arg}}}\text{--}\overset{\text{Z}}{\underset{|}{\text{Lys}}}\text{--}\overset{\text{OBzl}}{\underset{|}{\text{Asp}}}\overset{\text{OBzl}}{\underset{|}{\text{Glu}}}\text{--Thr--Ser--Tyr--)-OBzl}}{\overset{R}{|}}$$

1. TFA-anisole
2. Condensation

$$\text{Boc-(--}\overset{\text{MBzl (O)}}{\underset{|}{\text{Cys}}}\text{--}\overset{}{\underset{|}{\text{Met}}}\text{--}\overset{\text{Mts}}{\underset{|}{\text{Arg}}}\text{--}\overset{\text{Z}}{\underset{|}{\text{Lys}}}\text{--}\overset{\text{OBzl}}{\underset{|}{\text{Asp}}}\overset{\text{OBzl}}{\underset{|}{\text{Glu}}}\text{--Thr--Ser--Tyr--)-OBzl}$$

1. 1M TFMSA-thioanisole/TFA (or MSA-cresol)
2. Reduction of Met(O), if necessary.

$$\text{H-(--}\overset{\text{SH}}{\underset{|}{\text{Cys}}}\text{--Met--Arg--Lys--Asp--Glu--Thr--Ser--Tyr--)-OH}$$

Mts= mesitylene-2-sulfonyl.

Fig. 1. Principle of the organo-sulfonic acid deprotecting procedure for protein synthesis.

RESULTS AND DISCUSSION

Role of thioanisole in acidolytic deprotection

Low recovery of methionine was noted, when Z-Met-OH was treated with TFMSA-TFA (1:3) or MSA in the presence of anisole, because of the formation of a by-product, S-methylmethionine sulfonium compound.[9] The sulfur atom of methionine trapped the methyl group of anisole. This reaction seems to

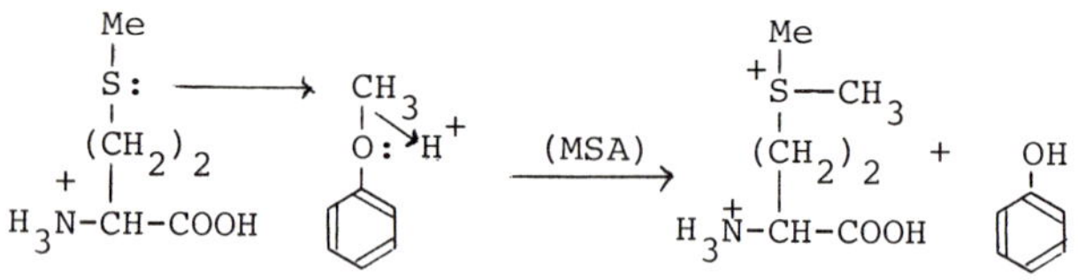

Fig. 2. Role of methionine (sulfur source) in the facilitated cleavage of aromatic ethers.

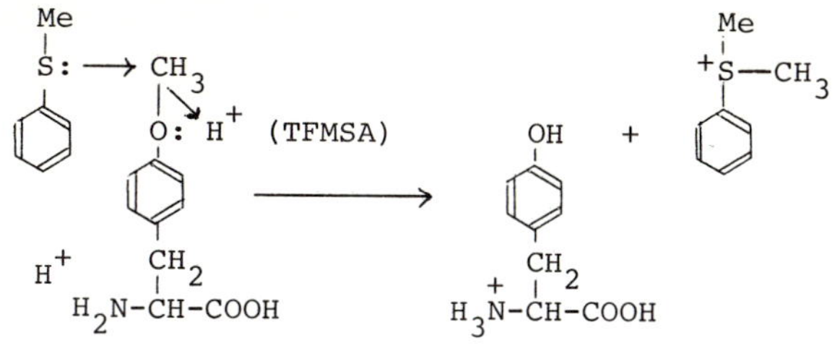

Fig. 3.   Deprotection of O-methyltyrosine by TFMSA-thioanisole.

proceed not by the simple acidolysis, but by a sort of S_N2 type substitution reaction explainable by the "Hard-Soft concept,"[10] i.e., interaction between [H^+ (a hard acid)-$\underline{O}$Me (a hard base)] and [S (a soft base)-$\underline{C}H_3$ (a soft acid)] as shown in Fig. 2.

The above result seems to open a new concept for the acidolytic deprotecting procedure:

(i)   The facilitated cleavage of aromatic ethers can be achieved by these acids in the presence of sulfur compounds, such as methionine or possibly thioanisole. This idea was soon applied to the synthesis of N-methylenkephalinol,[11] O-Methyltyrosine was employed and this methyl group was removed from Boc-N-Me-Tyr(Me)-Gly-Gly-Phe-Met-Ol by TFMSA in the presence of thioanisole as a sulfur source (soft nucleophile) without forming the 3-alkyltyrosine derivative (Fig. 3).

(ii)   Based on the "Hard-Soft" concept, we found that the Moc (methyloxycarbonyl) group could be cleaved by MSA with an aid of sulfur compounds within 5 hr at 5° (Fig. 4). In this instance, dimethylsulfide was more effective than thioanisole.[12] Different from the Z group, Moc-amino acids could be converted to corresponding acid chlorides and thus, the amide bond formation by the acid chloride method was possible. MSH-release inhibiting factor was synthesized by this procedure.[13] Z(OMe)-Lys(Moc)-OH was prepared. This compound will be a useful compound, like Lys(For), for the synthesis of relatively small peptides.

(iii)   Kiso et al., a student in our laboratory,[14] found that thioanisole is an effective accelerator for the TFA cleavage of the Z group and proposed the "Push-Pull"

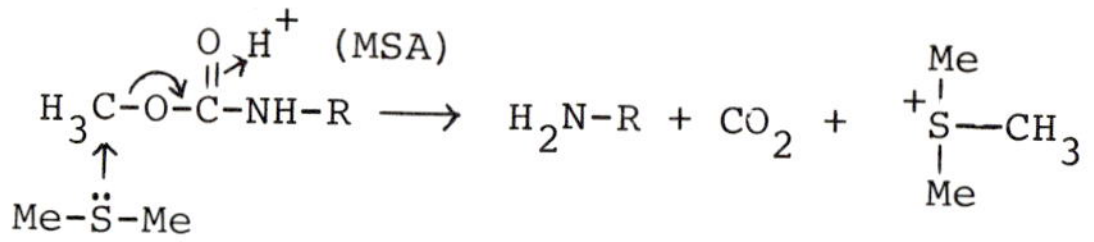

Fig. 4. Cleavage of the Moc group by MSA-dimethylsulfide.

mechanism, based on the "Hard-Soft" concept, to explain these
phenomena (Fig. 5). The promoting effect on this cleavage
reaction of the nucleophile (sulfur source) was in the order;
thioanisole 〉 dimethylsulfide 〉 ethanedithiol. No such
effect was found on phenol and anisole. The complete
cleavage of the Z group from Lys(Z) (0.1 mmol) was achieved
by TFA (27 mmol) in the presence of thioanisole (5 mmol) at
25° for 3 hr, while in TFA-anisole, the cleavage of the Z
group was incomplete even after 27 hr at 25°.

We confirmed that the benzyl ester from Glu(OBzl) and Asp-(OBzl) and even
the Arg-protecting group, Mts, could be removed by TFA-thioanisole, though
complete removal of these groups required 24 hrs treatment. A role of thioanisole, as an accelerator in acidolytic deprotection by TFA and TFMAS/TFA
is apparent.

Met-containing peptides can be synthesized without protecting its sulfur
atom by the MSA or TFMSA procedure, if anisole was replaced by m-cresol or
thioanisole or other scavengers which has no property of alkyl donor. It
should be mentioned that Met(O) can be reduced in nearly 80% during the treatment with 1M TFMSA-thioanisole/TFA at room temperature for 60 min. In the

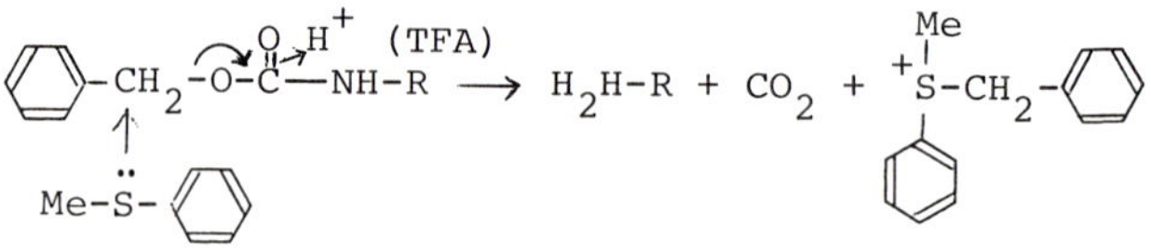

Fig. 5. Cleavage of the Z group by TFA-thioanisole.

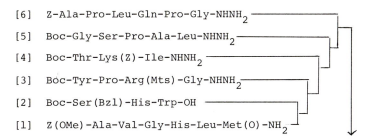

[6] Z-Ala-Pro-Leu-Gln-Pro-Gly-NHNH$_2$

[5] Boc-Gly-Ser-Pro-Ala-Leu-NHNH$_2$

[4] Boc-Thr-Lys(Z)-Ile-NHNH$_2$

[3] Boc-Tyr-Pro-Arg(Mts)-Gly-NHNH$_2$

[2] Boc-Ser(Bzl)-His-Trp-OH

[1] Z(OMe)-Ala-Val-Gly-His-Leu-Met(O)-NH$_2$

H-Ala-Pro-Leu-Gln-Pro-Gly-Gly-Ser-Pro-Ala-Leu-Thr-Lys-Ile-
   Tyr-Pro-Arg-Gly-Ser-His-Trp-Ala-Val-Gly-His-Leu-Met-NH$_2$

Fig. 6. Synthetic scheme of GRP (chicken)

thioanisole-mediated deprotection, we decided to use m-cresol as an additional
scavenger to cancel the possible alkylating property of S-alkyl-thioanisole
sulfonium compound.[15]

   (iv) Synthesis of gastrin releasing peptide was achieved using this
      thioanisole-mediated TFMSA deprotecting procedure. As shown
      in Fig. 6, a heptacosapeptide amine corresponding to the
      entire amino acid sequence of chicken gastrin releasing
      peptide (GRP) was synthesized, like porcine GRP,[7] by assem-
      bling six peptide fragments followed by deprotection with
      a new deprotecting system discussed above, 1M TFMSA-
      thioanisole in TFA. The deprotected peptide was purified
      by ion-exchange chromatography on CM-cellulose followed by
      partition chromatography on Sephadex G-25. The yield we
      obtained in the final deprotecting and purification steps was
      33%.[16] Thus, we could offer one of examples demon-
      strating the usefulness of this new deprotecting procedure
      for the practical peptide synthesis.

Synthesis of Cys-containing peptides

    (i) Cysteine sulfoxide. Cys(Bzl) can not be cleaved completely
      by HF and even by MSA within 60 min at 20°, while Cys(MBzl)
      is cleaved by MSA, 1M TFMSA/TFA, like HF.

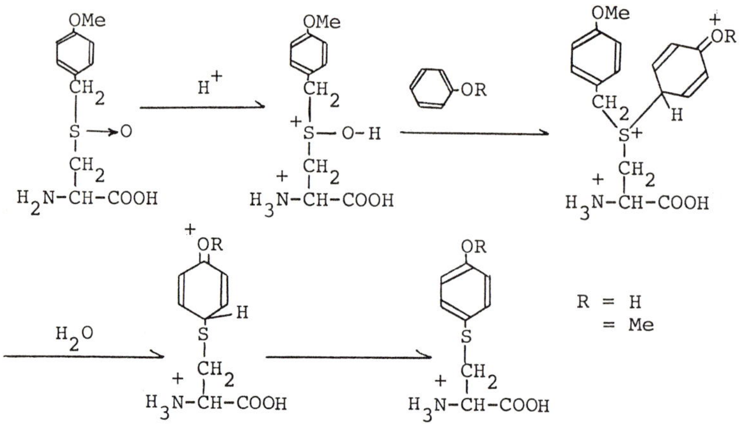

OMe

CH_2

$S \rightarrow O$

CH_2

$H_2N-CH-COOH$

H^+

OMe

CH_2

$^+S-O-H$

CH_2

$H_3N-CH-COOH$

OR

OMe

CH_2

S^+

CH_2

$H_3N-CH-COOH$

^+OR

H

H_2O

^+OR

H

S

CH_2

$H_3N-CH-COOH$

OR

S

CH_2

$H_3N-CH-COOH$

R = H
= Me

Fig. 7. Behavior of S-substituted cysteine sulfoxide under deprotecting conditions in peptide synthesis.

We wish to describe here the sulfoxide of Cys(MBzl), which has been hitherto unknown in the literature. This problem is not serious for the synthesis of small Cys-containing peptides, but has to be taken into consideration, when we engage in the synthesis of relatively large peptides.

In the course of the synthesis of RNase A,[3] we hydrolyzed protected inter- mediates containing Cys(MBzl) and Tyr residues by 6N HCl in the presence of phenol, expecting a better recovery of tyrosine[17] and found an unidentified peak on the short column of an amino acid analyzer with the retention time of 26 min. This compound, S-p-hydroxyphenylcysteine, was found to be derived from the sulfoxide of the Cys(MBzl) residue, (Fig. 7). H-Cys(Bzl)(O)-OH was stable to HF and MSA, but H-Cys(MBzl)(O)-OH was found to be converted quantitatively to S-p-methoxyphenylcysteine by MSA-anisole or by HF-anisole.[18] H-Cys(Acm)(O)-OH gave the identical product, but not quantitatively, when treated with MSA or HF in the presence of anisole.[19]

These results indicated that the sulfoxide, if it once formed, should be reduced before deprotection, otherwise satisfactory recovery of cysteine from Cys(MBzl) could not be expected. Thiophenol in organic solvents was found effective to reduce the sulfoxide in protected peptides, except for Cys(Acm)(O). Boc-Cys(Acm)(O)-OH was converted by thiophenol to S-acetamidomethyl-phenyl sulfide and N^α-Boc-S-phenylthiocysteine (Fig. 8). At present, we have no suitable way of reducing the sulfoxide of Cys(Acm), if it once formed during

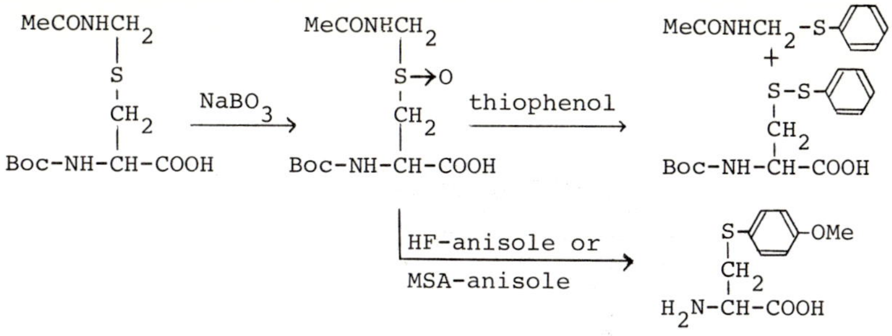

Fig. 8. Behavior of Cys(Acm) (0).

the synthesis. Cys(Tri) resisted to oxidation with hydrogen peroxide or
sodium metaperiodate, presumably by the steric hindrance of three phenyl
residues. The occurrence of the sulfoxide in synthetic peptides can be
estimated by hydrolysis with 6N HCl in the presence of phenol followed by
identification of S-p-hydroxyphenylcysteine by an amino acid analyzer.

(ii) EGF was selected as a model peptide for our present synthesis.
This peptide contains not only Cys residues, but also acid-
labile Trp residues, which offer a certain difficulty from
the synthetic viewpoint. The protected tripenta-
contapeptide ester corresponding to the entire amino acid
sequence of EGF[8] was synthesized by assembling relatively
small 15 peptide fragments as shown in Fig. 9. Side
reaction at the Trp residue, i.e., indole-alkylation
observable during the TFA deprotection, was suppressed
by the use of anisole containing ethanedithiol.[20] All
protecting groups were removed by treatment with 1M TFMSA-
thioanisole/TFA. After reduction with dithiothreitol
followed by air-oxidation, the crude product was purified
by gel-filtration on Sephadex G-25 and then by ion-exchange
chromatography on DEAE cellulose. The product here
obtained still contains a slight impurity, but exhibited a
sharp cross precipitation band with the natural EGF
antisera. When tested in rats, synthetic peptide
inhibited the histamine-stimulated gastric secretion.

338

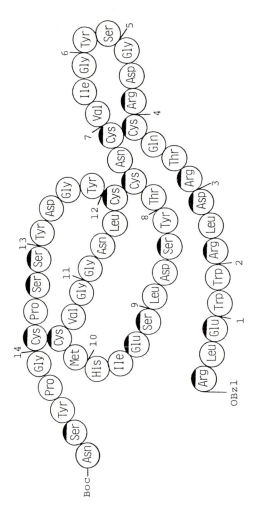

Protected amino acids employed: Arg(Mts), Glu(OBzl), Cys(MBzl), Ser(Bzl), Asp(OBzl).

⬤— Positions of fragment condensation.

Fig. 9. Synthetic scheme of EGF.

CONCLUSION

In order to overcome difficulties which exist in the solution method for
protein synthesis, various improvements have been made. A new concept of
thioanisole-mediated deprotection procedure was discussed and a hitherto
unknown side reaction of the sulfoxide of Cys was elucidated. Two model
peptides, gastrin releasing peptide (chicken GRP) and epidermal growth factor
(EGF) were synthesized, using the above thioanisole-mediated deprotection procedure.

REFERENCES

1. Yajima, H., et al. (1974) J.C.S. Chem. Comm., 107-108.
   Yajima, H., et al. (1975) Chem. Pharm. Bull. (Japan), 23, 371-374.
2. Nishimura, O. and Fujino, M. (1976) Chem. Pharm. Bull. (Japan), 24,
   1568-1575.
3. Yajima, H. and Fujii, N. (1981) J. Am. Chem. Soc., 103, 5867-5871.
4. Yajima, H., et al. (1978) J.C.S. Chem. Comm., 482-483.
5. Yajima, H., et al. (1980) Chem. Pharm. Bull. (Japan), 28, 3140-3142.
6. Yajima, H., et al. (1978) Chem. Pharm. Bull. (Japan), 26, 3752-3757.
7. McDonald, T. J., et al. (1979) Biochem. Biophys. Res. Commun., 90, 227-233.
   McDonald, T. J., et al. (1980) FEBS Lett., 122, 45-48.
8. Savage, C. R. Jr., Hash, J. H., and Cohen, S. (1973) J. Biol. Chem., 238,
   227-234.
9. Irie, H., et al. (1976) J.C.S. Chem. Comm., 922-923.
   Irie, H., et al. (1977) Chem. Pharm. Bull. (Japan), 25, 2929-2934.
10. Pearson, R. G. (1966) Science, 151, 172-177.
    Pearson, R. G. and Songstad, J. (1967) J. Am. Chem. Soc., 89, 1827-1836.
11. Kiso, Y., et al. (1979) J.C.S. Chem. Comm., 971-972
12. Irie, H., et al. (1980) Chem. Letters, 875-878.
13. Celis, M. E., Taleisnik, S., and Walter, R. (1971) Proc. Natl. Acad. Sci.
    USA, 68, 1428-1433.
    Nair, R. M. G., Kastin, A. J., and Schally, A. V. (1971) Biochem. Biophys.
    Res. Comm., 43, 1376-1381.
14. Kiso, Y., Ukawa, K. and Akita, T. (1980) J.C.S. Chem. Comm., 101-102.
15. Lundt, B. F., et al. (1978) Int. J. Peptide Protein Res., 12, 258-268.
16. Akaji, K., et al. (1981) Int. J. Peptide Protein Res., 18, 180-194.
    Akaji, K., et al. (1981) Chem. Pharm. Bull. (Japan), 29, 3080-3082.
17. Iselin, B. (1962) Helv. Chim. Acta, 45, 1510-1515.
18. Funakoshi, S., et al. (1979) Chem. Pharm. Bull. (Japan), 27, 2151-2156.
19. Yajima, H., et al. (1980) Chem. Pharm. Bull. (Japan), 28, 1942-1945.
20. Ogawa, H., et al. (1978) Chem. Pharm. Bull. (Japan), 26, 3144-3149.

ASYMMETRIC INDUCTION IN PEPTIDE SYNTHESIS

KUNG-TSUNG WANG
Institute of Biological Chemistry,
Academia Sinica, Taipei, Taiwan, R.O.C.

ABSTRACT

Asymmetric induction in peptide synthesis are studied. The enzymatic method gives almost complete induction while the chemical method gives only partial induction. The effect of the solvents, the concentration, the temperature and the protecting groups are studied using Asp-OH (OBzl) and Phe-COOCH$_3$. The highest asymmetric yield are 63% L-L-dipeptide and 37% L-D-dipeptide using Cbz-protection and acetonitrile as solvent at 0°C. Dicyclohexylcarbodiimide are used as coupling reagent.

INTRODUCTION

Most of the peptide synthesis are using chiral pure isomers of protected amino acids to obtain a chiral pure peptide. The racemization of amino acid derivatives during the synthetic process is one of the problems to obtain pure biological active polypeptides. Very little work was done on the peptide bond formation using racemic amino acids because it is useless to have a mixture of diastereoisomeric peptides.

Recently, we succeeded in the separation of protected dipeptides by HPLC and we are able to analyze the diastereoisomeric peptides prepared from racemic amino acid (Fig. 1). It is interesting to note that only Lichrosorb Si60 (5 μM) column (E. Merck Co.) with dichloromethane as eluent can resolve the diastereo-isomers. Now, we have an analytical tool with which to study the asymmetric induction in peptide bond formation.

We classify the asymmetric induction into two categories, namely, biological method and chemical method. One can say also enzymatic and nonenzymatic method. We have prepared several Cbz-protected dipeptides by reverse reaction of proteolytic enzyme papain. The reaction gave pure chiral protected dipeptides in good yield.[1] But except for CBZ-protected amino acids, the other protecting groups gave only poor yield; possibly the solubility of the product is higher. It is only recently that we succeeded in obtaining a high yield of Boc-protected dipeptide by salting out effect (Fig. 2, Table 1). The purpose of these syntheses were to obtain a Boc-protected dipeptide for fragment solid phase synthesis. It is single peak in HPLC and our estimate is that there is less

342

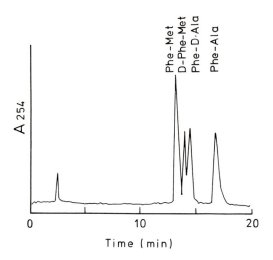

Fig. 1. The separation of Cbz- and TMP-protected dipeptides.
Column: Lichrosorb Si 60 (5 μm); Eluent: CH_2Cl_2; Flow rates: 1.0 ml/min.

TABLE 1

PAPAIN-CATALYZED DIPEPTIDE SYNTHESIS

Substrates	Product	Yield(%)	m.p.
Boc-Leu + Phe-OTMB	Boc-Leu-Phe-OTMB	88	142.5-143.5
Boc-Leu + DL-Phe-OTMB	Boc-Leu-Phe-OTMB	87	142.5-144
Boc-Ala + Phe-OTMB	Boc-Ala-Phe-OTMB	72	134-135
Boc-DL-Ala + Phe-OTMB	Boc-Ala-Phe-OTMB	70	133-135
Boc-Met + Phe-OTMB	Boc-Met-Phe-OTMB	76	148.5-150
Boc-DL-Met + Phe-OTMB	Boc-Met-Phe-OTMB	87	149-150
Boc-Ser(Bzl) + Phe-OTMB	Boc-Ser(Bzl)-Phe-OTMB	85	98-99
Boc-Thr(Bzl) + Phe-OTMB	Boc-Thr(Bzl)-Phe-OTMB	90	141-142
Boc-Cys(Bzl) + Phe-OTMB	Boc-Cys(Bzl)-Phe-OTMB	86	134.5-135.5
Boc-Lys-(Z-Cl) + Phe-OTMB	Boc-Lys(Z-Cl)-Phe-OTMB	90	104-105

Biological method

$$\text{Z-D,L-Phe + L-Ala-OTMB} \underset{\longleftarrow}{\overset{\text{papain}}{\longrightarrow}} \text{Z-L-Phe-L-Ala-OTMB}$$

$$\text{Boc-D,L-Met + L-Phe-OTMB} \underset{\longleftarrow}{\overset{\text{papain}}{\longrightarrow}} \text{Boc-L-Met-L-Phe-OTMB}$$

Fig. 2. Papain-catalyzed Boc-protected dipeptide synthesis.

than 0.1% of D-L-dipeptide contamination. The general procedure for these reactions are, dissolve the protected amino acid in buffer solution and after addition of an appropriate amount of enzyme, incubated at 37°C for 18 hours. The precipitates were filtered, washed with 0.1 N H_2SO_4, 0.1 N Na_2CO_3 and water successively. After drying at room temperature, it was recrystallized from 95% ethanol. We had developed a catalytic hydrogen transfer method to deprotect the trimethylbenzyl group from Boc-protected dipeptides. The products in Table 1 are the inexpensive sources of chiral pure Boc-dipeptides for fragment solid phase synthesis of proteins.

Very little work was cited in the references for the chemical asymmetric induction in the peptide synthesis. Recently, Benoiton et al.[2] showed the asymmetric induction by the reaction of racemic oxazolone derivatives of leucine to L-N^ε-Z-lysine methyl ester. The best results are 50% optical excess of L-L-dipeptide.

We studied the asymmetric induction of CDDI mediated coupling of Cbz-L-Asp-OH to D,L-phenylalanine methyl ester (Fig. 3). The products were analyzed by HPLC. The effects of concentration, solvent, temperature, and protecting groups on the asymmetric induction were studied and the highest diastereoisomeric yield is 63% L-L dipeptide and 37% L-D dipeptide.

MATERIALS

Papain was purchased from E. Merck, Germany. Its potency was described as 3.5 m Anson-E/mg. The enzyme was used without further purification.

D,L-Methionine, D,L-phenylalanine, benzyloxycarbonyl chloride (Cbz-Cl) and S-butyloxycarbonyl-4,6-dimethyl-2-mercaptopyrimidine were from Protein Research Foundation, Japan. Other amino acids were purchased from Kyowa Hakko Kogyo Co., Japan.

The N-benzyloxycarbonyl (Cbz-) amino acids were synthesized by the method compiled by Fletcher and Jones[3] and N-butyloxycarbonyl (Boc-) amino acids were synthesized by the established method.[4]

344

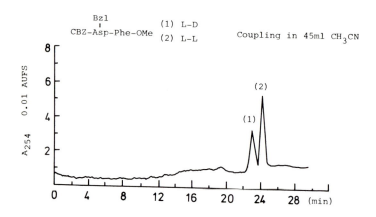

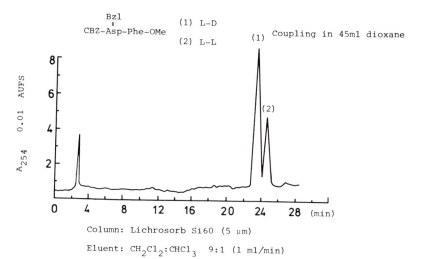

Column: Lichrosorb Si60 (5 μm)

Eluent: CH₂Cl₂:CHCl₃ 9:1 (1 ml/min)

Fig. 3. The assymetric induction of DCCI mediated reaction

The HPLC system included shimadzu LC-3A, pump, a shimadzu LC injector, and Waters Model 440 u.v. detector. The column is Lichrosorb Si60 (5 μm), E. Merck, Germany. LC grade solvents are purchased from Alps Chemicals, Taipei.

METHODS

Enzymic reaction:

(I) Benzyloxycarbonyl-protected peptides. To a solution of benzyloxycarbonyl and trimethylbenzyl-protected L-form amino acids (1 mmol each) or DL-form amino acids (2 mmol) in ethanol/McIlvaine buffer (1:1 v/v, 10 ml), pH 6.0, were added 5 ml of papain solution (500 mg papain). The reaction mixture was incubated at 37°C for 18h. The colloidal precipitate was collected by filtration and recrystallized from ethanol.

(II) t-Butyloxycarbonyl-protected peptides. Substrates (1 mmol for protected L-form amino acids or 2 mmol for protected DL-form amino acids) were dissolved in ethanol/acetate buffer (1:2 v/v, 15 ml), pH 4.5, followed by addition of 5 ml papain solution (750 mg papain) and 500 mg of sodium chloride. The reaction mixture was incubated at 37°C for 24 h. The colloidal precipitate was collected by filtration and recrystallized from ethyl acetate/hexane. Separation of diastereoisomeric dipeptides by HPLC.

A Lichrosorb Si60 (5 μm), column were used to separate the protected diastereoisomeric dipeptides with CH_2Cl_2 as eluent.

Chemical reaction:

The mixture containing N-Cbz-β-Bzl-L-Asp (1.00 mmole), DL-Phe-OMe·HCl (2.00 mmole) and triethylamine (2.00 mmole) in 50 ml solvent was stirred at chosen temperature for an hour and dicyclohexylcarbodiimide (1.10 mmole) was added to the solution. The solution was stirred continuously for further twelve hours. The dicyclohexylurea was filtered off and the solvent evaporated under vacuum. The residue was dissolved in 25 ml ethyl acetate and washed with $NaHCO_3$, H_2O, 0.2 N H_2SO_4 and H_2O, 2 ml x 3 successively. The solution was dried over anhydrous Na_2SO_4 and evaporated under vacuum to get the crude product, which was analyzed by HPLC and calculated the two diastereomers relative contents.

RESULTS AND DISCUSSION

The concentration effects are shown in Table 2. We found no significant difference in the L-L/L-D ratio, so we used 50 ml solvent for all other reactions.

The second effect is temperature that low temperature is favored for asymmetric induction (Table 3). At 25°C, no significant induction was observed

TABLE 2

CONCENTRATION EFFECT, AT 5°C.

	Acetonitril		Dioxane		Methanol	
	L-D	L-L	L-D	L-L	L-D	L-L
5 ml	38	62	58	42	49	51
15 ml	50	60	67	33	49	51
30 ml	41	59	65	35	48	52
45 ml	38	62	67	33	45	55
135 ml	41	59	66	34	44	56

but at 0°C, acetonitrile gave more L-L isomer and dioxane gave more L-D isomer at 25°C but gave more L-L isomer at 0°C.

The third and most important is solvent effect (Table 4). We found that dioxane favored the L-D isomer formation. Toluene also favored the formation of L-D isomer.

The fourth effect we studied was protecting groups (Table 5). Little induction was found using Fmoc- and Boc- protecting groups. But the oldest protecting group, Cbz- group, showed the highest induction. The best result is 63% L-L isomer and 37% L-D isomer. This protecting group is preferred for more L-L dipeptide formation.

TABLE 3

TEMPERATURE EFFECT

Solvent	25°C		0°C	
	L-D	L-L	L-D	L-L
Acetonitrile	48	52	37	63
Dioxane	57	43	65	35
Chloroform	53	47	60	40
Methanol	47	53	44	56
Ethyl Acetate	52	48	43	57

TABLE 4

SOLVENT EFFECT COUPLING AT 0°C, 50 ml solvent

	L-D	L-L	Yield (%)
EtoAc	43	57	87
MeOH	44	56	72
CCl_4	66	34	91
TOLUENE	63	37	84
THF	56	44	91
DMF	45	55	95
DIOXANE	65	35	87
$CHCl_3$	60	40	81
CH_2Cl_2	52	48	85
EtoH	62	38	95
CH_3CN	37	63	85

TABLE 5

SYNTHESIS OF FMOC-VAL-PHE-OTMB SOLVENT EFFECT COUPLING AT 0°C, 50 ml SOLVENT

Solvent	Yield (%)	L-L	L-D
DMF	85	60	40
CCl_4	96	60	40
EtOH	88	57	43
MeOH	86	56	44
THF	92	53	47
TOLUENE	90	50	50
CH_3CN	92	49	51
EtoAc	96	46	54
DIOXANE	93	46	54
$CHCl_3$	90	44	56

348

CONCLUSIONS

1. Biological method is completely stereospecific.

2. Chemical method is partially stereospecific.

3. The manipulation of reaction condition need further studies for the goal of complete stereoselectivity in chemical method.

4. The carbobenzoxy (Cbz) group is preferred for more L-L dipeptides.

5. Acetonitrile is preferred for more L-L dipeptides.

ACKNOWLEDGEMENTS

The work was supported by National Science Council, R.O.C. (NSC 71-0203-B001-02).

REFERENCES

1. Chou, S. H., et al. (1978) J. Chinese Chem. Soc., 25, 215-218.
2. Benoiton, N. L., Kroda, K., and Chen, F. M. F. (1981) Tetrahedron Lett., 22, 3361-3364.
3. Fletcher, G. A. and Jones, J. H. (1972) Int. J. Peptide Protein Res., 4, 347.
4. Nagasawa, T., et al. (1973) Bull. Chem. Soc. Jap., 46, 1269-1272.

Biochemical and Biophysical Studies of Proteins and Nucleic Acids,
Lo, Liu, and Li, eds.

BINDING OF SYNTHETIC CLUSTERED LIGANDS TO THE
Gal/GalNAc LECTIN ON ISOLATED RABBIT HEPATOCYTES[†]

YUAN CHUAN LEE,[*] R. REID TOWNSEND,[*] MARK R. HARDY,[*]
JÖRGEN LÖNNGREN,[+] AND KLAUS BOCK[**]
[*]Department of Biology and The McCollum-Pratt Institute, The Johns Hopkins
University, Baltimore, Maryland 21218; [+]The Department of Organic Chemistry,
Arrhenius Laboratory, University of Stockholm, S-106 91 Stockholm, Sweden;
and [**]The Department of Organic Chemistry, The Technical University of
Denmark, DK-2800 Lyngby, Denmark

ABSTRACT

A series of synthetic oligosaccharide analogs of natural N-acetyl-lactosamine
type glycans were tested for their ability to inhibit the binding of labeled
natural ligands (asialoorosomucoid and asialotriantennary glycopeptide) to
isolated rabbit hepatocytes at 2°C. Inhibitory potency (which can be related
to affinity constant) decreased dramatically in the order of tetraantennary,
triantennary, biantennary, and monoantennary. The range of concentrations
required for 50% inhibition of labeled ligand binding extended from about 1 mM
for the monoantennary oligosaccharide to about 1 nM for triantennary oligosaccharides, even though the absolute Gal concentration increased only three-
fold. The tetraantennary undeca-oligosaccharide was only a slightly better
inhibitor than the best of the triantennary oligosaccharides. The branching
configurations of the lactosamine units were found to contribute significantly
to the binding affinity. Correlation of the distances between terminal
galactose residues and ligand affinity suggests that there are three binding
sites on the lectin spaced by 15 and 23 Å.

INTRODUCTION

It is now recognized that lectins from both plants[1] and animals[2,3] can bind
multiple sugars of a single oligosaccharide structure resulting in greater
ligand specificity as well as tighter binding. In most cases, a single type
of monosaccharide is responsible for binding; however, recently two different
sugars, fucose and mannose, have been shown to be involved in the binding of
oligosaccharides to pea and lentil lectin.[4] The hepatic lectin on the surface
of mammalian hepatic parenchymal cells binds Gal/GalNAc-terminated oligosaccharides and can effectively discriminate between clusters of one, two, and

---

[†]Contribution 1213 from the McCollum Pratt Institute, The Johns Hopkins
University, Baltimore, MD 21218.

three galactosyl residues. This has been demonstrated using either synthetic cluster glycosides[2] or asialo-glycopeptides derived from glycoproteins.[3] We have extended these observations by studying the binding of a series of synthetic oligosaccharides which resemble N-acetyl-lactosamine type oligosaccharides.[5] By using synthetic oligosaccharides, we have been able to investigate the effect of the following on ligand binding: (1) the number of Gal residues per cluster; (2) the composition of branch structures; and (3) the mode of branching as determined by the linkage configuration to the branching mannose. These studies revealed that a synthetic ligand with three Gal residues as the terminal sugars possesses an affinity similar to that reported for such high affinity ligands as asialo-orosomucoid and asialo-fetuin.

The relative affinities of the oligosaccharides was computed from the concentration of ligand required for 50% inhibition of binding of labeled ligand to rabbit hepatocytes at 2°C. We found an hierarchy of binding affinities of the oligosaccharides in the order of tetraantennary > triantennary $\gg$ biantennary $\gg$ monoantennary. It was found that the binding constant of a tetraantennary ligand ($K_d = 10^{-9}$ M) was one million-fold greater than that of a monoantennary ($K_d = 10^{-3}$ M) while the total galactose concentration was increased only 4-fold, clearly demonstrating the binding synergism of clustered Gal residues. The relative affinity was also greatly influenced by the linkage positions on the mannose from which extends the galactose terminated branches. It was also found that distances between Gal residues of oligosaccharides in their preferred conformations could be correlated with ligand affinity and yielded information on the spatial arrangements of the Gal binding sites.

EXPERIMENTAL PROCEDURES

Materials. Hepes was obtained from Research Organics (Cleveland, OH). Silicone oil (DC 550) was from Accumetric (Elizabethtown, KY) and light mineral oil was from Barre Drug Co. (Baltimore, MD). [^{125}I]NaI (carrier free) in 0.1 M NaOH was from Amersham Corp. (Arlington Hts., IL).

Synthetic oligosaccharides. The structures of synthetic oligosaccharides are shown in Figs. 2 and 3. The synthesis of Galβ(1,4)GlcNAcβ(1,6)Man, HEPTA, PENTA-2,4 and PENTA-2,6 and NONA I have been reported.[6-9] Other oligosaccharides were prepared by similar methods (Lönngren et al. unpublished results).

BI-GP   TRI-GP

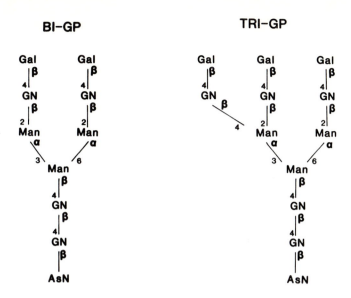

Fig. 1. Structures of naturally derived bi- and triantennary glycopeptides. The biantennary glycopeptide was from human fibrinogen,[10] and the triantennary glycopeptide was from human α-1-protease inhibitor.[5]

Asialo-glycopeptides. Asialo-biantennary glycopeptide (Fig. 1) was prepared from human fibrinogen as described.[10] Triantennary glycopeptide (Fig. 1)* was prepared from α-1-protease inhibitor by a modification[5] of the procedure previously described.[11] Tyrosine was coupled to asialo-triantennary glycopeptide as previously described[2] and the Tyr-asialo-triantennary glycopeptide was radio-labeled by a modified Chloramine-T method[12] using a 2-fold molar excess of [125I]NaI over glycopeptide. The resulting 125I-Tyr-asialo-triantennary glycopeptide had a specific activity of approximately 1 x 10[9] cpm/nmol.

    The exact concentrations of oligosaccharides or asialo-glycopeptides in aqueous stock solutions were determined by analysis of component neutral sugars by automated liquid chromatography.[13] The stock solutions were appropriately diluted with modified MDE.[14]

352

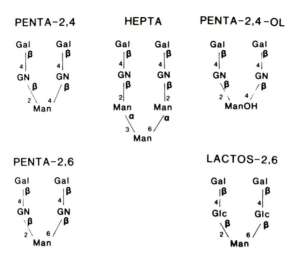

Fig. 2. Structures of synthetic biantennary oligosaccharides. Methods for preparation of these oligosaccharides are given in "Experimental Procedures."

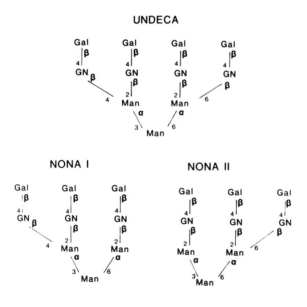

Fig. 3. Structures of synthetic tri- and tetraantennary oligosaccharides. Methods for preparation of these oligosaccharides are given in "Experimental Procedures."

Inhibition assay. The inhibition assays were carried out at 2°C in capped
12 x 75 mm polystyrene tubes. Each tube contained approximately 2.5 x 10^6
rabbit hepatocytes, prepared by an established procedure,[25] inhibitor of known
concentration, and 0.8 nM of ^{125}I-Tyr-asialo-triantennary glycopeptide or
0.25 nM ^{125}I-asialoorosomucoid in a total volume of 1 mL of modified MDE. The
tubes were rotated end over end at 4 rpm for 2 h at 2°C, after which time,
200 µL samples in duplicate were taken and pipetted into 400 µL microfuge tubes
containing approximately 150 µL of silicone/mineral oil mixture (4:1 v/v) at
2°C. The tubes were immediately centrifuged in an Eppendorf microfuge (Model
5412) for 10 s. The cell pellet was counted for radioactivity in a Packard
PRIAS auto-gamma counter.

Nonspecific binding was defined as the amount of ^{125}I-Tyr-asialo-triantennary
glycopeptide or ^{125}I-asialoorosomucoid bound to the cells when the incubation
was carried out in the presence of a 2000-fold molar excess of Gal_{44}-AI-BSA.[15]
The $[I]_{50}$ was determined using a computer curve-fitting program (ALLFIT) based
on a logistic equation as previously described.[16] The data used for the
ALLFIT program were also analyzed with the SCAFIT program[17] to obtain K_d
(dissociation constant of ligand-receptor complex) and R (receptor concentration). The ALLFIT and SCAFIT programs were obtained from Biomedical Computer
Technology Information Center (Vanderbilt Medical Center, Nashville, TN).

RESULTS AND DISCUSSION

In our previous studies with Gal-BSA neoglycoproteins, we found that the
binding affinity of neoglycoproteins to rabbit liver plasma membranes increased
exponentially with the number of Gal residues coupled to BSA.[18,19] This
"cluster effect" was also demonstrable in the binding of synthetic cluster
galactosides to isolated hepatocytes.[2] An analogous trend was also reported
using the isolated human hepatic lectin.[3] With the availability of synthetic
branched oligosaccharides with unique structures, it became possible to analyze
fine structural requirements for binding. We present data to relate the number
of Gal-terminated side chains and their mode of branching in the oligosaccharide
to ligand affinity.

As shown in Table 1, the $[I]_{50}$ values for mono-, bi-, and triantennary
oligosaccharides were found to be approximately 1 mM, 1 µM, and 1 nM, respectively, showing a dramatic (1 x 10^6-fold) enhancement of inhibitory potency
as the number of branches increased (3-4-fold).

The $[I]_{50}$ value of the monoantennary ligand, Galß(1,4)GlcNAcß(1,6)Man, was
similar to that of Gal, approximately 1 mM,[2] indicating that GlcNAc and Man in

TABLE 1

COMPARISON OF $[I]_{50}$ AND K_d VALUES USING EITHER ^{125}I-ASOR OR ^{125}I-Tyr-TRI-GP AS LABELED LIGAND

Compounds	^{125}I-ASOR[a]		^{125}I-Tyr-TRI-GP[b]	
	$[I]_{50}$	K_d	$[I]_{50}$	K_d
	(µM)	(µM)	(µM)	(µM)
Gal-GlcNAc-Man	N.D.	N.D.	821	283
HEPTA	49.9	41.3	20.4	13.2
LACTOS-2,6	6.2	4.63	3.6	2.31
PENTA-2,6	4.5	5.45	2.8	1.75
PENTA-2,4-OL	1.5	5.13	1.9	1.33
PENTA-2,4	0.27	1.33	0.25	0.168
BI-GP	2.4	2.17	1.8	
	(nM)	(nM)	(nM)	(nM)
NONA II	145 ± 40.6	197	111	81.3
Tyr-TRI-GP	N.D.	N.D.	6.2	3.45
NONA I	7.4 ± 2.3	8.61	2.0	1.85
UNDECA	3.4 ± 1.0	4.72	1.3	0.877
TRI-GP	5.8 ± 1.9	10.1	4.3	2.40

[a] ^{125}I-ASOR (2.5×10^{-10} M) was used as labeled ligand.
[b] ^{125}I-Tyr-TRI-GP (8×10^{-10} M) was used as labeled ligand.

the chain are not contributing to the binding. Addition of another N-acetyl-lactosamine chain to the Man to form biantennary structures yielded ligands with $[I]_{50}$ values of 10^{-5} to 10^{-7} M (Fig. 5, Table 1). Structures with three and four lactosamine branches exhibited even greater inhibitory potency up to an $[I]_{50}$ of 10^{-9} M (Table 1, Fig. 4).

The inhibition by ligands of biantennary structure are shown in Fig. 5. Desialylated biantennary glycopeptide (BI-GP), and two synthetic pentasaccharides, PENTA-2,6 and LACTOS-2,6, had similar $[I]_{50}$ values, about 10^{-6} M (Table 1). By comparison of the structures of these pentasaccharides, it is clear that the replacement of GlcNAc in PENTA-2,6 with Glc (resulting in LACTOS-2,6) did not significantly change the inhibitory potency. It should be

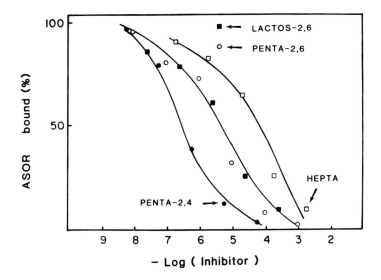

Fig. 4. Inhibition of binding of ^{125}I-asialoorosomucoid by rabbit hepatocytes with synthetic biantennary oligosaccharides. The concentration of the labeled ASOR was 2.5×10^{-10} M in s suspension of 2×10^{6} cells per ml.

pointed out that these pentasaccharides had the same affinity as desialylated biantennary glycopeptide in which each of the β-N-acetyl-lactosamine chains is spaced by an additional Man from the branching Man (Fig. 2). A synthetic heptasaccharide (HEPTA) corresponding to the same structure as found in the natural biantennary glycopeptide minus the GlcNAcβ(1,4)GlcNAcβAsn, gave an unexpected result. The synthetic heptasaccharide was less inhibitory (∼20-fold) than asialobiantennary glycopeptide.

As mentioned above, PENTA-2,6, having two branches β-linked to reducing terminal Man at the 2 and 6 positions, showed a similar affinity toward the lectin as the asialo-biantennary glycopeptide. However, PENTA-2,4, an analog of PENTA-2,6 having the β-N-acetyl-lactosaminyl unit at positions 2 and 4 of the terminal Man, was a 10-fold stronger inhibitor than asialo-biantennary glycopeptide. It should be noted that this pentasaccharide unit is a part of one natural triantennary glycopeptide (Fig. 1), and thus may be a major structural determinant in the tighter binding ($[I]_{50} = 7.4 \times 10^{-9}$ M) displayed by this triantennary glycopeptide (see discussion below). Reduction of the terminal Man (PENTA-2,4-OL) is accompanied by a 5-fold increase in the $[I]_{50}$

356

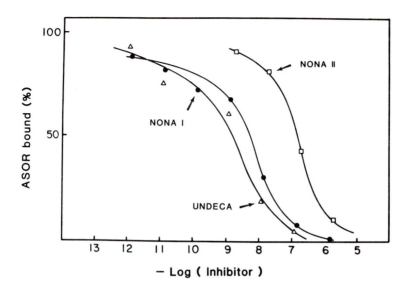

Fig. 5. Inhibition of binding of ^{125}I-asialoorosomucoid by rabbit hepatocytes with synthetic tri- and tetraantennary oligosaccharides. The concentration of the labeled ASOR was 2.5 x 10^{-10} M in a suspencion of 2 x 10^{6} cells/ml.

(Table 1), presumably because of the changes in conformation resulting from reduction.

Figure 5 shows inhibition of ^{125}I-asialoorosomucoid binding to hepatocytes by triantennary and tetraantennary ligands. NONA I possesses the terminal branched structure contained in the triantennary oligosaccharides isolated from α-1-acid glycoprotein[20] and fetuin.[21] Its inhibitory potency was similar to that found for asialo-triantennary oligosaccharide from α-1-protease inhibitor. Another triantennary ligand, NONA II (Fig. 3), was a 20-fold worse inhibitor than NONA I. However, NONA II is a 40-fold better inhibitor than PENTA-2,6, which is a part of NONA II.

Undecassacharide (UNDECA) containing four N-acetyl-lactosamine branches represents the terminal branched structure of the tetraantennary glycopeptides isolated from human α-1-acid glycoprotein.[20] UNDECA was only slightly more inhibitory than NONA I or asialo-triantennary glycopeptide from α-1-protease inhibitor. An additional β-N-acetyl-lactosamine 6-linked to the branching Man in NONA I thus seems to have little additional effect on inhibitory potency. It is possible that the binding requirement of the cell-surface receptors is largely fulfilled by the triantennary configuration of NONA I. However, in view of the difference in inhibitory potency between NONA I and NONA II, it is also possible that tetraantennary oligosaccharides of different structural design from UNDECA may show increase in binding affinities greater than 10^{-9} M.

The relationships of the binding activities described above to the conformation of the oligosaccharides is extremely interesting. The preferred conformations of oligosaccharides depend on the conformations of the individual monosaccharides and the torsion angles at the glycosidic linkages. The conformation of the complex N-acetyl-lactosamine type carbohydrate portions of glycoproteins have been determined[22] using computerized molecular modeling corrected for the exo-anomeric effect (HSEA-calculations).[23,24] The theoretical calculations have been corroborated by [1]H- and ^{13}C-nuclear magnetic resonance investigations[22] performed on synthetic oligosaccharides[24] derived from these structures.

These studies indicated that the oligosaccharides adopt conformations in solution with the N-acetyl-lactosamine units widely separated in space. The HSEA-calculations allow determination of the distance between the terminal β-D-galactosyl groups (i.e., between the C-4s) in these structures. For example, for the UNDECA oligosaccharide in its most preferred conformation, the Gal-Gal distances (A, B, C, D, E, and F) are shown in Fig. 6. Distance A (∼15 Å) is the same as the Gal-Gal distance in PENTA-2,4, whereas distance D (∼25 Å) is identical to that in HEPTA. Distance C (∼23 Å) corresponds to that in PENTA-2,6 (Fig. 6).

A minor proportion of the latter two oligosaccharides can also be present in alternate conformers (i.e., *gauche-gauche* conformers instead of *gauche-trans* conformers at the C-5--C-6 bonds of the branching D-mannose residues). For example, the Gal-Gal distance in PENTA-2,6 can be 14 Å in the alternate conformation. For HEPTA, however, either conformer results in ∼25 Å Gal-Gal distance.

These data are in good agreement with the binding results of the hepatic lectin. Since PENTA-2,4 binds better than other oligosaccharides with two terminal β-D-galactosyl residues, two Gal-binding sites in the lectin are most likely to be 15 Å apart. PENTA-2,6 binds better than HEPTA because a minor population of PENTA-2,6 can be a conformer with 14 Å between the terminal β-D-galactosyl groups. The same type of argument holds for the relative binding strengths of NONA I and NONA II if the "biantennary" parts of the "triantennary" structure can be regarded to be contributing to the binding independently. From these results, it may be speculated that the hepatic

358

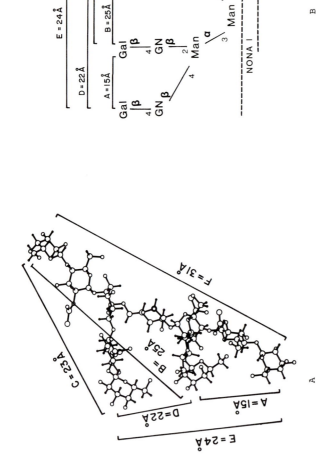

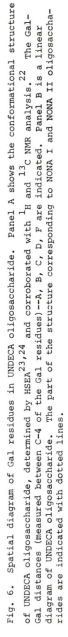

A

B

Fig. 6. Spatial diagram of Gal residues in UNDECA oligosaccharide. Panel A shows the conformational structure of UNDECA oligosaccharide, determined by HSEA[23,24] and corroborated with ^{1}H and ^{13}C NMR analysis.[22] The Gal–Gal distances (measured between C-4 of the Gal residues)--A, B, C, D, F are indicated. Panel B is a linear diagram of UNDECA oligosaccharide. The part of the structure corresponding to NONA I and NONA II oligosaccharides are indicated with dotted lines.

TABLE 2

ENERGETICS OF LIGAND BINDING BY RABBIT HEPATOCYTES

Ligand	$\Delta\Delta G$ (Kcal/mol)[a]
Gal-GlcNAc-Man	0
HEPTA	0.42
LACTO-2,6	1.37
PENTA-2,6	1.53
BI-GP	1.62
PENTA-2,4-OL	1.68
PENTA-2,4	2.71
MONA II	3.20
TYR-TRI-GP	4.93
TRI-GP	5.13
NONA I	5.27
UNDECA	5.68

[a] Using the monoantennary trisaccharide as a reference point, enhancement of binding was calculated using the relationship $\Delta\Delta G = RT\ln \frac{K_d}{K_{d'}}$, where $K_{d'}$ is the apparent dissociation constant for the receptor ligand complex of Gal-GlcNAc-Man and K_d is that of the ligand being considered. The greater $\Delta\Delta G$ values indicate stronger binding. The K_d values obtained from inhibition of labeled TYR-Tri-GP (Table 1) were used in these calculations.

lectin has three Gal binding sites the distances of which are about 15 and 25 Å apart.

The enhanced inhibitory power of the clustered oligosaccharides can be expressed in thermodynamic terms as shown in Table 2. When the monovalent Gal-GlcNAc-Man is used as a reference compound, binding of HEPTA, PENTA-2,6, PENTA-2,4, NONA II, and NONA I are stronger by 0.42, 1.53, 2.71, 3.20, and 5.27 Kcal/mol, respectively.

Our results presented here demonstrate the advantages of using synthetic oligosaccharides of defined structure. An interesting question of how the lectin on the hepatocyte surface discriminates the fine structural features of oligosaccharides and glycopeptides will be our next concern. Whether this is a property of a single lectin molecule or is related to aggregation of lectin molecules on the cell surface awaits further investigation.

360

ACKNOWLEDGEMENTS

We would like to thank Dr. Theresa Kuhlenschmidt for generously supplying Gal_{44}-AI-BSA. Yuan Chuan Lee is supported by United States Public Health Service-National Institutes of Health Research Grants AM 9970 and CA 21901. R. Reid Townsend is supported by United States Public Health Service-National Institutes of Health Training Grant HL06188 and by the Daland Fund of the American Philosophical Society. Mark R. Hardy is supported by a Fellowship from The Johns Hopkins University and Jörgen Lönngren is supported by the Swedish Natural Science Research Council.

REFERENCES

1. Hammarström, S. (née Dillner), et al. (1982) Proc. Natl. Acad. Sci. USA, 79, 1611-1615.
2. Connolly, D. T., et al. (1982) J. Biol. Chem., 257, 939-945.
3. Baenziger, J. and Maynard, Y. (1980) J. Biol. Chem., 255, 4607-4613.
4. Kornfeld, K., Reitman, M. L., and Kornfeld, R. (1981) J. Biol. Chem., 256, 6633-6640.
5. Lee. Y. C., et al. (1983) J. Biol. Chem., 258, 199-202.
6. Arnarp, J. and Lönngren, J. (1980) J. Chem. Soc. Chem. Comm., 788, 1000-1002.
7. Arnarp, J. and Lönngren, J. (1981) J. Chem. Soc., Perkin Trans. I, 7, 2070-2074.
8. Arnarp, J., Haraldsson, M., and Lönngren, J. (1981) Carbohydr. Res., 97, 307-313.
9. Arnarp, J., Haraldsson, M., and Lönngren, J. (1982) J. Chem. Soc., Perkin Trans., I, 8, 1841-1844.
10. Townsend, R. R., et al. (1982) J. Biol. Chem., 257, 9704-9710.
11. Hodges, L. C., Laine, R., and Chan, S. K. (1979) J. Biol. Chem., 254, 8208-8212.
12. Greenwood, F. C., Hunter, W. M., and Glover, J. S. (1963) Biochem. J., 89, 114-123.
13. Lee, Y. C. (1972) Methods Enzymol., 28, 63-72.
14. Schnaar, R. L., et al. (1978) J. Biol. Chem., 253, 7940-7951.
15. Lee, Y. C., Stowell, C. P., and Krantz, M. J. (1976) Biochemistry, 15, 3956-3963.
16. DeLean, A., Munson, P. J., and Rodbard, D. (1978) Am. J. Physiol., 235, E97-E102.
17. Munson, P. J. and Rodbard, D. (1980) Anal. Biochem., 107, 220-239.
18. Krantz, M. J., et al. (1976) Biochemistry, 15, 3963-3968.
19. Lee, R. T. and Lee, Y. C. (1980) Biochemistry, 19, 156-163.
20. Fournet, B., et al. (1978) Biochemistry, 17, 5206-5214.
21. Nilsson, B., Nordén, N. E., and Svensson, S. (1979) J. Biol. Chem., 254, 4545-4553.
22. Bock, K., Arnarp, J., and Lönngren, J. (1983) Eur. J. Biochem., 129, 171-178.
23. Lemieux, R. U., et al. (1980) Can. J. Chem., 58, 631-653.
24. Thøgersen, H., et al. (1982) Can. J. Chem., 60, 44-57.
25. Connolly, D. T., et al. (1983) Biochem. J., 214, 421-431.

THE CHEMISTRY AND FUNCTION OF GONADOTROPINS
OF PIKE EEL AND CHINESE CARPS

FORE-LIEN HUANG AND TUNG-BIN LO
Institute of Biological Chemistry, Academia Sinica and Institute of Biochemical
Sciences, National Taiwan University, Taiwan, Republic of China

ABSTRACT

Gonadotropins (GTHs) of pike eel, carp, silver carp and big-head carp were
isolated from pituitary glands of respective fishes by 40% ethanol-6% ammonium
acetate, pH 5.1 extraction and were purified by DEAE-cellulose chromatography
and electrophoresis. These hormones were biologically characterized by the
stimulation of androgen production in carp testes. The amino acid composition
of these hormones were similar to each other. Two dissimilar subunits (pSI
and pSII) of pike eel GTH were dissociated in propionic acid and separated
by hydrophobic interaction chromatography on phenyl-Sepharose CL-4B. pSI
showed no biological activity but pSII exhibited very low GTH activity. Two
subunits reassociated in 0.05 M phosphate buffer, pH 7.4 to restore GTH
activity. The results of radioimmunoassay of various piscine GTHs showed that
the biological activity and immunogenic cross-reactivity were well correlated,
and the species specificity of mammalian, marine fish and freshwater fish GTHs
was well defined. The androgens produced in carp testes by piscine GTHs were
also analyzed by high pressure liquid chromatography and it was found that the
major androgens were 11-ketotestosterone, 11β-hydroxy-androstenedione and 11β-
hydroxytestosterone.

INTRODUCTION

The reproductive processes of vertebrates are closely regulated by gonadotropins (GRHs) secreted from anterior pituitary glands. In mammals, there are
two types of GTHs, follitropin (FSH) and lutropin (LH). Both of them have been
well studied, including their chemical structure, immunological properties
and physiological functions. In comparison, study on fish GTH is not so
extensive as those of mammalian GTHs. The available data concerning the
chemical structure, immunological properties and physiological function of
fish GTH are limited. In Taiwan, large quantity of pituitary glands from
several species of fishes can be collected. We therefore carry out a series
of work on fish GTH. In this study, GTHs of four species of fishes, pike eel
(*Muraenesox cinereus*), carp (*Cyprinus carpio*), silver carp (*Hypophthalmichthy*

molitrix) and bighead carp (*H. nobilis*) are investigated. In this paper, their chemical properties, immunological properties and preliminary physiological function (steroidogenesis) are presented.

BIOASSAY OF FISH GTH

In our study, four bioassays were used to determine the GTH activity.

^{32}P-phosphate uptake by 1-day old chicken testes. The procedures were essentially following those of Breneman et al.[1] The minimal effective dose was 1 µg for ovine LH (oLH) and 2 µg for pike eel GTH (pGTH).

Induction of catfish ovulation. The procedures were the same as previously described.[2] The minimal effective dose of pGTH was 40 µg.

Stimulation of testosterone production by dispersed rat Leydig cells. The procedures were essentially the same as those described by Moyle and Ramachandran.[3] The minimal effective dose of oLH was 5 ng whereas for pGTH it was 0.5 mg.

Stimulation of androgen production by carp testis *in vitro*. This method was recently developed in our laboratory.[4] The minimal effective dose of carp GTH (cGTH) was 25 ng.

Although the sensitivity of these four bioassays was different, yet the relative activities of various preparations of pGTH assayed by these four methods were well correlated. Because method four had the highest sensitivity, therefore it was used routinely for bioassay of fish GTH.

CHEMISTRY OF FISH GTH
Isolation and purification of fish GTH

Fish pituitary glands were either preserved by freezing or in acetone. They were sonicated in 40% ethanol-6% ammonium acetate, pH 5.1 and then extracted as described by Hartree.[5] Crude GTH was precipitated by adjusting ethanol concentration up to 80%. After desalting by gel filtration on a column of Sephadex G-25, it was fractionated by chromatography on a DEAE-cellulose column. pGTH was eluted by 30 to 80 mM whereas carp, silver carp and bighead carp GTH (cGTH, sGTH and bGTH, respectively) were eluted by 200 to 300 mM ammonium bicarbonate. pGTH was further fractionated by preparative disc gel electrophoresis and five fractions were obtained.[6]

Recently, the isolation and purification of GTHs from fish pituitary glands have been reported by several authors. From these data, it was found that the GTH content in fish pituitary glands was unusually high, e.g., 4.2% in carp,[7] 7.6% in sturgeon[8] and 8% in pike eel.[6] In general, the GTH content in mammalian pituitary glands was rather low, e.g., 0.6% of LH[9] and 0.003% of FSH[10] in ovine pituitary glands.

The electrophoretic properties of fish GTH

The disc gel electrophoretic pattern of pGTH was complicated. There were six bands, denoted as G1 to G6. All of them, except G1 (not assayed), had GTH activity. Among them G4 had the highest and G6 the lowest activity. They all showed practically identical SDS gel electrophoretic patterns with 2 bands. Their molecular weights were estimated to be 10,500 and 15,000, respectively.[6] These results suggested that pGTH was composed of isohormones and that each isohormone was composed of two nonidentical subunits.

The disc gel electrophoretic patterns of sGTH and bGTH were similar to each other. They all consisted of two electrophoretic zones. The zone with slower mobility (Rf = 0.36) had higher GTH activity than the zone with faster mobility (Rf = 0.56). The SDS gel electrophoretic patterns of sGTH and bGTH were also similar to each other. All of them consisted of two bands and their molecular weights were estimated to be 14,000 and 16,000, respectively.[11] These results also suggested that sGTH and bGTH were composed of isohormones and each iso-hormone was composed of two nonidentical subunits.

The chemical compositions of fish GTH

The amino acid and carbohydrate compositions of pGTH, sGTH and bGTH are listed in Table 1. The amino acid composition of GTHs from various species of fishes were similar to one another. Two characteristic features were observed, high content of half cystine and proline as in mammalian LH and FSH, and more acidic residues than basic residues.

Isolation of pGTH subunits

The pGTH was dissociated in 1 M propionic acid at a concentration of 4 mg/ml for 24 hr at 35° C. After lyophilization, the dissociated pGTH was subjected to hydrophobic interaction chromatography on a phenyl-Sepharose 4B column pre-equilibrated with 0.1 N acetic acid. The elution was started with 0.1 N acetic acid (unretarded fraction) and followed by 0.1 M ammonium acetate (retarded fraction). When these two fractions were subjected to SDS gel electrophoresis, they corresponded to the two bands of the native pGTH. The unretarded fraction showed faster mobility and was denoted as subunit I (pSI) whereas the retarded fraction showed slower mobility and was denoted as sub-unit II (pSII).[12]

The pSI had no GTH activity but pSII had 3% GTH activity as compared to intact pGTH. However, when pSI and pSII were co-incubated in 50 mM phosphate buffer (pH 7.4) at 25°C for 24 hr, 75% GTH activity was restored.[12] These

TABLE 1

CHEMICAL COMPOSITION OF PISCINE AND HUMAN GTHs

Compositions	cGTH[d]	pGTH[e]	sGTH[f]	bGHT[f]	hLH[g]	hFSH[h]
Amino acids[a]						
Lys	6.6	5.2	5.1	4.6	4.2	5.5
His	2.8	2.6	2.7	2.6	3.1	2.5
Arg	4.2	5.8	3.9	3.3	6.8	3.8
Asx	9.9	9.7	9.0	9.4	6.1	6.4
Thr	7.5	7.4	8.1	8.6	7.7	8.5
Ser	6.1	8.6	6.6	6.7	6.7	8.1
Glx	8.0	8.5	10.7	10.0	7.9	9.7
Pro	9.0	8.4	11.4	10.6	10.3	6.8
Gly	3.3	3.8	3.5	3.7	5.9	4.7
Ala	3.3	3.4	3.4	3.4	4.3	5.1
½ Cys	10.4	8.9	7.5	7.9	10.5	8.9
Val	9.4	7.2	9.6	9.7	8.4	6.8
Met	2.4	1.5	1.6	1.6	2.3	5.5
Ile	2.4	3.9	3.0	2.6	2.8	3.0
Leu	6.6	6.1	6.3	6.5	6.1	6.4
Tyr	5.2	5.1	4.6	4.8	2.9	4.7
Phe	2.8	3.6	3.1	3.1	3.0	3.4
Trp	_[b]	-	-	-	0.5	0.4
Carbohydrate[c]						
Neutral sugar	12.5	14.6	-	-	17.0	7.3
Amino sugar	12.0	5.9	-	-	14.3	5.3
Sialic acid	0.35	1.0	-	-	3.1	5.0
N-Terminal	Tyr	Tyr	-	-	Val	Ala
	Ser	Ser	-	-	Ser	Asn

[a]Residue per 100 residues.

[b]Not determined.

[c]g per 100 g protein.

Data taken from (reference number in parenthesis) d (17); e (6); f (11); g (30; h (31).

results indicate that the pSI and pSII thus isolated are the real and only subunits of pGTH.

The reassociation rate of pSI and pSII to form functional pGTH molecule was fast. The plateau was reached within 1 hr of incubation, while 60% of plateau level was achieved in 15 min of incubation. Moreover, the same extent of reassociation was also achieved at 2°C, 15°C, 25°C, or 35°C after 24 hr of incubation.[13] Similar results had also been observed in cGTH.[14] The reassociation of subunits of human chorionic gonadotropin[15] and human LH[16] were slower and temperature dependent.

The amino acid composition of pGTH subunits

As shown in Table 2, the amino acid composition of pSI and pSII were different from each other. Both of them had high contents of proline, half-cystine, and aspartic acid/asparagine.[13] In comparison with other GTHs, the amino acid composition of pSI and pSII were close to those of the corresponding cGTH subunits[17] but different from those of the corresponding mammalian GTH subunits.[18,19]

IMMUNOLOGICAL PROPERTIES OF FISH GTH AND ITS SUBUNITS

The antisera of cGTH, pGTH, pSI and pSII were induced in albino rabbit and the corresponding antiserum was denoted as AS-cGTH, AS-pGTH, AS-pSI and AS-pSII, respectively. When AS-cGTH was tested, its cross-reactivities to the following GTHs and subunits were: cGTH, 100%; sGTH, 152%; bGTH, 146%; pGTH, 9.8%; pSII, 13.8%; pSI, oLH and oFSH, 0%. In comparison, the cross-reactivities of AS-pGTH to the following GTHs and subunits were: pGTH, 100%; pSII, 37%; pSI, 3.7%; sGTH, 0.9%; cGTH, 0.5%; bGTH, 0.3%; oLH and oFSH, 0%.[20] These results indicate that the degree of cross-reactivity of GTH antiserum to heterologous GTHs was related to their phylogenic position.

The cross-reactivities of AS-pSI and AS-pSII to pGTH, pSI and pSII were also investigated. pSII slightly (5.8%) but pGTH fully competed with pSI in binding to AS-pSI. On the other hand, pSI could not but pGTH highly (70%) competed with pSII in binding to AS-pSII. The variability of the reactivity of GTH to the antisera of its subunits was also found in other types of GTH. The cGTH reacted slightly more to antiserum of subunit II than to that of subunit I.[21] On the other hand, bovine FSH reacted more to the antiserum of bovine FSH α subunit than to that of β subunit.[22]

THE ACTION OF FISH GTH ON THE STIMULATION OF ANDROGEN PRODUCTION IN CARP TESTIS

Stimulation of androgen production in testis is one of the well-established functions of LH. On the other hand, study of fish GTH on adrogen production has not yet been undertaken. Recently, we have found that fish GTHs can

TABLE 2

THE AMINO ACID COMPOSITION OF PISCINE AND OVINE GTH SUBUNITS

Amino acids[a]	psI[b]	csI[c]	oLHα[d]	oFSHα[e]	psII[b]	csII[c]	oLHβ[d]	oFSHβ[e]
Lys	6.3	10.4	9.6	10.5	3.8	3.4	1.9	7.8
His	3.0	3.1	2.9	3.2	1.8	2.6	2.8	3.1
Arg	4.7	4.2	2.9	4.5	5.8	4.3	6.6	4.0
Asx	9.6	11.4	6.7	1.2	9.6	8.6	4.7	10.0
Thr	6.9	6.2	9.6	6.2	7.3	8.6	4.7	9.5
Ser	6.5	6.2	5.8	5.7	9.4	6.0	4.7	7.5
Glx	10.5	6.2	9.6	12.6	9.1	9.5	5.7	11.0
Pro	7.9	6.2	7.7	5.6	11.1	11.2	18.9	6.7
Gly	3.9	3.1	4.8	4.7	2.6	3.4	5.7	4.9
Ala	4.7	5.2	7.7	8.3	2.7	1.7	6.6	8.3
½ Cys	11.2	10.4	9.6	4.9	9.1	10.3	9.4	8.0
Val	5.0	8.3	5.8	5.5	8.6	10.3	7.6	6.3
Met	1.7	3.1	3.9	0.3	1.6	1.7	1.9	0.3
Ile	3.8	2.1	1.9	2.7	4.1	2.6	3.8	3.6
Leu	5.8	5.2	1.9	9.0	6.7	7.8	11.3	4.8
Tyr	3.6	5.2	4.8	2.7	4.2	5.2	0.9	3.8
Phe	5.1	3.1	4.8	4.2	2.8	2.6	2.8	3.8

[a]Residues per 100 residues.
Data taken from (reference number in parenthesis) b (13); c (17); d (18); e (19).

stimulate carp testis to produce androgen *in vitro*,[4] therefore, we use this
newly developed system to study the action of fish GTHs on androgen production
in carp testis.

The effect of cAMP on androgen production

cAMP can mimic the action of cGTH to stimulate carp testis to produce
androgen *in vitro*. At lower concentration of cGTH, the action of cAMP and cGTH
were additive when they were coexisting in the incubation medium. The effect
of cAMP was gradually decreased as the concentration of cGTH was increased and
finally disappeared when maximal dose of cGTH was present.[23] These results
suggested that the action of cGTH may also be via cAMP as its secondary
messenger as found in the case of mammalian LH.[24-26]

Dependence of the action of cGTH on the *de novo* RNA and protein synthesis to stimulate androgen production

Both inhibitors, actinomycin D (RNA synthesis inhibitor) and cycloheximide
(protein synthesis inhibitor), can inhibit the action of cGTH on the stimulation of androgen production in carp testis.[23] The inhibitory effect of actinomycin D was dependent on the time interval between the addition of cGTH and
actinomycin D. Its inhibitory effect was gradually decreased as the time
interval between the addition of cGTH and actinomycin D and finally disappeared
if the time interval was lagged to three hrs. On the other hand, the inhibitory effect of cycloheximide did not become lessened as the time interval
between the addition of cGTH and cycloheximide was increased. These results
suggested that the expression of the action of cGTH was dependent on the *de novo*
protein synthesis and that the protein thus induced under the action of cGTH
was very labile. However, the RNA for this protein may be more or less stable
and hence may be accumulated. These phenomena were also observed in the case
of mammalian LH,[24,25,27,28]

Qualitative and quantitative analysis of androgen production in carp testis under the action of pGTH

The analyses of androgens of our previous experiments were made by radioimmunoassay (RIA). The antiserum was prepared against testosterone-17-
hemisuccinate. This antiserum cross-reacted strongly to androstenedione (153%)
but weakly to 11-ketotestosterone (6%) as compared to testosterone (100%).
Unless testosterone was the major androgen, otherwise our results obtained by
RIA will not reflect the change of total androgen content. In order to get the
information of the qualitative and quantitative change of androgens under the

action of pGTH, we developed a high pressure liquid chromatography system for androgen analysis. The system consisted of a reverse phase C_{18} column and a solvent system of water:methanol:tetrahydrofuran = 50:40:10. By this system, 6 androgens, adrenosterone (KA), 11-ketotestosterone (KT), 11β-hydroxy-androstenedione (OHA), 11β-hydroxytestosterone (OHT), androstenedione (A) and testosterone (T), can be well separated and quantitated.[29]

The carp testicular androgen content was 0.8 μg/100 g. KT was the major androgen (68%), following T (19%) and A (9%). After *in vivo* pGTH treatment (1 μg/g body weight, 4 hr), the androgen content was increased to 8 μg/100 g, about 10-fold increase. The major androgen was still KT (40%), yet 11β-hydroxylated androgen became very prominent (OHA, 16%; OHT, 27%). In addition, the *in vitro* pGTH treatment also greatly enhanced the androgen production. For *in vitro* experiments, 20 g tissue was incubated in Krebs Ringer bicarbonate buffer, pH 7.4, supplemented with 0.1% bovine serum albumin, 0.03% chick ovomucoid and 1 mM theophylline at 25°C for 4 hr in the presence and absence of pGTH (1 mg). After incubation, the androgen present in the incubation medium was analyzed. In the absence of pGTH, only 0.4 μg androgen was present in the medium. The KT was the major androgen (45%), following T (27%) and then A (17%). When pGTH was added, 10 μg androgen was present in the medium. The androgen composition was also greatly changed: OHT, 43%; OHA, 27%; and KT, 22%.[29]

The above results indicated that KT was the major carp testicular androgen. After pGTH treatment, KT production was enhanced. In addition, OHA and OHT were also increased from barely detectable levels to very prominent levels. These results suggested that OHA and OHT may be the intermediates of KT biosynthesis. By a series of work, we elucidated the major pathway of KT biosynthesis in carp testis *in vitro* as T → OHT → KT.

CONCLUDING REMARKS

In this paper, some preliminary experimental results of chemical, immuno-logical and endocrinological characterizations of fish GTHs are presented. Although they are far from completion, several interesting evidences are observed. Fish GTHs consist of two nonidentical subunits as in mammalian hormones but there exists an apparent species-specificity both endocrino-logically and immunologically. Fish GTHs do stimulate androgen production as mammalian hormones do but the major androgen in carp testis is 11-keto-testosterone rather than testosterone. For further studies of fish GTHs, obviously more efforts are needed. Elucidation of molecular structures, hybridization of subunits of GTHs from various species, pathway of steroido-genesis in fish testis are important to realize the molecular evolution of GTHs and to manipulate the fish reproductive problems. They are in progress in our laboratory.

ACKNOWLEDGEMENTS

Thanks are due to Ms. Yea-sha Chang, Messrs. Chang-jen Huang, Geen-dong Chang, Te-chong Liu, Tse-chong Wu and Wen-juin Hsieh for their technical assistance. This work was supported in part by grants from the National Science Council, Republic of China.

REFERENCES

1. Breneman, W. R., Zeller, F. J., and Greek, R. O. (1962) Endocrinology, 71, 790-798.
2. Huang, F. L., et al. (1974) J. Fish. Soc. Taiwan, 3, 41-50.
3. Moyle, W. R. and Ramachandran, J. (1973) Endocrinology, 93, 127-134.
4. Huang, F. L. and Chang, Y. S. (1980) Proc. Natl. Sci. Council, ROC, 4, 392-400.
5. Hartree, A. S. (1966) Biochem. J., 100, 754-761.
6. Huang, F. L., et al. (1981) Int. J. Peptide Protein Res., 18, 69-78.
7. Burzawa-Gerard, E. (1971) Biochimie., 53, 545-552.
8. Burzawa-Gerard, E., Goncharov, B. F., and Fontaine, Y.A. (1975) Gen. Comp. Endocrinol., 27, 289-295.
9. Papkoff, H., et al. (1965) Arch. Biochem. Biophys., 111, 431-438.
10. Scherwood, L. D., Grimek, H. J., and McShan, W. H. (1970) J. Biol. Chem., 245, 2328-2336.
11. Chen, C. T. (1982) Master Thesis, Graduate Institute of Biochemical Science, National Taiwan University, pp. 1-89.
12. Lo, T. B., Huang, F. L., and Chang, G. D. (1981) J. Chromatogr., 215, 229-233.
13. Huang, F. L., et al. (1982) Proc. Natl. Sci. Council ROC, 6, 30-36.
14. Fontaine, Y. A. and Burzawa-Gerard, E. (1978) in Structure and Function of the Gonadotropin, McKerns, K. W., ed., New York: Plenum Press, pp. 361-380.
15. Aloj, W. M., et al. (1973) Arch. Biochem. Biophys., 159, 497-504.
16. Ingram, K. C., Aloj, W. M., and Edelhoch, H. (1973) Arch. Biochem. Biophys., 159, 596-605.
17. Jolles, J. E., et al. (1977) Biochimie, 59, 893-898.
18. Bahl, O. P. (1973) in Hormonal Proteins and Peptides, Li, C. H., ed., Vol. 1, New York: Academic Press, pp. 171-200.
19. Papkoff, H. and Ekblad, M. (1970) Biochem. Biophys. Res. Commun., 40, 614-621.
20. Chang, Y. S., Huang, F. L., and Lo, T. B., unpublished data.
21. Burzawa-Gerard, E. and Kerdelhue, B. (1978) Ann. Biol. Anim. Bioch. Biophys., 18, 773-780.
22. Workewych, J. and Cheng, K. W. (1979) Endocrinology, 104, 1075-1082.
23. Chang, Y. S. and Huang, F. L. (1982) Gen. Comp. Endocrinol., 48, 147-153.
24. Dufau, M. L., Catt, K. J., and Tsuruhara, T. (1971) Biochim. Biophys. Acta., 252, 572-579.
25. Cigorraga, S. R., Dufau, M. L., and Catt, K. J. (1978) J. Biol. Chem., 253, 4297-4304.

26. Cooke, B. Z., et al. (1979) Biochim. Biophys. Acta., 583, 320-331.
27. Dufau, M. L., Mendelson, C., and Catt, C. J. (1974) J. Clin. Endocrinol. Meta., 39, 610-613.
28. Mendelson, C., Dufau, M., and Catt, C. J. (1975) Biochim. Biophys. Acta., 411, 222-230.
29. Huang, F. L., et al. (1982) Proc. Natl. Sci. Council ROC, in press.
30. Sairam, M. R. and Li, C. H. (1975) Biochim. Biophys. Acta., 412, 70-81.
31. Rathnam, P. and Saxena, B. B. (1971) J. Biol. Chem., 246, 7087-7094.

ISOLATION AND PROPERTIES OF CHUM SALMON PROLACTIN

HIROSHI KAWAUCHI, KEN-ICHI ABE, AKIYOSHI TAKAHASHI, TETSUYA HIRANO[*]
SANAE HASEGAWA,[*] NOBUKO NAITO[**] AND YASUMITSU NAKAI[**]
School of Fisheries Sciences, Kitasato University, Sanriku, Iwate 022-01;
[*]Ocean Research Institute, University of Tokyo, Nakano, Tokyo 164; and
[**]School of Medicine, Showa University, Shinagawa, Tokyo 142, Japan

ABSTRACT

A highly purified prolactin (PRL) was isolated from the chum salmon pituitary by extraction with acid acetone, gel filtration on Sephadex G-25, and ion-exchange chromatographies on DEAE-cellulose and CM-Sephadex C-25 with a yield of 1 mg/g of wet tissue. It was 100 times more potent than that of ovine PRL in sodium-retaining activity for hypophysectomized *Fundulus*. The salmon PRL emerged as a single and symmetrical peak on Sephadex G-100 with Ve/Vo = 2.0. Polyacrylamide gel electrophoresis revealed only one band at pH 4.3, whereas no band was seen at pH 7.5. The isoelectric point was estimated to be 10.3 by gel electric focusing. The circular dichroism spectrum of the salmon PRL was similar to that of *Tilapia* PRL, showing an α-helix content of 50%. The salmon PRL had a molecular weight of 23,400 daltons by gel filtration and 22,300 daltons by SDS gel electrophoresis, with a single amino-terminal residue, isoleucine, and a single carboxyl terminal residue, half-cystine. In the sequence comparison with that of human PRL and growth hormone, the clusters of invariant residues were found in both terminal regions, although the disulfide at amino-terminal of mammalian PRLs was missing. An antibody raised to the highly purified salmon prolactin reacted only with PRL cells in the pituitary of chum salmon.

INTRODUCTION

It has been suggested that prolactin (PRL) and growth hormone (GH) have evolved from a common ancestral molecule, based mainly on structural similarities between the mammalian hormones.[1] PRLs are known to exhibit multiple actions among vertebrates, whereas GH is exclusively concerned with stimulation of animal growth.[2,3] In order to obtain better insight into the evolutionary history of these hormones and the great diversity of their functions, it is essential to isolate and characterize PRLs and GHs of lower vertebrates.

In fish, *Tilapia* PRL has now been purified and extensively characterized physicochemically.[4] However, little is known about the primary structure.

This paper describes the purification procedure, and physicochemical and biological properties of the chum salmon PRL, together with partial amino acid sequence.

MATERIALS AND METHODS

Purification. Whole pituitary glands were taken from mature female chum salmon, *Oncorhynchus keta*, caught in Tsugaruishi River, Iwate, Japan during the month of December.

Routinely 50 g of the pituitaries were extracted with 150 ml of acid acetone (conc. HCl:acetone, 1:28, v/v) at 0 C for 1 hr and the residue was re-extracted with 100 ml of 80% acetone at 0 C for 1 hr. The extract was precipitated by addition to acetone (3 l.) prechilled for 24 hr at 4 C. The precipitate was submitted to gel filtration on Sephadex G-25 (medium) column (5.5 x 68 cm) in 0.1 N acetic acid. The unretarded fraction was lyophilized and dissolved in water at pH 9.0. The solution was introduced into a column of DEAE-cellulose, equilibrated with 0.05 M ammonium acetate pH 9.0. An adsorbed fraction was successively subjected to ion-exchange chromatography of CM-Sephadex C-25 (1.7 x 20 cm), equilibrated with 0.05 M ammonium acetate, pH 4.6. Elution was performed with a gradient formed by passing 0.2 M ammonium acetate, pH 9.0 through a mixing flask containing 300 ml of the starting buffer.

Physicochemical characterization. The purified PRL was examined by slab gel electrophoresis at pH 4.3 and 7.5 in 7.5% polyacrylamide gel[5] stained with 0.25% Coomassie brilliant blue R 250[6] and by gel isoelectric focusing in 5% polyacrylamide gel at 2% ampholine (pH 3.5-10 and pH 7-11),[7] stained with 0.04% Coomassie brilliant blue G 250 in 3.5% $HClO_4$.[8] The isoelectric point was estimated by gel isoelectric focusing with ampholine (pH 7-11). Protein samples containing pI-marker proteins (Oriental Yeast Co.) were focused in duplicate disc gels. The molecular weight was estimated by SDS slab gel electrophoresis[9] and by exclusion chromatography on Sephadex G-100 using 0.1 M Tris-HCl, pH 8.2.

Circular dichroism spectra were taken in a Cary model 60 spectropolarimeter, equipped with a model 6002 circular dichroism attachment in 0.1 M Tris-Cl, pH 8.2. The content of α-helix was estimated as described by Bewley et al.[10]

An absorptivity value of the salmon PRL, $E_{280 \text{ nm}}^{0.1\%}$ was determined to be 0.306 based on weight, assuming a 15% moisture content. Protein concentrations were determined from absorption spectra after correction for light scattering as described by Beaven and Holiday.[11]

Amino acid sequence analysis. For amino acid analysis, the protein was hydrolyzed for 22 hr at 110 C with constant boiling HCl. The analysis was carried out with an LKB automated amino acid analyzer (Type 4400). Amino-terminal analysis was performed by the dansyl method[12] and carboxyl-terminal residue was determined by amino acid analysis of hydrazinolysate of performic acid oxidized PRL, according to Akabori et al.[13] The amino acid sequence of the intact hormone was determined by the fluorescein-isothiocyanate as described previously.[14] The fluorescein-thiohydantoin amino acids were identified by thin layer chromatography on polyamide sheets.

For further determination of the amino acid sequence, the protein was treated with cyanogen bromide, followed by performic acid oxidation. The resulting peptides were fractionated by gel filtration on Sephadex G-50 in 0.1 N acetic acid and Sephadex G-75 in 0.1 N acetic acid, by droplet countercurrent chromatography on DCC-A (Tokyorika Co.) with n-butanol:pyridine:0.6 M ammonium acetate (pH 6.8) (5:3:10, v/v), and high voltage paper electrophoresis in formic acid-acetic acid, pH 1.9 at 2,000 V. Each fragment was digested by trypsin, chymotrypsin and thermolysin, and subjected to high performance liquid chromatography on reverse phase column C 18 with 10 mM ammonium acetate (pH 4.0) and isopropanol. The amino acid sequence of these peptides were determined by the dansyl Edman procedures.[15]

Biological characterization. Sodium-retaining activity of the salmon PRL was examined using hypophysectomized *Fundulus*. Ovine PRL (NIAMDD o-PRL-14) was used as a reference standard. The fishes were given intraperitoneal injections of saline (0.9% NaCl) or PRLs. They were sacrificed 24 hr after the injection. Blood samples were taken from the caudal artery following the method described by Jozuka and Adachi.[16] Aliquots of plasma (1 μl each) were diluted to 1 ml with deionized water and Na concentration was determined by atomic absorption spectrophotometer (Hitachi 208).

Immunohistochemical characterization of PRL cells. Antisera were raised against the salmon PRL in rabbits. The antigen (1 mg) was dissolved in 0.9% NaCl (1 ml) and emulsified with complete Freund's adjuvant (1 ml). Seven injections, 1 mg each, were given subcutaneously into the foot pads and the back at intervals of three weeks. They were bled two weeks after the last injection, and the sera were lyophilized. An immunoglobulin (IgG) fraction was prepared from the antisera by ammonium sulfate fractionation. The anti-salmon PRL rabbit IgG was used for immunocytochemical identification of PRL secreting cells in the chum salmon pituitary with unlabelled peroxidase anti-peroxidase method as described by Sternberger et al.[17] Specificity of the

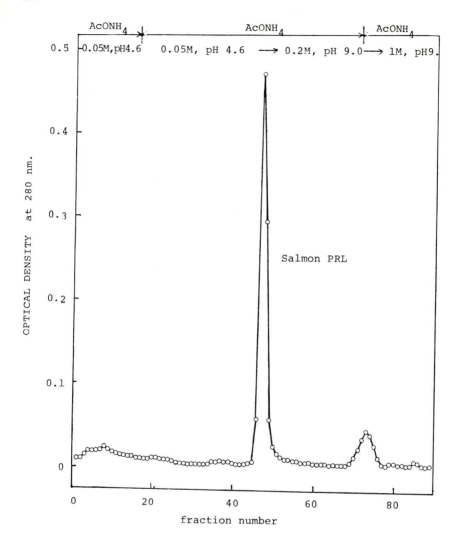

Fig. 1. Ion-exchange chromatography of the salmon PRL fraction on CM-Sephadex C-25. Initially, acid acetone extract of the salmon pituitary was chromatographed on Sephadex G-25. The protein fraction was subjected to ion-exchange chromatography of DE-52, equilibrated with 0.05 M ammonium acetate. The unadsorbed fraction was chromatographed on a column of CM-Sephadex C-25 (1.7 x 20 cm). Elution was performed with a gradient formed by passing 0.2 M ammonium acetate, pH 9.0 through a mixing flask containing 300 ml of the starting buffer.

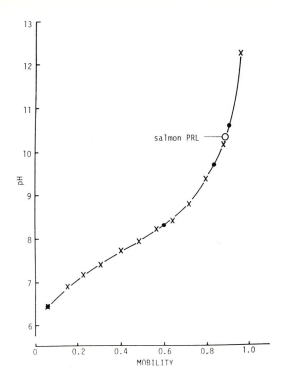

Fig. 2. Estimation of the isoelectric point of the salmon PRL. The pI-marker proteins consisted of cytochrome C from horse and the acetylated cytochrome C; pI, 6.4, 8.3, 9.7, and 10.6 (Oriental Yeast Co.).

immunocytochemical staining was confirmed by replacing the antiserum with normal rabbit serum or with antiserum absorbed with the salmon PRL.

RESULTS

   Isolation. Salmon prolactin was isolated as one of the predominant components from an acid acetone extract of the pituitary. Peptide fractions containing melanotropins and endorphins were separated effectively by gel filtration of Sephadex G-25. Salmon prolactin was separated from a large amount of N-terminal peptides of salmon proopiocortins[18] by ion-exchange chromatography of DEAE-cellulose at pH 9.0. The unadsorbed fraction consisted mainly of prolactin, and purified by ion-exchange chromatography of CM-Sephadex C-25 as shown in Fig. 1. When subjected to gel filtration on Sephadex G-100, the salmon PRL emerged as a single symmetrical peak with Ve/Vo = 2.0. A yield of 50 mg was obtained from 50 g (wet weight) of the pituitaries.

376

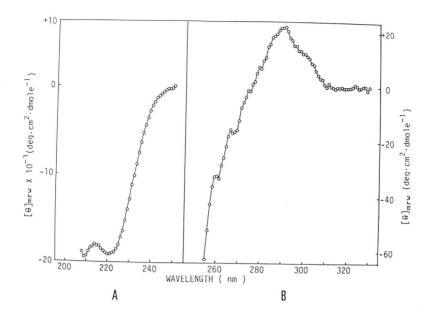

Fig. 3. Circular dichroism spectra of the salmon PRL in 0.1 M Tris-Cl, pH 8.2.
A. The amide band absorption. B. The side chain absorption.

Physicochemical properties. Polyacrylamide gel electrophoresis at pH 4.3
revealed only one intensely stained band, whereas no band was detected at pH
7.5. The isoelectric focusing of the salmon PRL exhibited a single band at
pH 3.5 - 10, and one intensely stained major band with a second faint band
running immediately behind it at pH 7-11. The isoelectric point of the main
band was estimated to be 10.3 (Fig. 2).

The molecular weight of the salmon PRL was determined to be 22,300 daltons
by SDS polyacrylamide gel electrophoresis and 23,400 daltons by exclusion
chromatography on Sephadex G-100.

The circular dichroism spectra are shown in Fig. 3. The salmon PRL exhibited
two negative bands at 221 and 209 nm, which are characteristic of α-helix.
An α-helix content was about 50% to the nearest 5%. The spectrum in the
region of side chain absorption exhibited a positive peak at 291 nm with a
shoulder around 300 nm, a crossover point at 276 nm and two weak, negative
shoulders near 267 and 261 nm.

TABLE 1

AMINO ACID COMPOSITION OF SALMON PRL COMPARED WITH *TILAPIA* AND OVINE PRLs

Amino Acid	Salmon PRL[a]	Tilapia PRL[b]	Ovine PRL[c]
Asp	21.2	16.4	22
Thr	8.0	9.4	9
Ser	25.8	21.6	15
Glu	13.7	17.4	22
Pro	13.4	10.8	11
Gly	9.5	8.1	11
Ala	6.2	9.6	9
Cys/2	3.7	4.2	6
Val	4.6	6.8	10
Met	9.1	5.4	7
Ile	10.0	9.3	11
Leu	26.4	24.5	23
Tyr	2.2	3.0	7
Phe	6.6	4.7	6
Trp	1.0	1.0	2
His	6.6	4.9	8
Lys	13.0	8.9	9
Arg	12.4	6.8	11
Total	193.4	172.8	199

[a]Values were calculated on the basis of the molecular weight of 22,300. Threonine and serine values were corrected for destruction. Tryptophan was determined by spectrophotometric method and by methane sulfonic acid hydrolysis.
[b]Taken from Farmer et al.[4]
[c]Taken from Li et al.[30]

Amino acid sequence. The amino acid composition of the salmon PRL was calculated on the basis of a molecular weight of 22,300 daltons. As shown in Table 1, the total number of the residues by this calculation was 194, which is comparable to that in ovine PRL but exceeds *Tilapia* PRL by 20 residues. Relatively high content of basic amino acids and low of Glu are consistent with the high value of isoelectric point. It is to be noted that the salmon PRL as well as *Tilapia* PRL consisted of 4 half-cystine residues which is in contrast with 6 residues in mammalian PRLs. As in the case for *Tilapia* PRL,

TABLE 2

AMINO ACID COMPOSITION OF CYANOGEN BROMIDE (CB) FRAGMENTS OBTAINED FROM SALMON PRL

Amino Acids	CB-1	CB-2	CB-3	CB-4	CB-5	CB-6	CB-7	CB-8	CB-9	CB-10
Asp	9.6	1.9		4.0	3.4	1.1		0.9		
Thr	2.2	1.0	1.0	0.9	1.5					
Ser	10.6	3.1		4.5	4.8	0.7			0.9	
Glu	8.3		1.0	1.3	2.4					
Pro	6.2		0.9	2.2	2.2				2.4	
Gly	3.3			3.7						
Ala	4.0				1.1	1.1	1.0			
Cys/2*	1.5	2.0	1.0							
Val	2.0	0.8								
Met**	+	+	−	+	+	+	+	+	+	+
Ile	4.8	0.6		2.9		1.0	1.1	1.4		
Leu	13.9	1.6		3.8	4.0	2.1	+	+		
Tyr	1.0			0.8						
Phe	0.4	1.7		3.0	1.0					
His	1.9	0.9		1.0	1.6					
Lys	6.1	2.8		2.0	2.4			1.0		
Arg	2.9	3.7	1.0	0.9	2.3		0.8		0.9	
Trp	+***									
Total****	(81)	24	5	33	28	7	4	4	5	1
NH$_2$-terminal residue	ND	Ser	Arg	Gly	Glu	Ile	Gly	Val	Pro	Met

The salmon PRL was treated with cyanogen bromide, followed by performic acid oxidation. The resulting peptides were fractionated on Sephadex G-50 and G-75 in 0.1 N acetic acid. The smaller peptides were separated by high voltage paper electrophoresis in formic acid-acetic acid, pH 1.9. The larger peptides were purified by droplet countercurrent chromatography with a solvent, n-butanol : pyridine : 0.6 M ammonium acetate (pH 6.8) (5:3:10, v/v).

* Determined after performic acid oxidation.
** Homoserine + homoserine lactone.
*** Detected by the ultraviolet absorption.
**** The values were obtained by sequence analysis except for CB-1.

relatively low content of tryptophan and tyrosine made the extinction coefficient low, 0.306.

Isoleucine was identified as a sole amino-terminal residue by the dansyl method. The amino-terminal sequence as determined by analysis of the intact hormone was H-Ile-Gly-Leu-Ser-Asp-Leu-Met-Glu-Arg-Ala-. Hydrazinolysis of the intact hormone gave no amino acid, whereas that of the performic acid oxidized hormone yielded cysteic acid. Thus, the COOH-terminal residue was determined to the half-cystine as are mammalian and *Tilapia* PRLs.

The salmon PRL gave 10 peptide fragments by cyanogen bromide cleavage and performic acid oxidation after fractionation using gel filtration, droplet countercurrent chromatography and high voltage paper electrophoresis. The amino acid composition of these peptides is summarized in Table 2 and the total residues are consistent with that of the intact hormone. By analyzing amino acid sequence of the fragments as well as the intact hormone, CB-6 was identified to be the amino-terminal peptide, followed by CB-5. It is evident that the salmon PRL lacks the amino-terminal disulfide loop of the mammalian PRLs. CB-3 was assigned to the carboxyl-terminal peptide, due to the presence of half-cystine residue and absence of methionine at the carboxyl-terminal. CB-2 exhibited significant sequence homology with the carboxyl-terminal portion of mammalian PRLs and GHs, especially of human PRL (172-192). The partial amino acid sequence is aligned with the structures of human PRL and GH in Fig. 4.

Biological properties. Figure 5 shows the sodium-retaining activity of the salmon PRL. When tested in hypophysectomized *Fundulus*, both salmon and ovine PRLs caused a significant increase in plasma sodium after 24 hr. The salmon PRL was 100 times more potent than that of ovine PRL; 20 ng/g of the salmon PRL gave a similar response with 2 μg/g of ovine PRL.

Immunohistochemical identification of PRL cells. Immunocytochemical staining of the chum salmon pituitary with anti-salmon PRL rabbit IgG, showed strong cross-reaction only in follicle-forming cells of the rostral pars distalis at a dilution of 1/4000 (Fig. 6). These cells correspond with the PRL cells of chum salmon identified with light and electron microscopy,[18] and distinguished from the GH cells in the proximal pars distalis. No specific reaction was observed between the antibody and the GH cells at all concentrations tested.

DISCUSSION

Extraction of pituitary with acid acetone has been employed for isolation of mammalian PRL as well as ACTH and β-LPH.[20] In contrast, fish PRLs have been

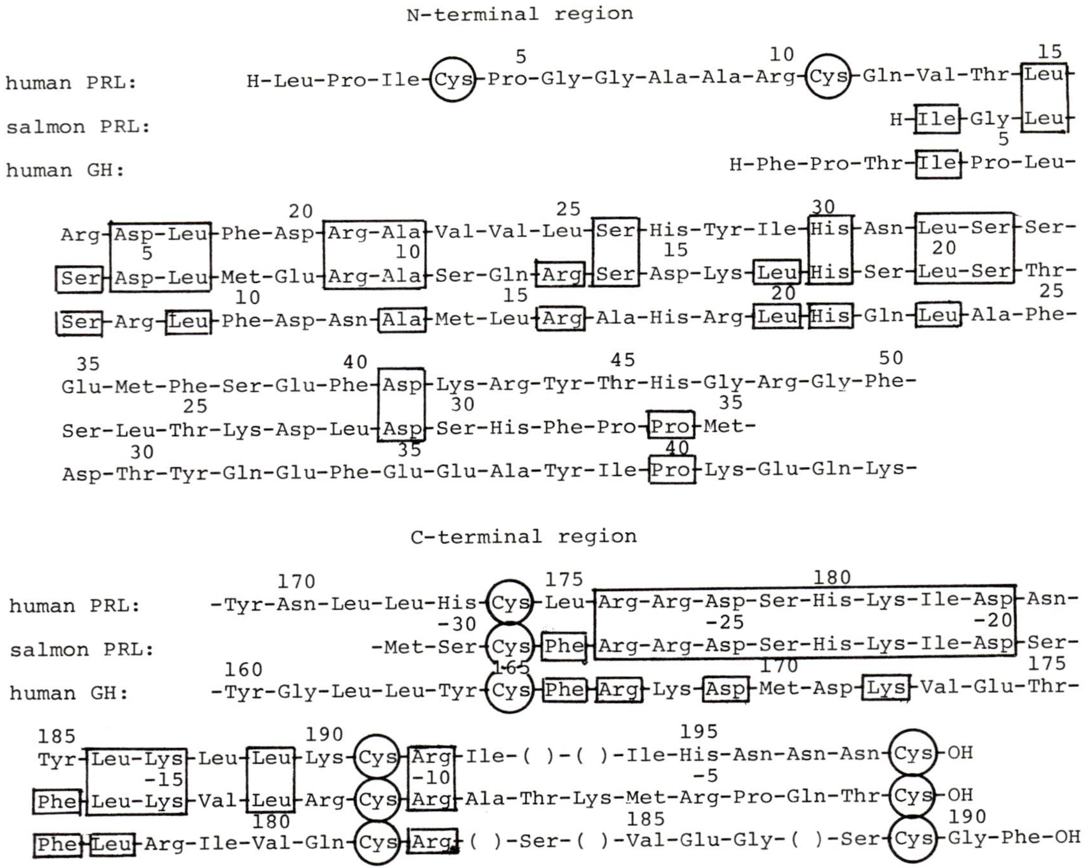

380

N-terminal region

human PRL: H-Leu-Pro-Ile-(Cys)-Pro-Gly-Gly-Ala-Ala-Arg-(Cys)-Gln-Val-Thr-Leu-
                               5   10   15

salmon PRL: H-Ile-Gly-Leu-
                5

human GH: H-Phe-Pro-Thr-Ile-Pro-Leu-

Arg-Asp-Leu-Phe-Asp-Arg-Ala-Val-Val-Leu-Ser-His-Tyr-Ile-His-Asn-Leu-Ser-Ser-
 5   10   15   20

Ser-Asp-Leu-Met-Glu-Arg-Ala-Ser-Gln-Arg-Ser-Asp-Lys-Leu-His-Ser-Leu-Ser-Thr-
      10   15   20   25

Ser-Arg-Leu-Phe-Asp-Asn-Ala-Met-Leu-Arg-Ala-His-Arg-Leu-His-Gln-Leu-Ala-Phe-
      5   25

Glu-Met-Phe-Ser-Glu-Phe-Asp-Lys-Arg-Tyr-Thr-His-Gly-Arg-Gly-Phe-
          25   30   35

Ser-Leu-Thr-Lys-Asp-Asp-Ser-His-Phe-Pro-Met-
      30   35   40

Asp-Thr-Tyr-Gln-Glu-Phe-Glu-Glu-Ala-Tyr-Ile-Pro-Lys-Glu-Gln-Lys-
          35   40   45   50

C-terminal region

human PRL: -Tyr-Asn-Leu-Leu-His-(Cys)-Leu-Arg-Arg-Asp-Ser-His-Lys-Ile-Asp-Asn-
                                  175   180

salmon PRL: -Met-Ser-(Cys)-Phe-Arg-Arg-Asp-Ser-His-Lys-Ile-Asp-Ser-
              -30   -25   -20
              165   170   175

human GH: -Tyr-Gly-Leu-Leu-Tyr-(Cys)-Phe-Arg-Lys-Asp-Met-Asp-Lys-Val-Glu-Thr-
          160   170   175

Tyr-Leu-Lys-Leu-Leu-Lys-(Cys)-Arg-Ile-( )-( )-Ile-His-Asn-Asn-(Cys)-OH
 185   -15   -10   -5   195

Phe-Leu-Lys-Val-Leu-Lys-(Cys)-Arg-Ala-Thr-Lys-Met-Arg-Pro-Gln-Thr-(Cys)-OH
        180   185   190

Phe-Leu-Arg-Ile-Val-Gln-(Cys)-Arg-Ser-( )-Val-Glu-Gly-( )-Ser-(Cys)-Gly-Phe-OH
                                       190

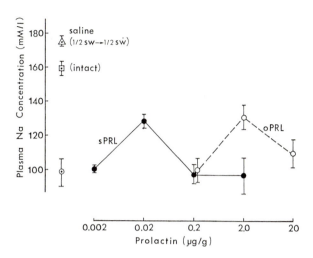

Fig. 5. Sodium-retaining activity of salmon and ovine PRLs in *Fundulus*.

prepared exclusively by extraction at neutral or alkaline conditions. In such conditions, however, several undesirable side reactions may take place: enzymatic degradation will occur at neutral pH, as well as deamidation and aggregation at alkaline pH. In addition, the chum salmon PRL was proved to be a strongly basic protein and may not be extractable efficiently in alkaline condition.

Most of the mammalian PRL preparations have revealed several bands on polyacrylamide gel electrophoresis, and some of the bands are assumed to be deamidated forms. The salmon PRL exhibited only one electrophoretic band at pH 4.3 and in ampholine (pH 3.5-10) but a very faint and slightly acidic band in ampholine (pH 7-11). The minor component could be a deamidated form as are the cases for mammalian PRLs. The isoelectric point of the salmon PRL (10.3) was

---

Fig. 4. Partial amino acid sequence of the salmon PRL in comparison with mammalian PRLs and growth hormones (GHs). The amino acid sequences are taken from human GH[28] and human PRL.[29]

382

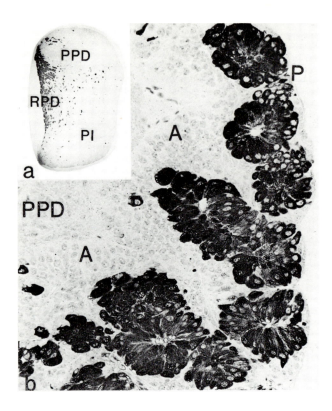

Fig. 6. a. Sagittal section of mature chum salmon (*O. keta*) pituitary stained with anti-salmon PRL at a concentration of 1/4000, counterstained with Mayer's hematoxylin. RPD, rostral pars distalis; PPD, proximal pars distalis; PI, pars intermedia. x 10. b. Higher magnification of the rostral pars distalis and proximal pars distalis (PPD) of the mature chum salmon pituitary stained with anti-salmon PRL at a concentration of 1/4000, counterstained with Mayer's hematoxylin. Immunoreactivity is evident in the PRL cells (P), whereas no reactivity is seen in the ACTH cells (A) or in the growth hormone and GHT cells in the PPD. x 300.

notably higher than those of mammalian PRLs; pI of ovine PRL is 5.6. According to Farmer et al.,[21] *Tilapia* PRL gave no electrophoretic band at pH 8.3, whereas two bands, one intensely stained major band with a second minor band running behind it were observed at pH 4.5.[4] It seems that the *Tilapia* PRL, as well as the salmon PRL, is a basic protein.

The molecular weight of the salmon PRL was estimated to be 22,300 daltons, which is 3,000 daltons larger than that of *Tilapia* PRL and similar to that of ovine PRL. Total amino acid residues were estimated to be 194, and amino acid analysis of ten cyanogen bromide fragments supports the estimation.

The salmon PRL possessed four half-cystine residues or two disulfide bridges, as in the case of *Tilapia* PRL. It has been speculated that the amino-terminal disulfide bond in mammalian PRLs is absent in *Tilapia* PRL, mainly because the amino-terminal tetrapeptide of the *Tilapia* PRL revealed no half-cystine residue. In addition, selective reduction of the amino-terminal disulfide bond of ovine PRL increased the biological activity of the hormone, when assayed on permeability change in *Gillichthys* urinary bladder.[22] As shown in the present studies, the maximum homology was found between the salmon PRL and human PRL and GH, when the amino-terminal residue of salmon was aligned with the 13th residue of human PRL. Three of the four half-cystine residues were identified at the carboxyl-terminal portion and aligned in the homologous positions with those of human hormones. Thus, the salmon PRL definitely lacks one disulfide loop at the amino-terminal portion. However, it seems also possible that the structure of the prohormone of the salmon PRL is similar with that of mammalian PRL, and that the processing of the prohormone could make the "phenotype" different.

The resemblance of primary structure between mammalian PRL and GH strongly suggested that they have evolved from a common ancestral molecule.[1] The sequence comparison (Fig. 4) seems to provide further evidence for a common ancestral molecule for PRL and GH.

Besides the similarity in the overall molecular features with the mammalian hormones, the protein obtained in the present study was identified as PRL by its immunohistochemical identification of the secreting cells and biological activity. Various hormone-secreting cells of the chum salmon pituitary have been identified by light and electron microscopy.[19] The PRL cells in the chum salmon are arranged in the form of follicles and form the bulk of the rostral pars distalis, whereas GH cells are found exclusively in the proximal pars distalis. In the present study, incubation of chum salmon pituitaries with anti-salmon PRL produced a strong reaction only in the follicular PRL cells, and no other cells in the pituitary gave this reaction.

It has been well established in many teleost species that PRL helps restore sodium balance in fresh water, maintaining the passive loss component through the gills at a low level. In the present study, the salmon PRL was 100 times as active as ovine PRL in increasing plasma sodium of hypophysectomized

Fundulus. Similarly, *Tilapia* PRL has been found approximately 100 times as active as ovine PRL in the *Tilapia* sodium-retaining bioassay.[4] These results indicate that molecular structure of fish PRL is considerably specialized for an osmoregulatory role.

When the structure of the salmon proopiocortin-related peptides are compared with those of mammalian hormones, it was found that amino acid residues which are essential for biological activity are conserved, i.e., His-Phe-Arg-Trp for melanotropin[23-25] and Tyr-Gly-Gly-Phe-Met for endorphin.[26,27] Among the clusters of invariant residues between salmon and human PRLs, the segment corresponding to human PRL 174-189 is the most conservative; 11 of 16 residues are invariant and five retain the physiological characteristics of the residues they replaced. Based on above observation, the core of invariant residues in PRL may be one of the responsible segments for the osmoregulatory activity.

REFERENCES

1. Bewley, T. A. and Li, C. H. (1971) Experientia, 27, 1368-1371.
2. Bern, H. A. (1875) Amer. Zool., 15, 937-948.
3. Clarke, W. C. and Bern, H. A. (1980) in Hormonal Proteins and Peptides, Li, C. H., ed., Vol. 8, New York: Academic Press, pp. 105-197.
4. Farmer, S. W., et al. (1977) Gen. Comp. Endocrinol., 31, 60-71.
5. Ornstein, L. (1964) Ann. N. Y. Acad. Sci., 121, 321-349.
6. Meyer, T. S. and Lamberts, B. L. (1965) Biochim. Biophys. Acta, 107, 144-145.
7. Wrigley, C. W. (1971) Methods Enzymol., 22, 559-564.
8. Reisner, A. H., Nemes, P., and Bucholtz, B. (1975) Anal. Biochem., 64, 509-516.
9. Weber, K. and Osborn, M. (1969) J. Biol. Chem., 244, 4406-4412.
10. Bewley, T. A., Brovetto-cruz, J., and Li, C. H. (1969) Biochemistry, 8, 4701-4708.
11. Beaven, G. H. and Holiday, E. R. (1952) Advan. Protein Chem., 7, 319-386.
12. Gray, W. R. (1967) Methods Enzymol., 11, 469-475.
13. Akabori, S., et al. (1965) Bull. Chem. Soc. Japan, 29, 507-518.
14. Muramoto, K., Kawauchi, H., and Tsuzimura, K. (1978) Agric. Biol. Chem., 42, 1559-1563.
15. Bruton, C. J. and Hartley, B. S. (1970) J. Mol. Biol., 52, 165-178.
16. Jozuka, K. and Adachi, H. (1979) Annot. Zool. Japan, 52, 107-113.
17. Sternberger, L. A., et al. (1970) J. Histochem. Cytochem., 18, 315-333.
18. Kawauchi, H., Takahashi, A., and Abe, K. (1981) Int. J. Peptide Protein Res., 18, 223-227.
19. Nagahama, Y. (1973) Mem. Fac. Fish. Hokkaido Univ., 21, 1-63.
20. Cole, R. D. and Li, C. H. (1955) J. Biol. Chem., 213, 197-201.
21. Farmer, S. W., et al. (1975) Life Sci., 16, 149-158.
22. Doneen, B. A., Bewley, T. A., and Li, C. H. (1979) Biochemistry, 18, 4851-4860.
23. Kawauchi, H. and Muramoto, K. (1979) Int. J. Peptide Protein Res., 14, 373-374.
24. Kawauchi, H., Adachi, Y., and Ishizuka, B. (1980) Int. J. Peptide Protein Res., 16, 79-82.

25. Kawauchi, H., Adachi, Y., and Tsubokawa, M. (1980) Biochem. Biophys. Res. Commun., 96, 1508-1517.
26. Kawauchi, H., Tsubokawa, M., and Muramoto, K. (1979) Biochem. Biophys. Res. Commun., 88, 1249-1254.
27. Kawauchi, H., et al. (1980) Biochem. Biophys. Res. Commun., 92, 1278-1288.
28. Li, C. H. (1972) Proc. Am. Phil. Soc., 116, 365-382.
29. Shome, B. and Parlow, A. F. (1977) J. Clin. Endocrinol. Metab., 45, 1112-1115.
30. Li, C. H., et al. (1970) Arch. Biochem. Biophys., 141, 705-737.

PROPERTIES OF CHEMICALLY MODIFIED SNAKE VENOM PHOSPHOLIPASES A$_2$*

C. C. YANG
Institute of Molecular Biology, National Tsing Hua University
Hsinchu, Taiwan 300, Republic of China

ABSTRACT

The relationship between the toxicity and the enzymatic activity of phospholipase A$_2$ isolated from snake venoms is poorly understood. Basic phospholipases A$_2$ are highly toxic *per se* or in complex with other subunits, while both acidic and neutral phospholipases A$_2$ are much less toxic. An approach to assess whether dissociation of toxicity and enzymatic activity of phospholipase A$_2$ can be achieved was made by chemical modifications of amino acid residues in the enzyme molecule. Basic *N. nigricollis* and acidic *N. naja atra* (Taiwan cobra) phospholipases A$_2$ have been chosen for the study because they possess a remarkable degree of homology and have differences in charge and toxicity.

Alkylation of one histidine residue at the active site in snake venom phospholipase A$_2$ with *p*-bromophenacyl bromide destroyed both enzymatic activity and lethal toxicity. By means of lysine modification, a dissociation between hydrolytic activity and pharmacological properties has been achieved. Carbamylation of lysine residues in the phospholipases A$_2$ from *N. naja atra* and *N. nigricollis* venoms did not affect the enzymatic activities, while the lethality and various pharmacological activities were drastically reduced or abolished. The effect of Ca^{2+} on the fluorescence emission intensity of ANS-Lys-modified *N. nigricollis* phospholipase A$_2$ and the ultraviolet difference spectra of lysine modified derivatives differ greatly from those of the native *N. nigricollis* phospholipase A$_2$ and become similar to those of the acidic *N. naja atra* phospholipase A$_2$.

Modification of carboxylate groups, on the other hand, causes loss of enzymatic activity but preservation of considerable pharmacological potency. *N. naja atra* phospholipase A$_2$ modified in the presence of Ca^{2+} ions at pH 5.5 yielded a protein with 12.6% of the native enzymatic activity, while 100% of the lethal potency and 58% antigenic activity were found. Although the carboxylated phospholipases A$_2$ modified in the absence of Ca^{2+} at both pH 3.5 and pH 5.5 have less than 2% of the original enzymatic activity, they have about 25% of the intraventricular lethality.

---

*This work was supported by the National Science Council, Republic of China.

It is concluded that the toxicity of pure phospholipases A_2 may be due to a direct effect which does not correlate with levels of phospholipid hydrolysis and that this direct effect is prominent in the relatively toxic phospholipase A_2 enzymes while it is less manifest in the less toxic phospholipase A_2 enzymes. There may be two distinct but perhaps overlapping active sites with histidine 48 and aspartic acid 49 being at the catalytic site and histidine 48 and lysines being at the pharmacologically active region.

INTRODUCTION

Snake venom phospholipases A_2 (PLA$_2$; EC 3.1.1.4) show considerable structural homology but differ greatly in enzymatic properties, toxicity and pharmacological actions.[1] The toxic PLA$_2$ are usually more basic than those of the less toxic enzymes.[2] Basic PLA$_2$ exhibits direct hemolytic activity, anticoagulant potency and myotoxic actions besides neurotoxicity. All presynaptically acting neurotoxins, e.g., notexin, crotoxin, β-bungarotoxin and taipoxin are basic PLA$_2$ per se or contain a basic PLA$_2$ as an indispensable part of their structures.[3,4] Although there is considerable evidence in favor of the essential role of PLA$_2$ activity in the involvement of toxicity,[5-7] no one has yet been able to pinpoint the structural features and physical and chemical properties that render these PLA$_2$ enzymes such potent toxins.

An approach to elucidating the relationship between enzymatic activity and toxicity could be made by chemical modifications of various amino acid residues in the PLA$_2$ enzyme molecules and by comparing the effects of modification on these two parameters. Since Volwerk et al.[8] demonstrated that the PLA$_2$ from porcine pancreas can be completely inactivated by alkylation of one histidine residue at the active site with p-bromophenacyl bromide, this procedure has been widely used for assessing the role of enzymatic activity in toxicity. Two cobra venom PLA$_2$ have been chosen for the study because they have a remarkable degree of homology while differing in charge and toxicity: the N. nigricollis enzyme is basic and relatively toxic while the acidic N. naja atra (Taiwan cobra) PLA$_2$ is less toxic.

The present study reports on a successful dissociation of enzymatic activity and toxicity achieved by progressive carbamylation of lysine residues in the PLA$_2$ of N. nigricollis and N. naja atra venoms and by carboxylate group modification using carbodiimide and semicarbazide.

MATERIALS AND METHODS

Naja nigricollis venom (Lot No. NGE45-IZ) of East African origin was
obtained from Miami Serpentarium Laboratories, Miami, Florida. *Naja naja atra*
venom was freshly collected and lyophilized. Potassium cyanate was purchased
from Fisher Scientific Company; N-ethylmorpholine, *p*-bromophenacyl bromide
and 1-ethyl-3-(3-dimethylaminopropyl)carbodiimide (EDC) from Sigma Chemical
Company. Sephadex G-25, G-100, CM-Sephadex C-25, SP-Sephadex C-25 and DEAE-
Sephacel were obtained from Pharmacia Fine Chemicals. Ammonium acetate,
ammonium bicarbonate, calcium chloride, semicarbazide, sodium cacodylate, sodium
deoxycholate and Tris(hydroxymethyl) aminomethane were obtained from E. Merck
(Darmstadt), 8-anilinonaphthalenesulfonate (ANS) ammonium salt from Pierce
Chemical Company. All other reagents were of analytical grade.

Isolation and purification of PLA$_2$

Three PLA$_2$ fractions (CMS-5, CMS-6 and CMS-9) were separated from *N. nigricollis* venom by chromatography on a CM-Sephadex C-25 column.[9] The isoelectric
points of CMS-5, CMS-6, and CMS-9 were 7.6, 8.3 and 10.6, respectively. CMS-9,
which is the most toxic and basic fraction, was further purified on a DEAE-
Sephacel column and was used in this study. The acidic PLA$_2$ from *N. naja atra*
venom was isolated and purified by successive chromatography on SP-Sephadex
C-25, DEAE-Sephacel and CM-Sephadex C-25 columns.[10] The major PLA$_2$ (pI 5.2)
which was used in this study, was separated from the minor enzyme (pI 4.8) on
a SP-Sephadex C-25 column. The purity of these enzymes was verified using disc
electrophoresis on a 7% polyacrylamide gel.[11]

Chemical modification of histidine residue in PLA$_2$

Modification with *p*-bromophenacyl bromide was performed at a molar protein:
reagent ratio of 1 : 10 in 0.025 M Tris-HCl buffer, pH 8.0. *p*-Bromophenacyl
bromide (125 µl, 40 mM in acetone) was added to 5 ml of PLA$_2$ (1.3 mg/ml) in
0.025 M Tris-HCl buffer, pH 8.0. Reaction was allowed to proceed at 30°C
for 40 min, and then the solution was acidified with glacial acetic acid to
pH 4.0 to stop the reaction. Excess reagents were removed by passing the
mixture through a column of Sephadex G-25 and the protein fraction was
lyophilized.

Carbamylation of lysine residues and separation of carbamylated derivatives

Carbamylation with KCNO was performed with 2 mmole KCNO and 70 mg (5 µmole)
PLA$_2$ in 0.5 M N-ethylmorpholine acetate buffer (pH 8.0) according to the method

of Atark[12] and Karlsson et al.[13] The reaction proceeded for various time intervals (3 and 16 hr for *N. naja atra* and 40 hr for *N. nigricollis* enzymes) at 25°C.[10,14] Excess reagents were removed by passing the mixture through a column of Sephadex G-25 equilibrated with 0.1 M ammonium acetate buffer (pH 4.0) and the protein fractions were lyophilized.

Carbamylated *N. naja atra* PLA$_2$ (3 or 16 hr, 160 mg) was dissolved in 0.05 M ammonium acetate buffer (pH 6.8) and fractionated by DEAE-Sephacel column chromatography. Eight peaks were obtained and each peak was revealed as a single band by disc electrophoresis.[10] Carbamylated *N. nigricollis* PLA$_2$ (40 hr, 180 mg) was dissolved in 0.05 M ammonium acetate buffer (pH 5.0) and the clear supernatant was chromatographed on an SP-Sephadex C-25 column.[14] Five peaks, A, B, C, D, and E, were eluted and the last three were used in the present study.

Guanidination with O-methylisourea

Guanidination was performed with O-methylisourea[15] and gave rise to products free of unmodified enzyme. Three μmoles of PLA$_2$ were incubated in 2 ml of 0.5 M O-methylisourea-HCl solution, adjusted to pH 10.8 with 6 M NaOH, and the reaction allowed to proceed at 4°C for 72 hr. The mixture was then passed through a column of Sephadex G-25 and the protein fraction lyophilized. Guanidinated *N. naja atra* PLA$_2$ had an isoelectric point of 5.7 (native, 5.2) and *N. nigricollis* PLA$_2$ 9.8 (native 10.6). The number of lysines modified in the guanidinated *N. nigricollis* and *N. naja atra* PLA$_2$ was 9.4 and 4.5, respectively.

Modification of carboxylate groups

Carboxylate groups were modified using a water-soluble carbodiimide and semicarbazide. PLA$_2$ (5 mg/ml) was dissolved in 0.25 M cacodylate buffer (pH 5.5) containing 1 M semicarbazide. When the reaction was carried out at pH 3.5, 1 M semicarbazide was used as buffer. The reaction was started by the addition of solid EDC to a final concentration of 0.1 M. Addition of EDC (5 mg/ml) was repeated every 30 min to compensate for its decomposition in water. When no further change in enzymatic activity occurred, the reaction was terminated by the addition of solid sodium acetate (75 mg/ml) and the protein was desalted immediately on a column of Sephadex G-25 equilibrated with 0.02 M ammonium bicarbonate. In order to regenerate tyrosyl residues[16] the protein was treated with 1 M hydroxylamine for 3 hr at pH 7.5 and 25°C, followed by desalting and lyophilization. Modified proteins were purified on a SP-Sephadex C-25 column (2.8 x 35 cm) equilibrated with 0.05 M phosphate buffer (pH 7.5) and elution was performed with a linear NaCl gradient (0-0.8 M over 2400 ml) in the same buffer.[17]

Determination of PLA$_2$ activity

PLA$_2$ activity was determined by the titrimetric method described by de Haas et al.[18] with a slight modification. Titration was carried out with 0.001 M NaOH on a TTT1 Radiometer autotitrator at pH 8.0 and 25°C. An aqueous emulsion of egg yolk was used as substrate for which one egg yolk was homogenized with 300 ml water and supplemented with 2.7 mM deoxycholate and 20 mM CaCl$_2$. For each determination 10 ml of the substrate solution was used and a suitable amount of enzyme (0.2 to 2.0 µg) in 10 µl was added. There was a linear relationship between the activity and the amount of enzyme, and the curves were linear up to 3 min. One unit of enzyme activity was defined as a release of 1 µ equivalent of fatty acid per min.

Other tests

Amino acid analysis, ultraviolet difference spectroscopy, fluorescence measurements, assay of lethal toxicity and immunological procedures were performed essentially the same as described previously.[9]

RESULTS AND DISCUSSION

Chemical modification of the histidine residue in PLA$_2$

The basic *N. nigricollis* PLA$_2$ and acidic *N. naja atra* PLA$_2$ were subjected to chemical modification with *p*-bromophenacyl bromide at pH 8.0 and the modified enzymes lost both the enzymatic activity and toxicity (Table 1). The decrease in enzymatic activity was accompanied by a decrease in direct hemolysis (for basic *N. nigricollis* PLA$_2$), ability to hydrolyze red cell phospholipids, intraventricular LD$_{50}$ and CD$_{50}$ in rats, brain phospholipid hydrolysis and ability to block action potentials in eel electroplax.[19]

Amino acid analysis showed that only one His-residue was modified and the modified residue was identified to be His-47 in the sequence.[9,10] Chemical modification studies of PLA$_2$ from different sources revealed that the critical His-residue which was potentially reactive toward *p*-bromophenacyl bromide was located in the same or homologous position.[6,8,20-23]

The His-modified PLA$_2$ not only lost enzymatic activity and the ability to bind ANS, but its Ca^{2+}-induced difference spectrum also differs greatly from that of the native enzyme (Fig. 1). In spite of these differences, native and His-modified enzymes could be perturbed as a function of Ca^{2+} concentration.

TABLE 1

COMPARISON OF PHYSICAL AND BIOLOGICAL PROPERTIES OF NATIVE AND
His-MODIFIED PLA$_2$ FROM N. *nigricollis* AND N. *naja atra* VENOMS

Physical and biological properties	N. *nigricollis* PLA$_2$		N. *naja atra* PLA$_2$	
	Native	His-modified	Native	His-modified
PLA$_2$ activity (U/mg)[*]	1380	0.64	3400	0
Antigenic activity (%)	100	100	100	96
Ca^{2+} binding Kd (mM)	0.20[**]	0.29[**]	0.56[**]	0.55[**]
	(0.30)[+]	-	(0.59)[+]	-
ANS binding Kd (μM)				
with 10 mM Ca^{2+}	31.2	-	20.0	-
without Ca^{2+}	12.8	-	20.4	-
LD$_{50}$ (mg/kg mouse)	0.3	> 10	8	> 80

[*]
One unit of enzyme activity was defined as a release of 1 μ equivalent of fatty acid min^{-1}.
[**]From ultraviolet difference spectra.
[+]From fluorescence measurements.

The K$_d$ of the His-modified enzymes was similar to that of the native PLA$_2$ (Table 1) and the Scatchard plots reveal that there is only one kind of Ca^{2+} binding site for each protein. It is suggested that modification of His-47 does not disturb the metal ion-protein interaction.

It is interesting to note that Ca^{2+} exerts pronounced protection on the inactivation process by p-bromophenacyl bromide and the His-modified enzymes still possess the ability to bind Ca^{2+}. Verheij et al.[24] also observed that the His-48 methylated pancreatic PLA$_2$ shows total loss of enzymatic activity whereas the binding of Ca^{2+} remains intact. It is necessary that Ca^{2+} and p-bromophenacyl or methyl group do not bind to identical residues, although they may be close, and it could be inferred that His-47 or 48 are not directly

---

Fig. 1. Spectral changes of native and His-modified PLA$_2$ induced by Ca^{2+} in 0.025 M Tris-0.1 M NaCl buffer (pH 8.0) at 25°C. The sample cuvette contained 0.45 mg/ml of native or His-modified PLA$_2$ from N. *nigricollis* and N. *naja atra* venoms in the presence of Ca^{2+} (mM) as indicated. The cell in the reference position contained the respective enzyme (0.45 mg/ml) and spectra were recorded with a 0-0.1 absorbance scale.

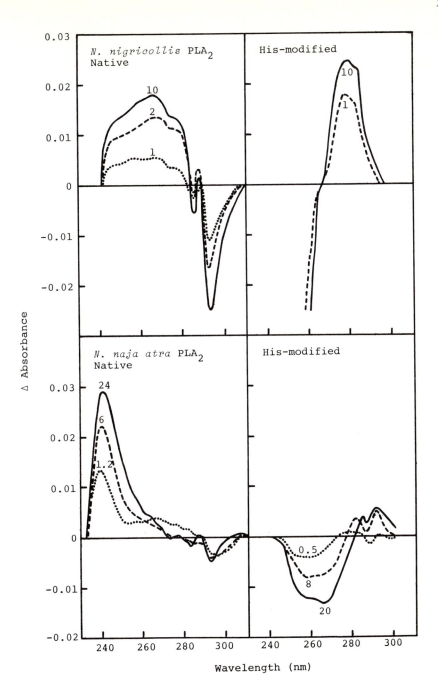

TABLE 2

ENZYMATIC ACTIVITY, ANTIGENICITY AND LETHALITY OF THE CARBAMYLATED
N. *naja atra* PLA$_2$ DERIVATIVES OBTAINED FROM DEAE-SEPHACEL COLUMN

Fractions	Number of Lys-residue modified[*]	pI	Enzymatic activity (%)	Antigenic activity (%)	LD$_{50}$[+] i.v. (mg/kg mouse)	LD$_{50}$[+] intraventricular (μg/rat)
	(out of 5)					
Native PLA$_2$	0	5.2	100	100	8.6	7.5
Fraction 1	1.1	4.7	97	97		
Fraction 2	1.2	4.7	88	96	> 26	95
Fraction 3	2.1	4.4	74	89	> 26	95
Fraction 4	3.1	3.7	74	88		
Fraction 5	3.2	3.7	74	88	> 26	> 225
Fraction 6	4.2	3.4	74	88		> 225
Fraction 7	4.3	3.4	59	76		
Fraction 8	5.0	3.0	59	72		

[*] The recovery factor of 0.19 was used to calculate the amount of homocitrulline formed as previously described.[25] $H = (5 - L/1 - 0.19)$ where H = moles of homocitrulline; L moles of lysine; 0.19 = the recovery factor of lysine during acid hydrolysis at 110°C for 24 hr.

[+] Condrea et al.[26]

involved in Ca^{2+} binding. The binding of Ca^{2+} can induce a conformational change which enables the enzyme to be more accessible to the substrate. The Ca^{2+}-induced conformational change not only makes the enzyme more active in performing its catalytic function, but also renders it less reactive toward p-bromophenacyl bromide.

Carbamylation of lysine residues in PLA$_2$

Modification of the enzymes with potassium cyanate at pH 8.0 resulted in selective carbamylation of Lys-residues. The carbamylated N. *naja atra* PLA$_2$ derivatives were separated on a column of DEAE-Sephacel and 8 fractions were obtained. The results of amino acid analysis showed that 1 to 5 Lys-residues were modified. As seen from Table 2, associated with modification of increasing number of Lys-residues were progressive decreases in pI values and marked decreases (3- to > 30-fold) in LD$_{50}$ values following intravenous and

TABLE 3

ISOELECTRIC POINT, ENZYMATIC ACTIVITY, ANTIGENICITY AND LETHALITY OF THE
CARBAMYLATED N. *nigricollis* PLA$_2$ DERIVATIVE OBTAINED FROM SP-SEPHADEX
C-25 COLUMN

Fractions	Number of Lys-residue modified*	pI	Enzymatic activity (%)	Antigenic activity (%)	LD$_{50}$[+] i.v. (mg/kg mouse)	intraventricular (μg/rat
	(out of 10)					
Native PLA$_2$	0	10.6	100	100	0.63	0.45
Fraction E	7.4	5.4	95	79	1.1	0.87
Fraction D	8.2	4.9	89	64	-	1.65
Fraction C	9.0	4.4	75	51	> 5.0	3.75
Fraction B	9.2	4.4	44	43		
Fraction A	10.0	3.9	15	40		

* See Table 2.
[+] Condrea et al.[26]

intraventricular injection. However, the decrease in enzymatic activity was
slight and antigenic specificity was unaffected. It is noteworthy that even
after four out of five Lys-residues had been modified (fraction 6), the enzyme
still possessed 74% of its enzymatic activity and 88% antigenic activity; however, the lethal potency decreased more than 30-fold. In addition, in contrast
to the native enzyme, fraction 5 had no effect on muscle contraction of the
phrenic nerve-diaphragm preparation even though phospholipid analysis of the
diaphragm showed that about the same percentage of phospholipids were hydrolyzed
as with the native enzyme.[26]

The carbamylated N. *nigricollis* PLA$_2$ derivatives were separated by chromatography on a column of SP-Sephadex C-25 and five fractions were obtained.
Amino acid analysis showed that the numbers of Lys-residues modified for
fractions A to E were 10, 9.2, 9.0, 8.2, and 7.4, respectively. The pI values
decrease with increasing carbamylation, converting the basic enzyme into an
acidic protein (Table 3). It is noteworthy that even after 9 out of 10 Lys-
residues had been modified (fraction C) and the pI of the enzyme decreased from
10.6 to 4.4, the enzyme still retains 75% of its enzymatic activity and 51%
antigenicity. However, the lethality decreased at least 8-fold and its direct

hemolytic and anticoagulant activities were lost,[26] revealing a clear dissociation between enzymatic activity and pharmacological properties.

There was also no correlation between toxicity or pharmacological properties and *in vivo* levels of phospholipid hydrolysis. Carbamylation abolished the direct lytic effect of *N. nigricollis* enzyme on guinea pig erythrocytes and its anticoagulant activity on rabbit plasma, while the phospholipid hydrolysis in the erythrocyte outer layer and in the rabbit plasma agreed with the high activities determined on purified substrates *in vitro*. This demonstrates unequivocally that the direct lytic property and the anticoagulant effect are not due to phospholipid hydrolysis in the preparations.[26]

As illustrated in Fig. 2, Ca^{2+}-induced ultraviolet difference spectra of *N. nigricollis* PLA_2 were characterized by a negative perturbation with a minimum at 293 nm and a positive peak at 268 nm. However, after 8 out of 10 Lys-residues were carbamylated, the Ca^{2+}-induced difference spectra differed greatly from that of the native enzyme and became similar to that of the acidic *N. naja atra* PLA_2. ANS, which has been used as a hydrophobic probe to study the conformational change of proteins, also showed a special affinity to the native and Lys-modified *N. nigricollis* PLA_2. It has been suggested that the hydrophobic pocket which interacts with ANS might be the site of the enzyme which interacts with the substrates.[9,19,27] As seen from Fig. 3, at pH 8.0, the emission-enhancing ability of the ANS-*N. nigricollis* PLA_2 complex decreased with increasing concentration of Ca^{2+} until the saturation level was reached. However, a reversed effect was noted for the Lys-modified fraction D and the less toxic acidic *N. naja atra* PLA_2. The affinity of the native *nigricollis* PLA_2 to ANS also decreased 2.5-fold as Ca^{2+} bound to the enzyme (Table 4). The different conformational changes induced by Ca^{2+} might be attributable to the charge properties of the enzyme which depends on the pH of the solution and the pI of the enzyme. When the pH of the buffer solution was lower than the pI of Lys-modified fraction D, it was found that not only the emission intensity decreased with increasing concentration of Ca^{2+} (Fig. 2, upper right), but also the dissociation constant of the ANS-enzyme complex was reduced in the absence of Ca^{2+} to one-half its original value in the presence of metal (Table 4). The same phenomenon was also observed for *N. naja atra* PLA_2[10] (Table 4), and the effects of Ca^{2+} on the interaction between ANS and Lys-modified fraction D or *N. naja atra* PLA_2, either at pH 8.0 or pH 3.0, were the same. Thus, by carbamylation of a toxic PLA_2 we created derivatives which resemble other less toxic native protein species.

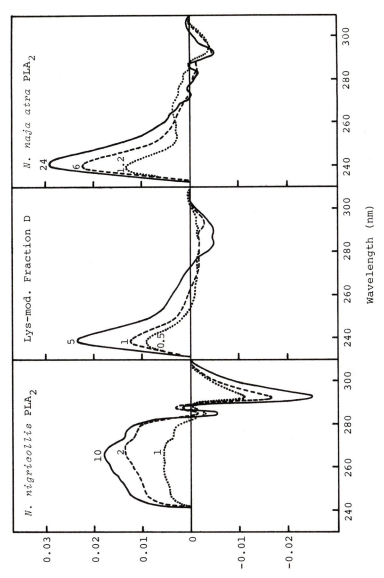

Fig. 2. Spectral changes of native and Lys-modified *N. nigricollis* PLA$_2$ induced by Ca^{2+} in 0.025 M Tris–0.1 M NaCl buffer (pH 8.0) at 25°C. The sample cuvette contained 0.45 mg/ml of native *N. nigri-collis* PLA$_2$, Lys-modified fraction D or *N. naja atra* PLA$_2$ in the presence of Ca^{2+} (mM) as indicated. The cell in the reference position contained the respective enzyme (0.45 mg/ml) and spectra were recorded with a 0–0.1 absorbance scale.

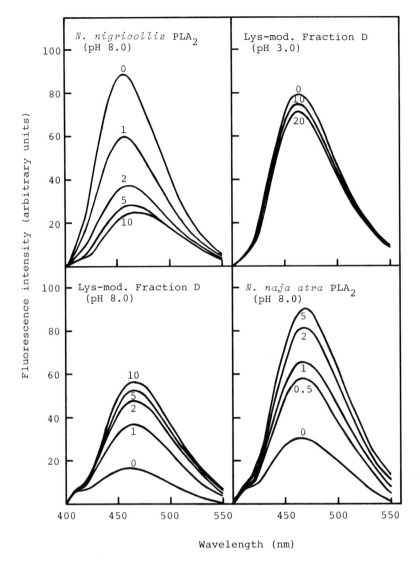

Fig. 3. Effect of Ca^{2+} on the interaction of native and Lys-modified *N. nigri-collis* PLA$_2$ with ANS at 25°C. The sample cuvette contained 0.175 mg native *N. nigricollis* PLA$_2$, 0.077 mg Lys-modified fraction D, 0.175 mg *N. naja atra* PLA$_2$ per ml 0.025 M Tris-0.1 M NaCl (pH 8.0) or 0.12 mg Lys-modified fraction D per ml 0.025 M acetic acid (pH 3.0) and 12.5 μM ANS in the presence of various concentrations of Ca^{2+} (mM) as indicated.

TABLE 4

COMPARISON OF PHYSICAL AND BIOLOGICAL PROPERTIES OF NATIVE AND
Lys-MODIFIED PLA$_2$ FROM *N. nigricollis* AND *N. naja atra* VENOMS

Physical and biological properties	*N. nigricollis* PLA$_2$		*N. naja atra* PLA$_2$	
	Native	Lys-modified (Fraction D)	Native	Lys-modified (Fraction 6)
PLA$_2$ activity (U/mg)[*]	1380	1226	3400	2500
Km (mM)	4.2	3.7	5.3	4.9
Antigenic activity (%)	100	64	100	88
Ca^{2+} binding K$_d$ (mM)				
at pH 8.0	0.20[**]	0.28[**]	0.56[**]	0.54[**]
	(0.30)[+]	(0.31)[+]	(0.59)[+]	(0.68)[+]
at pH 3.0	–	0.35[**]	0.46[**]	–
ANS binding K$_d$[+] (μM)				
at pH 8.0				
with 10 mM Ca^{2+}	31.2	18.0	20.0	16.9
without Ca^{2+}	12.8	16.0	20.4	17.9
at pH 3.0				
with 10 mM Ca^{2+}	–	18.0	25.0	–
without Ca^{2+}	–	9.6	12.0	–
LD$_{50}$ (mg/kg mouse)	0.3	1.0	8	> 80

[*] One unit of enzyme activity was defined as a release of 1 μ equivalent of fatty acid min^{-1}.

[**] From ultraviolet difference spectra.

[+] From fluorescence measurements.

Guanidination with O-methylisourea

In contrast to carbamylation of lysines with potassium cyanate which causes a marked decrease in the isoelectric point, guanidination with O-methylisourea, which converts lysines to homoarginines, does not markedly affect the charge on the protein. We can thus estimate the contribution of charge to the actions of the PLA$_2$. Guanidinated *N. nigricollis* PLA$_2$ had an isoelectric point 9.8 and *N. naja atra* PLA$_2$ 5.7, and the number of lysines modified was 9.4 (out of 10) and 4.5 (out of 5), respectively (Table 5). Modification of the lysines, either by guanidination or by carbamylation, had little effect upon enzymatic activity of either enzyme. In contrast, the intraventricular lethalities of both the

TABLE 5

PROPERTIES OF GUANIDINATED N. *nigricollis* AND N. *naja atra* PLA$_2$:
COMPARISON WITH CARBAMYLATED DERIVATIVES

PLA$_2$	Number of Lys-residues modified	pI	Enzymatic activity (%)	Intra-ventricular lethality[*] (%)
N. *nigricollis*				
Native	0	10.6	100	100
Guanidinated	9.4	9.8	100	13
Carbamylated				
Fraction E	7.4	5.4	95	57
Fraction C	9.0	4.4	75	13
N. *naja atra*				
Native	0	5.2	100	100
Guanidinated	4.5	5.7	100	67
Carbamylated				
Fraction 5	3.2	3.7	74	< 3

[*]Condrea et al.[26,28]

guanidinated and carbamylated N. *nigricollis* samples were equally decreased to about 13% of the native enzyme (Table 5).

It is also obvious that a decrease in the pI of the phospholipase alone was not responsible for the decrease in the intraventricular lethality of N. *nigricollis* PLA$_2$ by carbamylation, since the guanidinated enzyme showed a similar decrease in intraventricular lethality even though the pI of the enzyme was not significantly altered. In marked contrast, however, guanidination of the lysines in N. *nigricollis* PLA$_2$ only decrease intravenous lethality and anti-coagulant potency about 50%, whereas carbamylation of the lysines caused at least a 10-fold decrease in these parameters.[26,28] Also, the direct hemolytic potency of the N. *nigricollis* enzyme was unaltered by guanidination whereas direct hemolysis was lost following carbamylation. The conversion from a basic to an acidic phospholipase by carbamylation may thus have contributed to the greater loss in intravenous lethality, hemolytic activity and anticoagulant potency.[28]

Guanidination of 4.5 out of the 5 lysines in N. *naja atra* PLA$_2$ caused only a slight decrease in intraventricular lethality (33%) whereas carbamylation of three out of the five lysines caused more than a 30-fold decrease (Table 5).

TABLE 6

PROPERTIES OF NATIVE AND CARBOXYLATED *N. naja atra* PLA$_2$

Physical and biological properties	Native *N. naja atra* PLA$_2$	Carboxylated PLA$_2$		
		pH 5.5 + Ca^{2+}	pH 5.5	pH 3.5
PLA$_2$ activity (%)	100	12.6	2.0	1.5
Intraventricular lethality[*] (%)	100	100	25	25
Antigenic activity (%)	100	58	54	23
pI	5.2	8.0	8.0	9.4
Ca^{2+} binding Kd (mM)	0.84	1.43	-	-

[*] Rosenberg et al.[29]

It is obvious that carbamylation and guanidination of lysines do not induce
similar modifications of lethal potency. The isoelectric point of the carbamylated *N. naja atra* enzyme (3.7) is lower than that of the guanidinated enzyme
(5.7) and differs from that of the native enzyme (5.2). This difference in
charge, however, may not be a sufficient explanation for the very marked differences in properties which we have observed. This would agree with our
previous conclusion that the charge alone on the phospholipase molecule can not
be correlated directly with the toxicity of the enzyme. For example, by carbamylation of 7.4 out of the 10 lysines in *N. nigricollis* enzyme (Table 5,
fraction E), high toxicity was maintained even though the modified enzyme had
an isoelectric point of 5.4, showing that not only basic enzymes are toxic.

Modification of carboxylate groups

Carboxylate groups were modified using a water-soluble carbodiimide and semicarbazide at pH 3.5 and pH 5.5. Purification of the enzyme modified in the
presence of Ca^{2+} ions at pH 5.5 yielded a protein with 12.6% of the native
enzymatic activity, while 100% of the lethal potency and 58% antigenic activity
were found (Table 6) Although the carboxylated PLA$_2$ modified in the absence
of Ca^{2+} at both pH 3.5 and pH 5.5 have less than 2% of the original enzymatic
activity, they have about 25% of the intraventricular lethality (Table 6).
As seen from Fig. 4, the Ca^{2+}-induced difference spectra of carboxylated
PLA$_2$ at pH 5.5 + Ca^{2+} (Fig. 4b) showed a similar positive perturbation at
shorter wavelength to that of native enzyme (Fig. 4a), with a peak at 242 nm.
The peak could also be titrated as a function of Ca^{2+} concentration and the

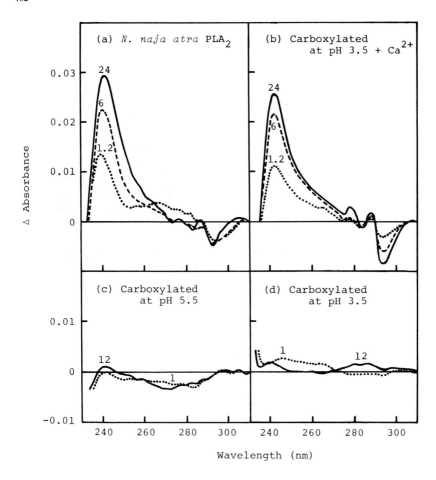

Fig. 4. Spectral changes of native and carboxylated N. *naja atra* PLA$_2$ induced by Ca^{2+} in 0.025 M Tris-0.1 M buffer (pH 8.0) at 25°C. The sample cuvette contained 0.45 mg/ml of native N. *naja atra* PLA$_2$ (a) and carboxylated PLA$_2$ at pH 3.5 + 1 M Ca^{2+} (b), at pH 5.5 (c), or at pH 3.5 (d) in the presence of Ca^{2+} (mM) as indicated. The cell in the reference position contained the respective enzyme (0.45 mg/ml) and spectra were recorded with a 0-0.1 absorbance scale.

binding data were plotted according to the Scatchard model. The dissociation constant of native PLA$_2$ for Ca^{2+} was 0.85 mM and that of carboxylated PLA$_2$ was 1.43 mM (Table 6), indicating that the binding ability did not change much after modification at pH 5.5 in the presence of Ca^{2+} ions. However, proteins modified

TABLE 7

CARBOXYLATE GROUPS NOT REACTING AT pH 3.5 and pH 5.5

Modification conditions	N. naja atra PLA$_2$	Bovine PLA$_2$[*]	
	Enzymatic activity (%)	Enzymatic activity (%)	Unmodified carboxylates
Unmodified	100	100	15
pH 3.5	1.5	0	1 (Asp-99)
pH 5.5	2.0	0	2 (Asp-99 & Asp-39)
pH 5.5 + Ca^{2+}	12.6	13.8	3 (Asp-99, Asp-39 & Asp-49)

[*] Fleer et al.[30]

in the absence of Ca^{2+} at both pH 3.5 and pH 5.5 lost their affinities for Ca^{2+} (Table 6) and the Ca^{2+}-induced difference spectra (Fig. 4c,d) changed greatly from that of native enzyme.

Our result is similar to the findings by Fleer et al.[30] on the modification of the carboxylate groups of the bovine PLA$_2$ (Table 7). They found that at pH 3.5, all but one (Asp-99) of the carboxylates were blocked. At pH 5.5, modification of 13 out of the total of 15 carboxylates occurred, leaving Asp-39 and Asp-99 unmodified. The modified protein showed complete lack of both Ca^{2+} binding and enzymatic activity. Modification of the enzyme at pH 5.5 in the presence of Ca^{2+}, protects Asp-49 from modification, yielding a protein with three unmodified carboxylates which is capable of binding Ca^{2+} ions and with 13.8% of its initial activity.

We also determined the Ca^{2+}-binding constants for native N. naja atra PLA$_2$ and carboxylated PLA$_2$ at several pH values by ultraviolet difference spectroscopy. The dissociation constants thus obtained were plotted against pH. The Hill plot showed that both native PLA$_2$ and carboxylated PLA$_2$ modified at pH 5.5 + Ca^{2+} had a slope of -0.97. It indicated that one titratable group was involved. The apparent pK value was 5.25. No titratable group was detected in carboxylated PLA$_2$ modified at pH 3.5 or pH 5.5 in the absence of Ca^{2+}. The result is in good agreement with the findings on bovine PLA$_2$[24,30,31] and on N. naja atra PLA$_2$ using pH dependence of the tryptophyl fluorescence.[32] It is clear that a single carboxylate group in both bovine and N. naja atra PLA$_2$ is involved in Ca^{2+} binding. The group is Asp-49 with an apparent pK value of 5.2.

PLA$_2$ from mammalian pancreas and from elapid venoms exhibit strong sequence homology. *N. naja atra* PLA$_2$ has 56% homology with bovine pancreatic PLA$_2$ and possesses the same number (15) of carboxylate groups as bovine PLA$_2$. It is surprising to notice that Asp-39, Asp-42, Asp-49, and Asp-99 are conserved in all known PLA$_2$ sequences and the Ca^{2+}-binding site is at Asp-49 of the molecules.

The modification of PLA$_2$ at pH 5.5 in the presence of Ca^{2+} yielded a protein with 12.6% of the native enzymatic activity, while 100% of lethal potency was recovered (Table 6). This indicates that the enzymatic activity and the lethal toxicity can be dissociated. This is the first report that a chemical modification of a PLA$_2$ can destroy enzymatic activity without destroying lethality at the same time.

By means of lysine modification, we were also able to demonstrate that the lethality and pharmacological effects of *N. nigricollis* and *N. naja atra* PLA$_2$ can be dissociated from the accompanying phospholipid hydrolysis. It is concluded that the toxicity of pure PLA$_2$ may be due to a direct, nonenzymatic, effect suggesting that there may be two distinct but perhaps overlapping active sites with His-48 and Asp-49 being at the catalytic site and His-48 and lysines being at the pharmacologically active region. Some consequences of this direct effect might confer upon the enzyme the ability to penetrate organized membranes, induce cardiotoxicity, induce hemolysis and interfere with coagulation. This direct effect is prominent in the relatively toxic PLA$_2$ enzymes while it is less manifest in the less toxic PLA$_2$ enzymes.

REFERENCES

1. Rosenberg, P. (1979) in Snake Venoms, Lee, C. Y., ed., Berlin: Springer, pp. 403-447.
2. Karlsson, E. (1979) in Snake Venoms, Lee, C. Y., ed., Berlin: Springer, pp. 159-212.
3. Eaker, D. (1978) in Versatility of Proteins, Li, C. H., ed., Proc. of Int. Symp. on Proteins, Taipei, New York: Academic Press, 413-431.
4. Lee, C. Y. (1979) Adv. Cytopharmac., 3, 1-16.
5. Strong, P. N., et al. (1976) Proc. Natl. Acad. Sci. USA, 73, 178-182.
6. Halpert, J., Eaker, D., and Karlsson, E. (1976) FEBS Lett., 61, 72-76.
7. Howard, B. D. and Truog, E. (1977) Biochemistry, 16, 122-125.
8. Volwerk, J. J., Pieterson, W. A., and De Haas, G. H. (1974) Biochemistry, 13, 1446-1454.
9. Yang, C. C. and King, K. (1980) Biochim. Biophys. Acta, 614, 373-388.
10. Yang, C. C., King, K., and Sun, T. P. (1981) Toxicon, 19, 645-659.
11. Gabriel, L. (1971) in Methods in Enzymology, Vol. 22, Grossman, L. and Moldave, K., eds., New York: Academic Press, pp. 565-578.
12. Stark, G. R. (1972) in Methods in Enzymology, Vol. 25, New York: Academic Press, pp. 579-584.
13. Karlsson, E., Eaker, D., and Ponterius, G. (1972) Biochim. Biophys. Acta, 257, 235-248.

14. Yang, C. C., King, K., and Sun, T. P. (1981) Toxicon, 19, 783-795.
15. Chervenka, C. H. and Wilcox, P. E. (1956) J. Biol. Chem., 222, 635-647.
16. Grouselle, M. and Pudles, J. (1977) Eur. J. Biochem., 74, 471-480.
17. Yang, C. C., Chen, S. F., and Fan, Y. C. (1982) Proc. 7th World Congress on Animal, Plant and Microbial Toxins, Brisbane, in press; Abstract in Toxicon (1982) 20, Suppl. No. 1, 70.
18. De Haas, G. H., et al. (1968) Biochim. Biophys. Acta, 159, 103-117.
19. Condrea, E., et al. (1981) Toxicon, 19, 61-71.
20. Yang, C. C. and King, K. (1980) Toxicon, 18, 529-545.
21. Roberts, M. F., et al. (1977) J. Biol. Chem., 252, 2405-2411.
22. Halpert, J. and Eaker, D. (1976) J. Biol. Chem., 251, 7343-7347.
23. Viljoen, C. C., Visser, L., and Botes, D. P. (1977) Biochim. Biophys. Acta, 483, 107-120.
24. Verheij, J. M., et al. (1980) Biochemistry, 19, 743-750.
25. Yang, C. C. and Chang, L. C. (1978) J. Chinese Biochem. Soc., 7, 63-77.
26. Condrea, E., et al. (1981) Toxicon, 19, 705-720.
27. Purdon, A. D., Tinker, D. O., and Spero, L. (1977) Can. J. Biochem., 55, 205-214.
28. Condrea, E., et al. (1982) Toxicon, in press.
29. Rosenberg, P., et al. (1982) Proc. 7th World Congress on Animal, Plant and Microbial Toxins, Brisbane, in press. (Abstract in Toxicon (1982) Suppl. No1. 1, 20, 52.)
30. Fleer, E. A. M., Verheij, H. M., and De Haas, G. H. (1981) Eur. J. Biochem., 113, 283-288.
31. Dijkstra, B. W., et al. (1978) J. Mol. Biol., 124, 53-60.
32. Teshima, K., et al. (1981) J. Biochem., 89, 13-20.

INTERACTION OF THE PHOTOAFFINITY LABEL, ARYLAZIDO-β-ALANYL NAD$^+$

WITH RABBIT MUSCLE GLYCERALDEHYDE-3-PHOSPHATE DEHYDROGENASE

SHIUAN CHEN, HARRY DAVIS, JOSEPH R. VIERRA AND RICHARD J. GUILLORY
Department of Biochemistry and Biophysics, John A. Burns School of Medicine,
University of Hawaii, Honolulu, Hawaii 96822

ABSTRACT

Arylazido-β-alanyl NAD$^+$ (A3'-O-{3-[N-(4-azido-2-nitrophenyl)amino]-propionyl} NAD$^+$) has been found to be an effective substrate for rabbit muscle glyceraldehyde-3-phosphate dehydrogenase. In the presence of 156 μM glyceraldehyde-3-phosphate, the apparent kinetic constants were evaluated at pH 7.0 as $k_{m,app}$ = 113.5 μM and V_{max} = 0.58 μmol analogue reduced·min^{-1}·mg protein^{-1} ($k_{m,app}$ = 39.4 μM and V_{max} = 1 μmol reduced·min^{-1}·mg protein^{-1} for NAD$^+$ at identical substrate concentrations). Enzymatic activity is completely inactivated when photoirradiation is carried out in the presence of a five-fold molar excess of analogue to enzyme. The photodependent inactivation was almost completely prevented by the presence of a five-fold molar excess of NAD$^+$ over that of the analogue. Experiments utilizing arylazido-[3-^{3}H]-β-alanyl NAD$^+$ demonstrate that the analogue attaches covalently to the enzyme following photoirradiation.

A study of the relationship of the photodependent inactivation to the stoichiometry of covalent binding of the NAD$^+$ analogue reveals that the association of but one mole of arylazido-β-alanyl NAD$^+$ per mole of enzyme results in 100% inactivation of enzymatic activity. The difference in the endogenous nucleotide level of the enzyme preparations is shown to be a significant factor with respect to the relative binding effectiveness of the NAD$^+$ analogue. It is concluded that arylazido-β-alanyl NAD$^+$, like NAD$^+$, binds to the enzyme in a negatively cooperative fashion.

INTRODUCTION

Previous work from this laboratory has shown that arylazido-β-alanyl NAD$^+$ acts as a substrate and a photodependent active-site labeling reagent for yeast alcohol dehydrogenase[1] and mitochondrial NADH-CoQ reductase.[1,2] This communication describes the analogue's substrate reactivity with, and its photodependent covalent binding to, rabbit muscle glyceraldehyde-3-phosphate dehydrogenase.

MATERIALS AND METHODS

All studies reported in this communication were performed with two preparations of crystalline rabbit muscle glyceraldehyde-3-phosphate dehydrogenase. The enzyme preparations were purchased as ammonium sulfate suspensions from the Sigma Chemical Company and used without further purification. Their specific activities as indicated by the supply house was 49 (preparation A) and 70 (preparation B) units per mg protein. When assayed according to the procedure described by Verlick[3] in which the assay mixture contains 10 mM EDTA, 10 mM sodium arsenate, 2 mM NAD^+, 0.7 mM DL-glyceraldehyde-3-phosphate, 4 mM cysteine and 30 mM sodium pyrophosphate, pH 8.4, the activity of the latter preparation was found to be 83.4 µmol NAD^+ reduced per min per mg protein at pH 8.6. All assays reported in this communication were carried out using a modification of the procedure of Eby and Kirtly.[4] Since the arylazido analogues are to a certain degree unstable in the presence of cysteine and at alkaline conditions,[5] the assay mixture was modified to contain 100 mM sodium phosphate buffer (pH 7.0) together with 0.32 mM DL-glyceraldehyde-3-phosphate, 0.2 mM NAD^+, 10 mM sodium arsenate and 10 mM EDTA in a total volume of 1.0 ml.

The NAD^+ was purchased from the Sigma Chemical Company and NADH from Boehringer; all other reagents were of highest analytical grade purity. Arylazido-β-alanyl-NAD^+ and arylazido-[3-^{3}H]-β-alanyl NAD^+ were prepared as previously described.[1] The radioactive analogue had a specific activity of 1.93×10^8 cpm·µmol^{-1}. Arylazido-β-alanyl $NADP^+$ was prepared as described by Chen and Guillory.[6]

The NAD^+ concentration of the two enzyme preparations was evaluated as follows. Four test tubes, containing 20, 40, 60, and 80 µl of enzyme suspension, were diluted to 1 ml with water and incubated in a boiling water bath for 3 minutes. Following centrifugation for clarification (6 min.) the supernatants from the 4 tubes were combined and the amount of protein in each of the precipitates determined using the procedure described by Lowry et al.[7] These values were used to evaluate graphically the protein concentration per unit volume of the enzyme preparation. The NAD^+ concentration of the preparation was evaluated from the increase in absorbance at 34 nm of an aliquot of the supernatant following reduction by yeast alcohol dehydrogenase (17 units) in the presence of 840 mM ethanol and 83 mM Tris buffer pH 8.0.

Binding studies were carried out by the procedure previously described.[8] Photoirradiation was carried out as previously described.[1] Special operations are detailed in the legends of the figures and tables.

RESULTS AND DISCUSSION

Arylazido-β-alanyl NAD$^+$ as a substrate for rabbit muscle glyceraldehyde-3-phosphate dehydrogenase

Figure 1 shows that arylazido-β-alanyl NAD$^+$ can act as a substrate for rabbit muscle glyceraldehyde-3-phosphate dehydrogenase. In the presence of 156 μM glyceraldehyde-3-phosphate, the apparent kinetic constants were evaluated at pH 7.0 as $k_{m,app}$ = 113.5 μM and V_{max} = 0.58 μmol of analogue reduced·min^{-1}·mg protein^{-1}. In comparison, the kinetic constants for NAD$^+$ measured at identical substrate concentrations were $k_{m,app}$ = 39.4 μM and V_{max} = 1.0 μmol NAD$^+$ reduced·min^{-1}·mg protein^{-1} (Fig. 1). The kinetic studies reveal that when arylazido-β-alanyl NAD$^+$ is used as a substrate the maximum enzymatic velocity is some 58% that of the maximum velocity in the presence of NAD$^+$. The substrate reactivity of the NAD$^+$ analogue indicates that the analogue could act as an active site directed reagent for glyceraldehyde-3-phosphate dehydrogenase. When an aliquot of the arylazido-β-alanyl NAD$^+$ enzyme mixture following deproteinization was applied to paper for chromatography and developed with the solvent system n-butanol/water/acetic acid 5/3/2 (v/v), no indication could be found for the presence of NAD$^+$ or NADH. This fact

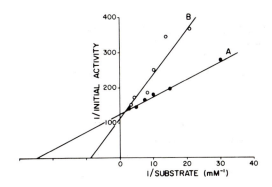

Fig. 1. Utilization of both NAD$^+$ and arylazido-β-alanyl NAD$^+$ as substrates for rabbit muscle glyceraldehyde-3-phosphate dehydrogenase. Enzymatic activity was assayed in 1 ml containing 100 mM phosphate buffer, pH 7.0, 17 mM Na$_2$HAsO$_4$, 10 mM EDTA, 156 μM glyceraldehyde-3-phosphate and 7.5 μg of enzyme preparation A at varying NAD$^+$ concentrations (A). When arylazido-β-alanyl NAD$^+$ was used in place of NAD$^+$ 15 μg enzyme was present (B).

TABLE 1

THE PHOTODEPENDENT EFFECT OF ARYLAZIDO PYRIDINE NUCLEOTIDE DERIVATIVES
ON RABBIT MUSCLE GLYCERALDEHYDE-3-PHOSPHATE DEHYDROGENASE

Additions	Percent Inhibition
1. _____ (Light Control)	0
2. Arylazido-β-alanyl NAD$^+$ (38 μM)	96
3. Arylazido-β-alanine (40 μM)	7
4. NAD$^+$ (37.5 μM)	0
5. Arylazido-β-alanine (40 μM) + NAD$^+$ (37.5 μM)	18
6. Arylazido-β-alanyl NADP$^+$ (41 μM)	23
7. Arylazido-β-alanyl NADP$^+$ (82 μM)	29

The irradiation mixture contained in 0.4 ml, 50 mM Tris HCl, pH 7.5, 200 μg
of rabbit muscle glyceraldehyde-3-phosphate dehydrogenase (preparation B) and
additional components as indicated in the Table. A 10 μl aliquot of the
mixture was used for enzymatic activity determination. The dark activity was
taken as 100%.

indicates that the observed enzymatic reduction of the analogue at pH 7 was
not due to NAD$^+$ formed by degradation of the analogue or as a contaminant
present in the reaction mixture.

Photodependent inactivation of rabbit muscle glyceraldehyde-3-phosphate dehydrogenase by arylazido-β-alanyl NAD$^+$

The specificity of the photodependent interaction of arylazido-β-alanyl NAD$^+$
with glyceraldehyde-3-phosphate dehydrogenase is shown in Table 1. Light
irradiation of the enzyme has no influence on the enzymatic activity under
our experimental conditions, while in the presence of 38 micromolar arylazido-
β-alanyl NAD$^+$ photoirradiation resulted in 96% inactivation. The presence of
40 μM arylazido-β-alanine, the photoreactive component of the NAD$^+$ analogue,
resulted in only a low level of 7% inhibition. Since NAD$^+$ at 37.5 μM did not
influence enzymatic activity under identical photoirradiating conditions it
is obvious that the photodependent inhibition due to arylazido-β-alanyl NAD$^+$
is not the result of a photosensitization effect due to NAD$^+$. A combination
of NAD$^+$ and arylazido-β-alanine did not significantly inhibit enzymatic
activity. The specific interaction of the arylazido-β-alanyl NAD$^+$ with muscle
glyceraldehyde-3-phosphate dehydrogenase is consequently judged to be a result
of the NAD$^+$ portion of the analogue directing the analogue towards the active
site and the arylazido-β-alanyl group providing a potential for photodependent

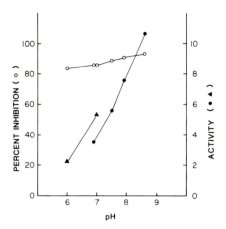

Fig. 2. Measurement of glyceraldehyde-3-phosphate dehydrogenase activity as a function of pH following photoirradiation at pH 7.5 in the presence of arylazido-β-alanyl NAD$^+$. The enzyme mixture contained in 0.4 ml, 50 mM Tris HCl pH 7.5, 36 μM arylazido-β-alanyl NAD$^+$, and 8.4 μM rabbit muscle glyceraldehyde-3-phosphate dehydrogenase (enzyme preparation B). Enzymatic activity was measured using a 10 μl aliquot of this mixture as described under Materials and Methods with either 100 mM phosphate buffer (pH 6 and 7) (▲) or pyrophosphate buffer (pH 6.9, 7.5, and 8.6) (●). The remaining enzyme mixture was then photoirradiated for 2 min at 10°C and the irradiated sample assayed as described above. Percent inhibition was calculated by taking the dark control at 100% activity (O). At pH 8.6 enzymatic activity was evaluated as 10.7 μmoles reduced min^{-1}.

insertion. Similar conclusions have been reached with respect to the analogue's interaction with yeast alcohol dehydrogenase[1] and the NADH-CoQ reductase of ox heart mitochondria.[1,2] This fact coupled to the substrate reactivity of the analogue makes it likely that the specific binding of the analogue occurs only at the substrate reactive site, rather than at a possible independent allosteric control site. A further demonstration of the specificity of the NAD$^+$ analogue for the muscle enzyme is found in the observation that arylazido-β-alanyl NADP$^+$, at 41 to 82 μM concentration, results in but 23 to 29% inhibition of enzymatic activity (Table 1). This is in sharp contrast to the interaction of these nucleotide analogues with yeast glyceraldehyde-3-phosphate dehydrogenase, in which case the NADP$^+$ analogue was shown to be as potent an inhibitor as the NAD$^+$ analogue.[9]

At 36 μM arylazido-β-alanyl NAD$^+$ the enzyme was inhibited to a similar extent (84-95%) when assayed from pH 6 to 8.6 in spite of a 5-fold difference

412

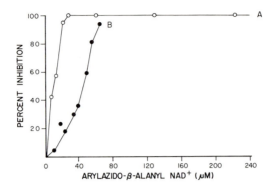

Fig. 3. Arylazido-β-alanyl NAD⁺-dependent photoinactivation of rabbit muscle
glyceraldehyde-3-phosphate dehydrogenase. (A) The irradiation mixture contained in 0.4 ml, 100 mM phosphate buffer, pH 7.0, 300 μg of rabbit muscle
glyceraldehyde-3-phosphate dehydrogenase (5.4 μM) (enzyme preparation A), and
varying concentrations of arylazido-β-alanyl NAD⁺ (7.0 to 223.0 μM). A 10 μl
aliquot of the mixture was taken for analysis of the dark control activity.
The remaining mixture was photoirradiated for 2 min at 10°C and the irradiated
sample then assayed for enzymatic activity. The dark control activity was
found to be stable and identical to the nonanalogue-treated preparation. This
value is taken as 100% activity. The assay was the same as that described in
Fig. 1. (B) The photoirradiated mixture, 0.24 ml, contained 50 mM Tris-HCl,
pH 7.5, 209 μg glyceraldehyde-3-phosphate dehydrogenase (enzyme preparation B)
and differing concentrations of arylazido-β-alanyl NAD⁺ (8.5 μM to 64 μM).
Enzymatic activity was measured prior to and following a 2 min photoirradiation.
The activity prior to photoirradiation was taken as 100%.

in specific activity between the extreme hydrogen ion concentrations (Fig. 2).
Consequently, it is reasonable to assume that the observed inhibitory data
collected at pH 7.0 represents a true photodependent inhibition due to a
specific interaction of the pyridine nucleotide analogue with the enzyme.

The powerful influence of arylazido-β-alanyl NAD⁺ on rabbit muscle
glyceraldehyde-3-phosphate dehydrogenase activity as illustrated in Fig. 3 is
surprising. When 5.4 μM glyceraldehyde-3-phosphate dehydrogenase (preparation
A) is subjected to a 2 min photoirradiation in the presence of 27 μM arylazido-
β-alanyl NAD⁺ in the final volume of 0.4 ml, enzymatic activity is completely
inactivated. This analogue concentration is but 5-fold in excess over the
enzyme concentration and only 28% of the K_m value evaluated for the analogue
acting as a substrate. Enzyme preparation A and B contained, respectively,
a 13- and 20-fold molar excess of NAD⁺. The analogue interaction with an
enzyme preparation containing a higher titer of endogenous NAD⁺ (preparation B)
is illustrated by plot B of Fig. 3. It is obvious that the endogenous

TABLE 2

NAD$^+$ PROTECTION OF ARYLAZIDO-β-ALANYL NAD$^+$ PHOTODEPENDENT

INACTIVATION OF RABBIT MUSCLE GLYCERALDEHYDE-3-PHOSPHATE DEHYDROGENASE

Additions	Percent Inhibition
I. Preparation A (300 μg)	
1. Arylazido-β-alanyl NAD$^+$	90
2. Arylazido-β-alanyl NAD$^+$ + NAD$^+$ (54 μM)	52
3. Arylazido-β-alanyl NAD$^+$ + NAD$^+$ (135 μM)	17
II. Preparation B (200 μg)	
1. Arylazido-β-alanyl NAD$^+$	96
2. Arylazido-β-alanyl NAD$^+$ + NAD$^+$ (188 μM)	42
3. Arylazido-β-alanyl NAD$^+$ + ADP$^+$ (212 μM)	96
4. Arylazido-β-alanyl NAD$^+$ + NMN (220 μM)	97
5. Arylazido-β-alanyl NAD$^+$ + ADP$^+$ (212 μM) + NMN (220 μM)	96

The experimental procedure was that outlined under Table 1 except for differing nucleotide and nucleotide derivative concentrations. In experiment I the concentration of arylazido-β-alanyl NAD$^+$ was 27 μM; in experiment II the analogue concentration was 38 μM.

nucleotide influences quantitatively the interaction of arylazido-β-alanyl NAD$^+$ with the enzyme. While nitrene formation from arylazido derivatives would be expected to result in direct covalent insertion of the analogue at its receptor site thus influencing natural ligand binding[10] kinetic considerations suggest that complete inactivation might be achieved only at infinite analogue concentrations.[11] The binding of pyridine nucleotide to rabbit muscle glyceraldehyde-3-phosphate dehydrogenase has been reported to involve negatively cooperative effects.[12,13] The low titer of analogue required to produce 100% photodependent inactivation of enzymatic activity is consistent with this suggestion in the case of the interaction of arylazido-β-alanyl NAD$^+$ with the enzyme (see below).

Active site reactivity for the NAD$^+$ analogue is further indicated by the ability of NAD$^+$ to effectively prevent the photodependent arylazido-β-alanyl NAD$^+$ inactivation of enzymatic activity. Photoirradiation of glyceraldehyde-3-phosphate dehydrogenase with 27 μM arylazido-β-alanyl NAD$^+$ in the presence of 135 μM NAD$^+$ (Table 2) afforded 73% protection against inactivation. Higher concentrations of arylazido-β-alanyl NAD$^+$ require higher concentrations of NAD$^+$ to protect against the analogue's photodependent inactivation of enzymatic activity. At 38 μM arylazido-β-alanyl NAD$^+$ a 5-fold excess concentration of NAD$^+$ (188 μM) provided but 54% protection against photodependent inactivation. The nucleotides ADP and NMN alone or in combination were not able to provide protection against the photodependent inhibition attributed to arylazido-β-alanyl NAD$^+$. This latter result indicates clearly that ADP and NMN have low binding affinity to the NAD$^+$ binding site in comparison to NAD$^+$ or the NAD$^+$ analogue. In addition it is clear that the protective ability of NAD$^+$ on the arylazido-β-alanyl NAD$^+$ photodependent inhibition cannot be ascribed to a light absorptive filtering effect.

The photodependent covalent binding of arylazido-β-alanyl NAD$^+$ to rabbit muscle glyceraldehyde-3-phosphate dehydrogenase

The photodependent covalent binding of the NAD$^+$ analogue to glyceraldehyde-3-phosphate dehydrogenase was most clearly indicated by studies utilizing the radioactive analogue. When 100 μg glyceraldehyde-3-phosphate dehydrogenase (preparation B) in 0.133 ml of 50 mM Tris-HCl, pH 7.5 was photoirradiated for 2 min in the presence of 23.3 μM arylazido-[3-^{3}H]-β-alanyl NAD$^+$, 0.39 mol NAD$^+$ analogue was bound per mol of enzyme as evaluated by the Sephadex G-50 syringe column centrifugation technique. The dark control, i.e., the enzyme analogue solution which was not subjected to photoirradiation, had but 15% (0.06 mol) of the analogue bound in comparison with the photoirradiated sample. When arylazido-β-alanyl NAD$^+$ was photoirradiated prior to mixing with the enzyme, or the enzyme was inactivated by treatment with sodium dodecyl-sulfate prior to mixing with the analogue again only 15% to 17% of the binding observed for the photoirradiated enzyme-cofactor mixture was observed. These controls demonstrate a specific light dependent covalent insertion of the analogue to the native enzyme.

The levels of radioactive analogue bound in the above described experiment is influenced by a number of factors. The crystallized rabbit muscle enzyme preparation is usually considered to have associated with it three to four NAD$^+$ molecules.[14] The enzyme preparation used in the above experiment had a 280 nm/260 nm absorptive ratio of 1.02 suggesting that there were indeed close to three NAD$^+$ molecules associated with each molecule of enzyme.[14] This enzyme preparation was shown to contain a 20-fold molar excess of NAD$^+$ over the enzyme concentration. Presumably the bound nucleotide would compete with the analogue for binding at the active site.

An additional consideration is that in this experiment the analogue concentration used (23.3 μM) is but 20.5% of the K_m value for the analogue acting as a substrate, a level at which one would not expect saturation of ligand

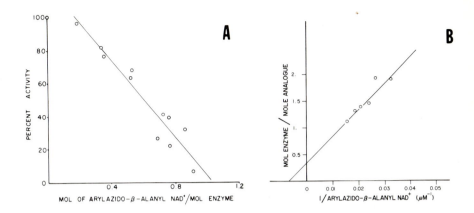

Fig. 4. The photodependent covalent binding of arylazido-[3-^{3}H]-β-alanyl NAD$^+$
to muscle glyceraldehyde-3-phosphate dehydrogenase. (A) The photoirradiated
mixture, 0.25 ml, contained 50 mM Tris-HCl, pH 7.5, 209 μg glyceraldehyde-3-
phosphate dehydrogenase (preparation B) and differing concentrations of
arylazido-[3-^{3}H]-β-alanyl NAD$^+$ (8.5 μM to 64 μM) specific activity 1.93 x
10^8 cpm/μmol. Enzymatic activity was measured prior to and following a 2 min
photoirradiation. The activity prior to photoirradiation was taken as 100%.
Following determination of enzymatic activity 50 μl of 5% sodium dodecylsulfate
was added to 200 μl of the enzyme mixture and the solution placed in a boiling
water bath for 3 min. Aliquots of the treated samples were then centrifuged
through syringe columns containing Sephadex G-50 (superfine) equilibrated
with 1% sodium dodecylsulfate. Radioactivity and the protein concentration of
the eluted solutions were then assayed. (B) A reciprocal plot generated
utilizing the data from Fig. 3A at analogue concentrations in excess of 31 μM.

binding sites. Figure 4A shows that binding of approximately one mole of
analogue per mole of enzyme results in 100% inactivation of enzymatic activity.
Koshland and his colleagues[13,15] have reported that binding of the first
molecule of NAD$^+$ influences the reactivity of the active site cysteine-149 in
all four subunits. Our result may suggest that the covalent binding of
arylazido-β-alanyl NAD$^+$ at one active site may modify the enzyme, limiting
nucleotide binding at the other sites.

In Fig. 4A it is observed that an apparent initial binding of about 0.17
mole of arylazido-β-alanyl NAD$^+$ per mole of enzyme does not measurably influence
enzymatic activity. This insensitivity at low analogue concentrations is
ascribed to one of two possible effects. There may be present in the enzyme
preparation a concentration of inactive enzyme amounting to about 17% of the

total which is still capable of binding arylazido-β-alanyl NAD$^+$. However, a much more plausible reason is that the apparently ineffective binding of 17% of the analogue is the result of nonspecific interaction with the enzyme. As indicated above 15% to 17% of maximal analogue binding is observed when arylazido-β-alanyl NAD$^+$ was photoirradiated prior to addition of the enzyme or when the enzyme was denatured prior to mixing with the analogue. When arylazido-β-alanine the photoreactive portion of the nucleotide analogue is included in the photoirradiation mixture, binding of arylazido-β-alanyl NAD$^+$ to the enzyme is decreased by about 13%. Consequently, we assume that our binding data is uncertain to the extent of 13% to 15% due to nonspecific binding of the arylazido-β-alanyl portion of the analogue. On the other hand the above studies clearly reveal that the interaction of arylazido-β-alanyl NAD$^+$ with muscle glyceraldehyde-3-phosphate dehydrogenase is a very specific reaction in which the binding of one mole of analogue results in complete inactivation of enzymatic activity.

Figure 4B is a reciprocal plot generated by utilizing the data from Fig. 4A at analogue concentrations in excess of 31 μM. At the lower analogue of concentrations, because of irreversible binding, the concentration of free analogue may be changing significantly; limiting the use of such data. Even at the higher analogue concentrations however the irreversible nature of the photodependent inhibition may render this approach somewhat uncertain.[11] Nevertheless, it is observed that while maximum inactivation of enzymatic activity occurs with the binding of but one mole of analogue per mole of enzyme, Fig. 4B indicates that maximum binding of the nucleotide analogue may occur at about 3 moles of analogue per mole of enzyme.

It is of interest that Bayne et al.[9] have shown that a total of 0.47 mol of arylazido-β-alanyl NAD$^+$ is bound covalently per subunit of an extensively dialyzed preparation of yeast glyceraldehyde-3-phosphate dehydrogenase. The analogue was in addition shown to undergo a substrate dependent reduction following its covalent binding to the enzyme.[9] A comparison of the arylazido nucleotide analogue binding to glyceraldehyde-3-dehydrogenase from these two sources shows that arylazido-β-alanyl NAD$^+$ binds to the yeast enzyme much more effectively than to the rabbit muscle enzyme. The difference in binding efficiency of the NAD$^+$ analogue may indicate a difference in the mode of coenzyme-enzyme interaction for the two enzymes. Indeed, the binding of pyridine nucleotide to rabbit muscle glyceraldehyde-3-dehydrogenase has been characterized by negative cooperativity[12,13] while binding to the yeast enzyme has been shown to have positive cooperativity.[16]

TABLE 3

NAD$^+$ AND NADH PROTECTION OF ARYLAZIDO-[3-^{3}H]-β-ALANYL NAD$^+$

PHOTODEPENDENT COVALENT LABELING OF RABBIT MUSCLE

GLYCERALDEHYDE-3-PHOSPHATE DEHYDROGENASE

Condition	Moles of Analogue Bound/20 μg
A	2.7×10^{-11}
B	1.4×10^{-11}
C	0.8×10^{-11}

Condition A, glyceraldehyde-3-phosphate dehydrogenase (preparation A) 200 μg (1.4 nmoles) in 0.27 ml, containing 0.2 M phosphate buffer pH 7, was photo-irradiated for 2 min in the presence of 2.49×10^{-9} mol (9.3 μM) arylazido-[3-^{3}H]-β-alanyl NAD$^+$. The irradiated mixture was dialyzed overnight against 2500 volumes of 1% sodium dodecylsulfate and then applied to a Sephadex G-50 column (1.2 x 23.5 cm) equilibrated with 1% dodecylsulfate. Condition B, the photoirradiated mixture contained the components of Condition A in addition to 1.5×10^{-8} mol NAD$^+$. Condition C, the photoirradiated mixture contained the components of Condition A in addition to 1.5×10^{-8} mol NADH.

In order to compare the effectiveness of NAD$^+$ and NADH as protective agents against the photodependent inactivation due to arylazido-β-alanyl NAD$^+$ the natural cofactors were tested under conditions which were known to result in a low level of photodependent analogue incorporation. As can be seen from Table 3 under conditions in which only about 10% of maximal labeling is accomplished (i.e., 0.09 mol of arylazido-β-alanyl NAD$^+$ per mole of enzyme), 56.4 μM NAD$^+$ provided 49% protection while NADH at the same concentration provided 69% protection. The greater effectiveness of NADH is consistent with its stronger binding to the enzyme.[17]

Glyceraldehyde-3-phosphate dehydrogenase was subjected to cyanogen bromide cleavage following photolabeling with arylazido-β-[3-^{3}H]- alanyl NAD$^+$. The cyanogen bromide cleavage products were then electrophoresed on sodium dodecylsulfate-urea acrylamide gels according to the procedure described by Swank and Munkres.[18] Analysis of the gel pattern revealed six major peptides of which the 25K peptide represents an incompletely digested fragment (Fig. 5). Radioactivity was associated exclusively with the 10K cyanogen bromide cleavage peptide. The amino acid sequence of the pig heart muscle enzyme is known to contain 9 methionine residues.[19] If the rabbit muscle enzyme is similarly constituted, the disposition of the methionine residues within the

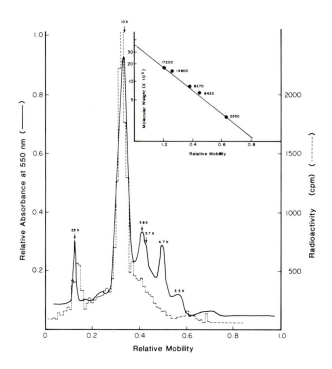

Fig. 5. The cyanogen bromide peptide pattern of arylazido-[3-^{3}H]-β-alanyl NAD$^+$ labeled glyceraldehyde-3-phosphate dehydrogenase. A mixture of (0.6 ml) containing 0.45 mg glyceraldehyde-3-phosphate dehydrogenase (enzyme preparation B), 10.7 nmol arylazido-[3-^{3}H]-β-alanyl NAD$^+$ (1.93 x 10^8 cpm/μmol), 16.4 nmol arylazido-β-alanine and 50 mM Tris HCl, pH 7.5 was photoirradiated for 2 min The photoirradiated mixture was then dialyzed in the dark at 5° against 1 liter of H$_2$O for at least 12 hours and then lyophilized. The lyophilized powder was

taken up in 1 ml 8 M urea containing 0.7 μmol-β-mercaptoethanol. After incubation at 45-55°C for 4 hours, iodoacetamide (0.8 μmol) was added and the incubation continued in the dark for an additional 15 min. The preparation was dialyzed against 1 liter of distilled water overnight (5°C) and then lyophilized. To the residue in 0.5 ml formic acid, 90 μmol CNBr was added and the mixture incubated at 37°C for 24 hours. The cyanogen bromide treated preparation was subjected to sodium dodecylsulfate-urea polyacrylamide gel

electrophoresis according to the procedure described by Swank and Munkres.[18] The migration pattern of cyanogen bromide peptides of myoglobin utilized as standards is indicated in the insert to the figure.

primary sequence and the limitations of the resolving power of the SDS-urea
gels would lead one to expect but five to six peptides clearly resolved under
the electrophoresis procedures described for Fig. 5.

CONCLUDING REMARKS

During the past ten years the preparation of three photoaffinity analogues
of pyridine nucleotides, the 3-diazoacetoxymethyl analogue of NAD^+,[20] the
3-azido NAD^+,[21] and the 3-diazirino NAD^+,[22] have been reported. Studies with
these analogues have tended to concentrate on the binding to particular
dehydrogenase enzymes without much attention being directed to their effects
on enzymatic activities either in the dark or following photoirradiation.
Studies correlating enzymatic activity with the interaction of such probes is
one of the principal, if not the only, reliable criteria attesting to the
specificity of interaction of such analogues with the true ligand binding site.
Such studies are consequently not only helpful but necessary in order to
assure that the binding of any analogue truly represents interaction with the
biologically relevant binding site.

Arylazido-β-alanyl NAD^+ has been shown to be an active site labeling reagent
for yeast alcohol dehydrogenase[1] and yeast alcohol glyceraldehyde-3-phosphate
dehydrogenase[9] as well as for a number of other dehydrogenase enzymes (see
Appendix) in addition to its reactivity with muscle glyceraldehyde-3-phosphate
dehydrogenase as reported above. Of all the enzymes tested arylazido-β-alanyl
NAD^+ appears to react most effectively with the muscle glyceraldehyde-3-
phosphate dehydrogenase enzyme. Thus the rabbit muscle glyceraldehyde-3-
phosphate dehydrogenase is singled out to illustrate the specificity of
interaction of the pyridine nucleotide photoprobes developed in this laboratory.

Arylazido-β-alanyl NAD^+ is a good substrate for the enzyme in the dark. On
the other hand in the presence of only 27 μM NAD^+ analogue (a 5-fold molar
excess over the enzyme concentration) the enzymatic activity is completely
inactivated following a 2 min period of light irradiation. The specificity
of this photodependent arylazido-β-alanyl NAD^+ inhibition is further illustrated
by the demonstration that the covalent association of but one mole of arylazido-
β-alanyl NAD^+ per mole of enzyme results in 100% inactivation of enzymatic
activity and that a single cyanogen bromide cleavage product, a peptide of
10K molecular weight is labeled by the analogue. In addition, the analogue,
similar to the natural substrate, binds to the enzyme in a negatively cooperative fashion. Finally, studies utilizing two enzyme preparations of differing
specific activity reveal that the endogenous nucleotide levels have a major
influence on the binding of the pyridine nucleotide probe. This photoprobe

is consequently anticipated to be an important tool in the study of the
allosteric activities of this complex enzyme.

A comparison of the relative reaction rates utilizing arylazido-β-alanyl
NAD^+ as a substrate for the muscle glyceraldehyde-3-phosphate dehydrogenase
and for yeast alcohol dehydrogenase[1] indicates as well that modification of
the ribose portion of NAD^+ has less of an effect on the interaction of the
modified substrate with glyceraldehyde-3-phosphate dehydrogenase than it does
with yeast alcohol dehydrogenase.[1] This is consistent with the finding that
$2'dNAD^+$ and $3'dNAD^+$ have substrate maximum velocity values of 4.3% and 0.8%
of that for NAD^+ for yeast alcohol dehydrogenase while the values are 31.2%
and 22.9%, respectively, for rabbit muscle glyceraldehyde-3-phosphate
dehydrogenase.[23] Suhadolnik et al.[23] suggest that the 2' and 3' hydroxyl group
on the adenosine portion of NAD^+ are not essential for coenzyme binding but
are required for proper orientation of the nicotinamide ring to produce a
productive complex.

REFERENCES

1. Chen S. and Guillory, R. J. (1977) J. Biol. Chem., 252, 8990-9001.
2. Chen S. and Guillory, R. J. (1979) J. Biol. Chem., 254, 7220-7227.
3. Velick, S. F. (1955) Meth. Enzymol., I, 401-406.
4. Eby, D. and Kirtley, M. E. (1976) Biochemistry, 15, 2169-2171.
5. Jeng, S. J. and Guillory, R. J. (1975) J. Supramol. Struc., 3, 448-468.
6. Chen, S. and Guillory, R. J. (1980) J. Biol. Chem., 255, 2445-2453.
7. Lowry, D. H., et al. (1951) J. Biol. Chem., 193, 265-275.
8. Chen, S. and Guillory, R. J. (1981) J. Biol. Chem., 256, 8318-8323.
9. Bayne, S., Sund, H., and Guillory, R. J. (1981) Biochimie, 63, 569-573.
10. Knowles, J. R. (1972) Acct. Chem. Res., 5, 155-160.
11. Guillory, R. J. (1979) Current Topics in Bioenergetics, 9, 267-414.
12. Conway, A. and Koshland, Jr., P. E. (1968) Biochemistry, 7, 4011-4022.
13. DeVijlder, J. J. M. and Slater, E. C. (1968) Biochim. Biophys. Acta,
    167, 23-34.
14. Fox, Jr., J. B. and Dandliker, W. B. (1956) J. Biol. Chem., 221, 1005-1017.
15. Teipel, J. and Koshland, Jr., D. E. (1970) Biochim. Biophys. Acta, 198,
    183-191.
16. Kirschner, K., et al. (1966) Proc. Nat. Acad. Sci. USA, 56, 1661-1667.
17. Bentley, P. and Dickinson, F. M. (1974) Biochem. J., 143, 11-18.
18. Swank, R. T. and Munkres, K. D. (1971) Anal. Biochem., 39, 462-477.
19. Harris, J. I. and Perham, R. N. (1968) Nature 219, 1025-1028.
20. Browne, D. T., Hixson, S. S., and Westheimer, F. H. (1971) J. Biol. Chem.,
    246, 4477-4484.
21. Hixson, S. S. and Hixson, S. H. (1973) Photochemistry and Photobiology,
    18, 135-138.
22. Standring, D. N. and Knowles, J. R. (1980) Biochemistry, 19, 2811-2816.
23. Suhadolnik, R. J., et al. (1977) J. Biol. Chem., 252, 4125-4133.

APPENDIX

THE INTERACTION OF ARYLAZIDO-β-ALANYL NAD(P)$^+$ WITH

SEVERAL NAD(H)- and NADP(H)-DEPENDENT ENZYMES

The interaction of several NAD(H)- and NADP(H)-dependent soluble enzymes
with the arylazido-β-alanyl photoaffinity analogues of NAD$^+$ and NADP$^+$ was
studied in order to assess the potential use of these probes for further
enzymatic studies. Table 1 lists the enzymes surveyed, the conditions employed,
and summarizes the major results of the study. In general, enzymes that were
photoinhibited by the analogues showed significant photodependent inhibition
at analogue concentrations of about 25 μM. In all cases photodependent
inhibition was significantly reduced by the presence of the natural coenzyme
during photoirradiation. Such protection is taken to indicate a competiton
for the active site between analogue and natural coenzyme. These studies
were limited to conditions in which the enzymes were not undergoing turnover
with respect to substrate activity. The titration curves illustrated in
Fig. 1 show that the maximum degree of photodependent inhibition varies from
enzyme to enzyme. Consequently, the different photoprobe reactivity is an
apparent characteristic of the individual enzymes and not a general characteristic of the photoprobes.

Chen and Guillory[1] have reported that yeast alcohol dehydrogenase can use
arylazido-β-alanyl NAD$^+$ as a coenzyme and was significantly inhibited by the
photoprobe upon photoirradiation in its presence. In the present study it
was observed that horse liver alcohol dehydrogenase was not photoinhibited by
either analogue. While this is an initially surprising observation it is
understandable on the basis that the yeast and liver enzyme have different
metabolic functions[2] and differ in subunit molecular weight, oligomeric form,
substrate specificity, and in 75% of sequence homology including some residues
necessary for catalysis.[3]

Except for sorbitol dehydrogenase and galactose dehydrogenase which were
photoinhibited by both analogues, the photodependent inhibition of the enzymes
studied was effected only by that analogue representing the natural coenzyme.
Thus, attachment of the arylazido-β-alanyl group does not effectively reduce
coenzyme specificity. With sorbitol dehydrogenase and galactose dehydrogenase
the analogue representing the natural coenzyme, arylazido-β-alanyl NAD$^+$,
exhibited the greater photodependent inhibition. In the case of galactose
dehydrogenase photodependent inhibition by both analogues was significantly
reduced by the natural coenzyme, NAD$^+$.

TABLE 1

PHOTODEPENDENT INHIBITION OF PYRIDINE NUCLEOTIDE DEPENDENT ENZYMES BY ARYLAZIDO-β-ALANYL NAD+ AND ARYLAZIDO-β-ALANYL NADP+

Enzyme	Coenzyme	Photo-irradiation mixture	% Inhibition with 25 µM analogue		Comments
			Arylazido-β-alanyl NAD+	Arylazido-β-alanyl NADP+	
Sorbitol dehydrogenase from sheep liver [EC 1.1.1.14] (Bergmeyer et al.[4])[a]	NAD+	13.3 µg enzyme in 0.2 ml of 0.1M phosphate pH 7	40	30	1) At 14.4 µM arylazido-β-alanyl NAD+, 144 µM NADH gave 72% protection from photoinhibition.
3-Hydroxybutarate dehydrogenase from *Rhodopseudomonas sphetoides* [EC 1.1.1.30] (Bergmeyer et al.[4])[b]	NAD+	25 µg enzyme in 0.2 ml of 0.1 M tris pH 7	10	0	1) At 59 µM arylazido-β-alanyl NAD+, 5.63 mM NAD+ gave 89% protection from photoinhibition. 2) At 137 µM arylazido-β-alanyl NAD+, 5.19 mM 3-hydroxybutarate gave 48% protection from photoinhibition.
6-Phosphogluconate dehydrogenase from yeast [EC 1.1.1.44] (Bergmeyer et al.[4])[b]	NADP+	4.7 µg enzyme in 0.2 ml of 0.1 M tris pH 7	0	59	1) At 9.2 µM arylazido-β-alanyl NADP+, 915 µM NADP+ gave 60% protection from photoinhibition.
Galactose dehydrogenase from *Pseudomonas fluorescens* [EC 1.1.1.48] (Bergmeyer et al.[4])[a]	NAD+	20 µg enzyme in 0.1 ml of 0.1 M phosphate pH 7	29	20	1) At 40 µM arylazido-β-alanyl NAD+, 2 mM NAD+ gave 91% protection from photoinhibition. 2) At 80 µM arylazido-β-alanyl NAD+, 1.6 mM NAD+ gave 76% protection from photoinhibition.

Glucose-6-phosphate dehydrogenase from Bakers' yeast [EC 1.1.1.49] (Bergmeyer et al.[4])[a]	6.6 μg enzyme in 0.4 ml of 0.1 M tris pH 7	NADP+	0	79
Dihydrofolate reductase from bovine liver [EC 1.5.1.3] (Osborn et al.[5])	0.16 μg enzyme in 0.2 ml of 10 mM phosphate pH 7	NADP+	–	39

1) At 3.1 μM arylazido-β- alanyl NADP+, 15.3 μM NADP+ gave 100% protection from photoinhibition.

2) With arylazido-β-alanyl NADP+ as coenzyme $K_M^{app} = 198$ μM, $V_{max}^{app} = 4.4$ μmol min^{-1} mg^{-1}, $k_i = 35$ μM.

3) Increasing concentrations of phosphate in the photoirradiation mixture resulted in decreasing amounts of photoinhibition.

The references cited refer to the assays used in assaying the respective enzymatic activities. Photoirradiation was carried out as previously described[1] except that the total photoirradiation period of two minutes was accomplished in 30-second intervals. The percentage of the photodependent inhibition was evaluated using enzyme activity measured following photoirradiation in the absence of analogue. All the enzymes were stable to photoirradiation under such conditions. The nonphotodependent inhibition was measured in the presence of analogue prior to photoirradiation and no significant inhibition was observed. Synthesis of the arylazido-β-alanyl NAD+ and NADP+ analogues was carried out as previously described.[1,6]

[a] Assay buffer was changed to phosphate pH 7.

[b] Buffer was changed to tris pH 7.

[c] The horse liver alcohol dehydrogenase [EC 1.1.1.1], rabbit muscle L-lactate dehydrogenase [EC 1.1.1.27], mitochondrial malate dehydrogenase [EC 1.1.1.37], and beef liver L-glutamate dehydrogenase [EC 1.4.1.3] were found not to be inhibited by arylazido-β-alanyl NAD+ at 840 μM, 27 μM, 160 μM, and 27 μM, respectively.

[d] The liver malic enzyme [EC 1.1.1.40], pig heart isocitrate dehydrogenase [EC 1.1.1.42] and beef liver L-glutamate dehydrogenase [EC 1.4.1.3] were found not to be inhibited by arylazido-β-alanyl NADP+ at 110 μM, 100 μM and 27 μM, respectively.

β-hydroxyacyl CoA dehydrogenase [EC 1.1.1.35] was photoinhibited 40% with 112 μM arylazido-β-alanyl NAD+ and 30% with 275 μM arylazido-β-alanyl NADP+.

[e] The mitochondrial malate dehydrogenase, the liver malic enzyme, and β-hydroxyacyl CoA dehydrogenase were provided by Dr. Ralph Bradshaw, Washington University, St. Louis.

424

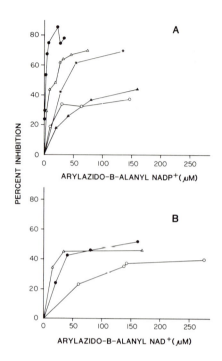

Fig. 1. Photodependent inhibition as a function of the concentration of arylazido-β-alanyl NAD+ and arylazido-β-alanyl NADP+ for a number of pyridine nucleotide dependent enzymes. The indicated concentrations (μM) of the photoprobes were included in the photoirradiation mixture and photoirradiation conducted as described in Table 1. (A) Photodependent inhibition by arylazido-β-alanyl NADP+ of glucose-6-phosphate dehydrogenase (●), 6-phosphogluconate dehydrogenase (Δ), dihydrofolate reductase (x), galactose dehydrogenase (▲) and sorbitol dehydrogenase (o). (B) Photodependent inhibition by arylazido-β-alanyl NAD+ of galactose dehydrogenase (●), sorbitol dehydrogenase (Δ), and 3-hydroxybutarate dehydrogenase (o).

Our studies have revealed as well that the reactivity of the photoprobes does not correlate readily with the AB stereospecificity of hydride transfer for the natural coenzyme. Enzymes photoinhibited by the NAD+ analogue consisted of both A- and B-type stereospecificities. The same was found with enzymes inhibited by the NADP+ analogue.

Similar conformations for the bound pyridine nucleotides are proposed for many of the enzymes we have studied. However, they do not all react with the photoprobes. This would indicate that the arylazido-β-alanyl adjunct on the nucleotide probes has an important influence on their interactions with some enzymes.

REFERENCES

1. Chen, S. and Guillory, R. J. (1977) J. Biol. Chem., 252, 8990-9001.
2. Jeffrey, J. (1980) in Dehydrogenases Requiring Nicotinamide Coenzymes, Jeffrey, J., ed., Base: Birkhauser Verlag.
3. Jornvall, H. (1977) in Pyridine Nucleotide-Dependent Dehydrogenases, Sund, H., ed., Berlin 30: Walter de Gruyter and Co.
4. Bergmeyer, H. U., Gawehn, K., and Grassl, J. (1974) in Methods of Enzymatic Analysis, Vol. 1, Bergmeyer, H. U., ed., New York: Academic Press.
5. Osborn, M. J. and Huennekens, F. M. (1958) J. Biol. Chem., 233, 969-974.
6. Chen, S. and Guillory, R. J. (1980) J. Biol. Chem., 255, 2445-2453.

Biochemical and Biophysical Studies of Proteins and Nucleic Acids,
Lo, Liu, and Li, eds.
427

CONCLUDING REMARKS

Robert L. Heinrickson
Department of Biochemistry
University of Chicago, Chicago, IL 60637, USA

The conference held in Taipei in 1982 was aptly named; we are on a new
frontier of research in molecular biology in which protein and nucleic acid
chemists will work together on a whole new set of problems hitherto beyond the
scope of our technical capacity. In 1906, Emil Fischer appealed to his fellow
scientists to apply their special skills in organic and physical chemistry to
the complex problems of biology and to develop the methods needed to conquer
the "impregnable fortress" of biomacromolecular structure. This conference
bears witness to that wedding of physical and biological science and to the
endless refinement in technology which has raised to such a sophisticated
level the state of molecular biology today.

Who would have thought a decade ago to include in the title of a conference
"Proteins, Nucleic Acids, and DNA Engineering?" These were very separate
enterprises and not to be confused with one another! But then, who would have
imagined at that time that nucleic acids would turn out to be easier to
characterize structurally than proteins? It is almost impossible to keep pace
with the astounding achievement in genetic engineering, microbiology, and in
the technology surrounding the manipulation of genetic materials. Just as the
protein chemist had become resigned to his fate of having sole responsibility
for the arduous task of unraveling protein sequences, the major burden for
total protein sequence analysis may rest more and more with the nucleic acid
chemist. Thus relieved on a particularly time and energy consuming chore,
the protein chemist can pursue the important and highly variegated features
of post-translational processing and the design of chemical modification studies
to probe structure-function relationships to proteins. Not that protein
sequence analysis is dead; far from it. The protein sequence determined from
analysis of nucleic acids requires some verification from the protein side.
Moreover, the new procedures from microsequencing of proteins will be crucial
for determining partial sequences appropriate for synthesis of oligonucleotide
probes which will in turn facilitate isolation and sequencing of related
genetic materials. Clearly, it is all beginning to come together now and the
prospects for the coming decade are most promising in regard to the potential
for gaining a much more detailed background of structural information with
which to interpret problems of biological interest.

This brings us to a consideration of that next higher arena of complexity at the organelle or cellular level. Many interesting papers were delivered based upon studies of membranes, mitochondria, and cell surface receptor proteins. Studies of this kind will benefit from the avalanche of structural information which will be forthcoming in the next years, but already important advances have been made in our understanding of membrane fluidity and the motion and disposition of proteins in membranes. One of the highlights of the conference was the plenary lecture by Dr. Ashwell in which some of his pioneering studies on uptake of glycoproteins by cell surface receptors were delineated. Studies of membrane ultrastructure and receptor proteins have shown ever increasing levels of sophistication and these will, in turn, set the stage for a more detailed knowledge of the molecular events of hormone action and cell-cell recognition phenomena which are at the heart of so many important processes of health and disease.

As regards polypeptide hormone action and the functions of peptides in general, it is clear that the field of peptide organic synthesis will continue to provide important contributions to our understanding. The overview of this rapidly growing area of research given by Dr. Li indicated expansion along many lines. New methods for group activation, protection, and deprotection are being added constantly to the endless repertoire of the peptide chemist and the synthetic products are increasing in reliability. The impact of reverse phase and ion exchange high performance liquid chromatographic methods in the purification of synthetic peptides has been enormous. Peptide synthesis will continue to use nature's blueprint as a basis for design of experiments to test the functional implications of amino acid substitutions. Another important direction in the 1980s will be peptide modeling studies in which the synthetic plan will be based upon secondary and not primary structural features of the peptide. Such work has already demonstrated that the function of many biologically active peptides is a consequence of their short range folding patterns and is sequence-independent. Thus, peptide synthesis will continue to play a vital role in understanding structure function relationships in proteins and peptides.

From the point of view of contributions from physical chemistry it was somewhat surprising not to find an x-ray crystallographer among the speakers at the conference in Taipei. X-ray crystallography has of course made a dramatic impact on our understanding of protein structure and continues to be an exceedingly important area of research. However, a field of inquiry which has been less in the limelight but which will nevertheless become increasingly important on the "new frontier," is the area of molecular dynamics. Physical chemists are applying with increasing success a number of elegant spectroscopic methods to the study of motion in proteins. Proteins and nucleic acids are not static entities but exhibit considerable dynamic character, part of which at least accounts for their functional capacities. The next dimension in our future understanding of structure function relationships will be in the area of dynamic measurements and the establishment of a rationale whereby, for example, the catalytic efficiency of an enzyme can be understood and predicted. The many talks at the conference on the subject of nuclear magnetic resonance spectroscopy bear witness to the growth in that field and the application of additional spectral methods will no doubt increase in the next decade.

Finally, we were privileged in the plenary lecture by Dr. Pestka to hear a most elegant presentation describing the attack on interferon by the new coalition of protein chemists and genetic engineers. It was an excellent example of how these manifold technical capabilities may be brought to bear on a problem of monumental interest relative to public health. Solving the problem required the skills of biochemists in purifying the interferon, protein chemists in the partial sequence analysis, and genetic engineers and nucleic acid chemists in cloning the gene and determining the structures of the interferons. Clearly, a great deal of work remains to be done in characterizing the many interferon types discovered in this work, but the potential for success in finding variants of importance is equally great.

Proteins are the workhorses of living systems, the final repositories in the transfer of genetic information. Their manifold functions are expressed in an equally diverse array of structural forms and their study and characterization remains of paramount importance. That importance was reflected in the Taipei conference for it is clear that American and Chinese colleagues alike are devoting considerable time and energy to questions of protein structure and function. The prospect before us is exciting indeed for the new tools of chemistry, physics, and genetics will serve as the basis for exploration of a fascinating new dimension of biological complexity.

434 434

434

Genes

leukocyte interferon, 29, 43

structural, chemical synthesis, 189-190

Glyceraldehyde-3-phosphate dehydrogenase, interaction with
arylazido-beta-alanyl NAD$^+$, 407-425

Glycoproteins, viral, at cell surface, 61-67

Gonadotropins of pike eel and Chinese carps, chemistry and
function, 361-370

Growth inhibition by leukocyte interferon, and antiviral
activity, 45-48

Hemoglobin, sickle, intracellular polymerization, 241

Hepatocytes

ligand binding to Gal/GalNAc lectin, 349-360

receptor-ligand complex, intracellular dissociation, 59

High pressure liquid chromatography (HPLC)

gonadotropins of pike eel and Chinese carps, 361-370

leukocyte interferon, 14-18

in protein chemistry, 296-298

Hydralazine reaction with amine oxidase, kinetic and
spectral studies, 317-329

Interferons, 11-58

action and assay, 12

amino acid sequence, 19-22, 31, 41, 43-45

bacterial expression of, 29-36

biological properties, 45-52

carbohydrate content, 19

classification, 13-14

clinical use, 53-54

cloning (recombinant), 23-29

crystallization, 52-53

hybrid molecules, 36-37, 48-52

monoclonal antibodies, 37-43

purification from human leukocytes and fibroblasts, 14-23

Isoniazid

reaction with amine oxidase, kinetic and spectrl studies,
317-329